The author of this book argues that while the structure, conduct, and performance of the world tin industry are subject to strongly competitive market forces, major intervention by international governments has exerted a controlling influence over the world tin market for the past sixty years. He analyzes these economic forces and political factors in the context of the Sixth International Tin Agreement and of continuing discussion over the proper structuring of the international tin market.

To Puey Ungphakorn

The World Tin Market

Political Pricing and Economic Competition

William L. Baldwin

Duke Press Policy Studies

Duke University Press Durham, N. C. 1983

Printed in the United States of America on acid-free paper

Library of Congress Cataloging in Publication Data

Baldwin, William Lee.
The world tin market

(Duke Press policy studies)
Bibliography: p.
Includes index.
1. Tin industry. I. Title. II. Series.
HD9539.T5B34 1983 338.2'7453 83-8888
ISBN 0-8223-0505-4

Contents

Tables

Figures

Acknowledgments

During the academic year 1978–79 I made visits of several months each to Thailand, Malaysia, and Great Britain, gathering information and collecting materials for this study. The work could not have been done without the consideration and generosity of a number of people who shared their knowledge of the tin industry with me and in some instances assisted me in meeting others who could be of help. I am deeply grateful to these informants not only for their time and willingness to be of assistance, but also for the candor with which most were willing to discuss sensitive problems and confidential matters. I have respected this frankness and openness by occasional resort to references such as "informed observers" or "knowledgeable individuals," and by combining figures given or shown to me in confidence in such a way as to avoid revealing the original information itself as well as the source.

In particular, I wish to express my appreciation to the following.

In Thailand: Sunt Rachdawong and Jumrus Atikul of the Department of Mineral Resources; Usoe Khaw, Benny Widyono, and Craig A. Emerson of the United Nations Economic and Social Commission for Asia and the Pacific; Peter E. Beal of Orion Associates; Robert F. Goninon of Thai Endeavor; and George E. Delehanty of Thammasat University and the Rockefeller Foundation.

In Malaysia: Yip Yat Hoong, K. C. Cheong, and Lee Hock Lock of the University of Malaya; Y. Mansoor Marican and Bruce R. Morris of Universiti Sains Malaysia; Alec Foo of Foo Nyit Tse and Brothers; Lim Keng Kay of K. K. Lim and Associates and the States of Malaya Chamber of Mines; Kam Cheng Eng of the Malayan Mining Employers' Association; Nazir Ariff and Khoo Kah Leng of Datuk Keramat Smelting; G. A. B. Galbraith of the Straits Trading Company; Datuk Wong Yoke-Meng and Mohd. Zarif bin Mohd. Zaman of the Ministry of Primary Industries; H. Hussin and Mohd. Zaki bin Haji Taib of the Mines Research Institute Malaysia; Abdullah Hasbi bin Haji Hassan of the South East Asia Tin Research and Development Centre; Fong Weng Phak of the Bank Negara Malaysia; Chung Tsu Tuan of the Department of Socio-Economic Research; and Stephen Chee of the International Committee on the Management of Population Research.

In Great Britain: Peter S. Lai, P. A. A. de Koning, Umar Yahya, Bernard C. Engel, and Neill L. Phelps of the International Tin Council; B. K. T. Barry of the International Tin Research Institute; Paul Pilkauskas of the United States Department of State; and William O. Sugg III of the United States Department of Commerce.

I am also grateful for assistance rendered by Laurence D. Stifel of the Rockefeller Foundation and Keith L. Harris of the United States Department of the Interior, and for perceptive comments and criticisms by my Dartmouth colleagues Alan L. Gustman, Lawrence G. Hines, Michael P. Mazur, John A. Menge, Salim Rashid, and John T. Scott.

The manuscript was edited with thoroughness and sensitivity by Sony Lipton, who also prepared the index. César A. Molina, James D. Wiltshire, Jr., and Evan C. Tepper, students at Dartmouth, provided conscientious research assistance. Elaine A. Vigneault typed the final manuscript with care and patience. My wife Marcia H. Baldwin read the entire manuscript and is responsible for a number of improvements in presentation.

I am indebted to Thammasat University, Bangkok, Thailand; Universiti Sains Malaysia, Penang, Malaysia; and the International Tin Council, London, England, for providing me with office and work space and for access to their library collections. My deepest obligation is to the Universiti Sains Malaysia, which gave me a temporary appointment as Visiting Researcher and provided housing for me and my family during our stay in Malaysia. I can only acknowledge, with warmest thanks, the hospitality, courtesy, and helpfulness shown to us by all members of the USM administration, faculty, and staff whom we met.

In addition to granting the sabbatical leave that made this project possible, Dartmouth College has provided financial support to the study through its Committee on Research.

The World Tin Market

Introduction

Nineteen eighty-one was a year of price gyrations, tensions, and high public visibility for the world tin industry. The year opened with spot tin selling on the London Metal Exchange (LME) at £6,292.50 per metric ton, down from a peak of £8,450 the previous March. By January 26, 1981, the price had fallen to £5,682. Through May, prices hovered around £6,000, and then in June, and more precipitously in July, began rising to reach a peak for the year of £8,555 in November. In early 1982, the price rose above £9,000 for a brief period.

These wide-swinging price movements were a manifestation of underlying political turmoil. Indeed, 1981 is an excellent year to illustrate the basic theme of this study: the performance of the world tin industry can best be understood as a process of political price determination imposed on a market that is structurally conducive to effective competitive processes.

The Fifth International Tin Agreement (ITA) was due to expire at the end of June 1981. The first of these five-year agreements had come into effect in 1956. By 1980 the repeatedly renewed tin pact was the oldest extant international commodity stabilization agreement, and widely recognized as the most successful. "The ITA represents 25 years of proven cooperation between producing and consuming countries," noted Datuk Leong Khee Seong, Malaysia's minister of primary industries.[1] But following a failure of negotiations in December 1980, the International Tin Council (ITC) announced in January 1981 that the fifth agreement would be extended until June 30, 1982, to allow the talks to continue under a less pressing deadline. The deadlock in the discussions was attributed primarily to the United States, the world's largest tin consuming country but a member of the ITC only since 1976, and to Bolivia, the producing nation whose economy was most dependent on tin and whose costs of production were the world's highest. The United States' delegation was pressing for larger contributions by member nations to an ITC buffer stock and for severe restrictions on the ITC's authority to impose export controls. The Bolivians, in addition to adamant opposition to the United States' proposals, were demanding that the sixth agreement contain a provision under which the ITC would gain a measure of control over sales of surplus tin from the United States' huge strategic stockpile of the metal.[2] In mid-January, a chill was cast on the pending negotiations when Secretary of State-designate Alexander M. Haig stated at his confirmation hearings that he was not "a fundamental enthusiast" for international commodity agreements.[3]

The ongoing activities of the ITC in implementing the provisions of the fifth agreement were also in disarray. Early in 1977 the council had depleted its buffer stock in a futile defense of the ceiling of its price support range, and had not been active on the tin market thereafter through 1980. This inactivity had been imposed on the ITC by a series of negative votes by consuming nation members, particularly the United States, West Germany, Japan, and the USSR, which had permitted the ceiling price to rise only from £3,240 at the beginning of 1976 to £5,390 at the end of 1980,[4] while LME prices rose from a 1976 low of £3,052.50 to the 1980 high of £8,450. At the ITC's April 1981 meeting, the consuming nations once again blocked an increase in the price range. An event common enough in other international organizations, but unprecedented in the ITC, followed: all of the producing nations' delegates walked out of the meeting.

As the year wore on, conditions deteriorated. At a June meeting of the ITA conference, delegates of all nations represented except the United States and Bolivia agreed on the text of a sixth agreement, and subsequently refused a request from the United States to reopen negotiations. In October, the United States announced its decision not to join the new pact. The sixth agreement was to come into effect only if signed by nations accounting for 65 percent of the production and 65 percent of the consumption of those participating in the conference.[5] At the end of the year neither Japan, accounting for approximately 18 percent of world consumption, nor the nations of the European Economic Community (EEC), accounting collectively for approximately 28 percent, had yet announced their decisions whether to sign the pact. In an editorial entitled "The New King-Makers," the leading trade magazine *Tin International* noted that the principal effect of the United States' withdrawal from the ITA would be to establish Japan as the most influential consuming member, and added, "Her geographical and political alignment with the ASEAN countries, which include the three major S.E. Asian tin-producing nations, could well presage a much better framework for understanding in the next ITA than has been seen during the chequered life of the current agreement."[6]

In April, despite the veto that month of an increase in the ITC's price range, the buffer stock manager began buying tin, as the price on the Penang market in Malaysia (the market on which the price range is based) fell to a three-year low and into the top of the band within which such buying was authorized. The Malaysian price was suffering not only from the generally depressed state of the world tin market, but also from a fall in the value of the British pound relative to the Malaysian ringgit. By June 30, the ITC's buffer stock held 3,865 metric tons of tin metal. At the same time, however, sales from the United States' strategic stockpile began to accelerate. In January 1980, President Carter had authorized sales of 30,000 long tons of stockpile tin, with a maximum of 10,000 tons to be sold in any one year. Under agreements made at the time the stockpile was acquired during the Second World War and the Korean War, and repeated in 1966, the United States was obligated to release surplus stocks "with due regard to the protection of tin producers, processors and consumers against avoidable disruption of their usual markets." At the end of 1980, only twenty-five of the authorized tons had been sold. But from January through July 1981, sales rose to 900 tons, followed by 260 tons in July, 400 in August, and 540 in September. Throughout the entire year, 5,920 tons were put on the market. Producing interests were incensed. Abdul Rahim Aki, president of the

States of Malaya Chamber of Mines, stated that the continuation of the sales after the ITC had entered the market as a buyer "makes the action highly irresponsible, and casts considerable doubt on the sincerity of the US in their assurances that the disposals would not adversely affect the market." Mining ministers of Bolivia, Malaysia, Thailand, and Indonesia protested the sales, the latter three in a joint statement noting, "Both the timing and method of releases and overhanging effect of the amount authorised for sale pose considerable bearish effects on an already weak tin market." The statement added that producers would be required to take "appropriate measures" unless the stockpile sales were suspended. *Tin International*, noting that in the 1966 reassurance the United States had made a commitment to the ITC "to moderate its tin sales programme if it should be inconsistent with the contingent operations of the buffer stock authorised under the ITA," called the unprecedented situation in which the buffer stock manager was buying and the United States government selling tin at the same time an "apparent absurdity."[7] Near the end of the year, *Tin International* commented editorially that the United States' action constituted "economic aggression," and quoted with approval one commentator's observation that the Reagan Administration appeared "simply bent on spurring its free market bronco right through the fabric of international understanding on tin."[8]

All of these events—the refusal of the United States to ratify the sixth agreement, the intransigence of the consuming members of the ITC in holding down the price range, and the accelerated sales of tin from the United States' strategic stockpile—appear to have contributed to a widely publicized climax for 1981. Datuk Seri Dr. Mahathir Mohamad, who had become prime minister of Malaysia on July 15, stated the following month, "If it is considered morally right for buyers to use stockpiles to lower prices, it is also morally right for sellers to use them to raise prices."[9] As tin prices rose, from £6,537.50 at the beginning of July to £8,455 by the end of November, both the trade and the general press began referring to a mystery buyer of impressive financial strength, who was forcing the price up, first by heavy purchases on the futures market from July through late November (estimated by some to be as high as 50,000 metric tons), followed by a sudden switch to the spot market on November 26.[10] While virtually all commentators noted that the manipulator was "believed connected with Malaysia,"[11] Malaysian governmental sources denied any official involvement. Yet on September 3, the *Business Times* of Malaysia reported, "The Prime Minister said that the consumers of tin should not take umbrage over reports circulating in London that some producer countries were buying tin in a bid to firm what would otherwise be a weak market. Datuk Seri Dr. Mahathir Mohamad asked: Why should consumers object when 'the kind of thing they do not like us to do' is being done on a continuing basis from their side."[12]

At its October meeting, the ITC did raise the floor and ceiling prices, by 6.85 percent. This increase was viewed, however, as little more than a gesture to the producing nations, which had been pressing for an increase of 28 percent. The delegates from the consuming nations now had a new argument—the ITC ought not to establish a floor price that would permit the mystery buyer to sell its tin holdings at a profit. The *Financial Times* of London commented, "Neither producers nor consumers are very happy about the compromise. Bolivia, in particular, was upset. However, Mr. Peter Lai, chairman of the Tin Council, claimed the way was now clear for ratification of the sixth International Tin Agreement, due to come into force next June when the present pact expires."[13]

In November, the *Daily Telegraph* of London noted that "the tin industry is in complete disarray with the Americans making sales from the strategic stockpile when the Malaysians are buying hand over fist to force tin to an all-time peak of over £8,650 a tonne when the world is in recession."[14] And on December 6, the *Observer* of London commented:

> If, and it has yet to be proved, the producers are behind the recent tin price support, and they manage to keep the market under control, it will be little short of a revolution in relationships in the tightly-knit world of commodity trading. Gone would be the days when the finance minister of a struggling Third World country dependent on its commodity exports would wait in his office for the latest prices on the Reuters news tape before knowing whether or not he could authorise the building of various projects.[15]

Late in the year, another possible explanation of the mystery buyer's activity on the LME surfaced. The *Far Eastern Economic Review* noted that Marc Rich and Company, a trading company serving as the New York agent for the Malaysia Mining Corporation, "may have persuaded Arab money that usury-free investment in tin would be reasonably secure," and would represent "a long-term investment with political benefits for a Muslim country."[16] In early 1982, the *New York Times* quoted George Kam of the Washington office of the Malayan Tin Bureau as stating that the mystery buyer was probably a private party. "Perhaps Malaysian," he speculated, "perhaps Indonesian, perhaps some Arabs, perhaps OPEC money—all that jazz. It's certainly not the Malaysian Government that's doing it."[17]

The objective of the mystery buyer is at least as obscure as its identity. If the purpose was an attempt to corner the market for manipulative profit, it was a failure in at least the short run. In late February 1982, when a number of traders were faced with obligations to deliver tin within a few days and concern was growing that the mystery buyer might succeed in cornering the spot market at the same time, the LME acted to limit the penalty for failure to deliver to no more than £120 per day. The spot price plummeted by £2,000, abetted, it was believed, by sales by the mystery buyer itself. The ITC intervened with buffer stock purchases to protect the floor price that had been set the previous October. In March, the LME's loss limit was removed. But the mystery buyer's objective might have been, as advocated for some time by Malaysia's Prime Minister Mahathir and Primary Industries Minister Leong, development of a producers' stockpile. Or, a long-term investor might still be able to profit from the adventure if it can hold tin until there is an end to the recession in the industrialized consuming countries. Finally, the objective may be related to an attempt by Malaysia to substitute a producers' cartel, already referred to by the press as TINPEC, for the ITC had the sixth agreement failed. Malaysia has recently indicated its intention to reduce its tin production by 25 percent after 1983.

The mystery buyer's most significant short-run effect was not on the tin market itself, in the opinion of informed observers, but rather on the future of the tin agreements. The squeeze was "highly technical," it was pointed out to me, in that the largest swings in price occurred on the spot market and were paid or received almost entirely by speculators and the mystery buyer as previous contracts for delivery came due and were closed out by paper transactions canceling the earlier ones. Industrial users of tin paid prices to dealers that were closer to the ninety-day forward price, which reached a peak of only around £8,000 at the same time that the spot price rose to £9,000. The publicity attending the mystery buyer's activities and conjectures as to his identity and motives may have had far more serious effects. There is concern that confidence in tin trading institutions around the world, particularly the Penang market for physical tin and the LME, has been eroded. The ITC has been made to look helpless in the face of such massive manipulation. Producers had earlier been antagonized by what they viewed as the intransigence of major consumers. The behavior of the mystery buyer raised comparable doubts among consumers as to the future intentions of the producing nations.

On the other hand, the ITC did step in, from February through April 1982, with buffer stock purchases, a call-up of member contributions to bolster its purchasing power, and finally export control, to stabilize the market. It may be that the ITC's success in countering the influence of the mystery buyer, coupled with the object lesson of what the alternatives to international agreement may be, will strengthen support for the Sixth ITA. In the words of one of my informants, some of the producing nations were "frightened to death" by what the mystery buyer was doing, and "totally confused."

Thus, in the spring of 1982 it appears that conflicting political forces are shaping the world tin market into a still unpredictable configuration. But crises, confrontations, and sweeping institutional changes are not new experiences for the tin industry. Indeed, other years have been more tumultuous than 1981. The current conditions have their roots in the past. For at least sixty years, the tin industry has been molded by a complex set of interactions between public and private interests. The remainder of this study explores such questions as: the extent to which a series of control schemes have been successful; the underlying structural conditions of the industry that may explain the degree of success, failure, or irrelevance; the potent and pervasive influence of fundamental market forces and private interests; and the national and international stakes at issue in public control of a worldwide market for a primary commodity.

1. The Devil's Metal: Economic Implications of the Geology and Technology of Tin

Tin and copper are two of the softer metals. Yet when they are combined in the proper proportions, the resulting bronze is remarkably hard—hard enough to be cast as gun metal. This hardening property of tin, inexplicable to the alchemists and those who worked the metal, became a mystery attributed to the influence of Satan and resulted in tin becoming known as *diabolus metallorum*, which has been translated both as "devil of the metals" and as "metal of the devil."[1]

One need not ascribe the hardening effect of tin on copper to black magic to note that the impact of a commodity on both its makers and its users is in large part dependent on nonhuman forces. The scope of private entrepreneurs and public agencies to shape an industry is constrained by the basic physical and technical characteristics of the commodity. It should suffice for purposes of this study merely to sketch the outlines of those underlying features of nature and technology which have clearly and significantly conditioned the economic structure, conduct, and performance of the modern tin industry; but a layman's understanding of these fundamentals of the industry does seem to be a necessary precondition to a meaningful economic investigation.[2]

Tin is often described as the scarcest and most valuable of the nonferrous base or common metals. It could, in fact, be regarded as semirare. Comparisons of total world production of the major nonferrous and rare metals put tin in a class by itself, as shown in table 1.1.

Similarly, tin is roughly an order of magnitude higher in price than the other nonferrous base metals, and an order of magnitude lower than silver, as shown in table 1.2.

These levels of production and price reflect, among other things, basic geological endowments. Table 1.3 indicates these endowments in the most fundamental sense, the average degree of concentration in the earth's crust expressed in parts per million, regardless of site, form, or difficulty of extraction.

Geographically, recoverable tin is highly concentrated in a few parts of the world. The richest deposits lie in a belt running from Yunnan in southern China through Laos, Burma, Thailand, and Malaysia to three Indonesian islands off the coast of Sumatra and south of Singapore. The known deposits in this belt are predominantly alluvial, although lode mines are being worked in all of these countries and are the principal source of ore in Yunnan. Offshore mining is becoming an important source of ore for Thailand and Indonesia. Throughout the present century, roughly two-thirds of the world's total tin output has come from this Asian belt. The second most important concentration of tin lies in the Bolivian altiplano—highlands in which most of the major mines

Table 1.1. World production of nonferrous base and rare metals, 1980

	1,000 metric tons
Aluminum	15,368.0
Copper	7,953.0
Zinc	5,806.0
Lead	3,518.0
Tin	248.0
Silver	10.6
Gold	1.2
Platinum	0.2

Source: *Minerals Yearbook*, 1980.

Table 1.2. Cash prices of nonferrous base and rare metals, April 1, 1982

	Price (US$ per lb.)	Price Index (tin = 100)
Lead	$ 0.28	0.04
Zinc	0.39	0.06
Copper	0.75	0.11
Aluminum	0.76	0.12
Tin	6.58	1.00
Silver	104.20	15.84
Platinum	4,608.48	700.38
Gold	4,768.90	724.76

Source: *New York Times*, April 2, 1982, p. D8.

lie between 11,500 and 16,000 feet above sea level. Virtually all of the Bolivian ore is lode ore, with veins averaging eight inches in diameter running through hundreds of feet of hard rock, requiring arduous and costly methods of mining. Until recently the altiplano has accounted for approximately one-fifth of the world's production. All other tin-bearing zones have been of far less significance than these two, although in the past two decades Russian, Australian, and Brazilian outputs have grown dramatically.

There are several areas that may be regarded as of second rank. Historically, two tin belts of Europe have been major ore sources, one running from Cornwall through Brittany and into the Iberian peninsula, and the other lying in central Europe in the Erzegebirge highlands of Saxony and Bohemia. The European deposits, worked for centuries, have now shrunk to relative insignificance, although there has been a recent resurgence of lode mining in Cornwall, with shafts being sunk to new depths in old mines. The major Australian deposits are located in Queensland, New South Wales, Tasmania, and the southwest of Western Australia. In Africa, large but unrelated deposits exist on the Jos

Table 1.3. Crustal concentration of nonferrous and precious metals

	grams per metric ton
Aluminum	83,000.0
Zinc	94.0
Copper	63.0
Lead	12.0
Tin	1.7
Silver	0.075
Platinum	0.046
Gold	0.0035

Source: Lee Tan and Yao Chi-Lung, "Abundance of Chemical Elements in the Earth's Crust and its Major Tectonic Units," *International Geology Review*, 12, 7 (July 1970): 782.

plateau of Nigeria, in a 600-mile stretch near the eastern border of Zaire running from the upper Congo River south to the Zambian border and spilling over into Rwanda, and in South Africa. In the early 1970s a major new source of tin was discovered and developed in the upper Amazon valley of Brazil, propelling that nation into the ranks of the principal tin producers. Finally, there are deposits of unknown richness in eastern Siberia and the far eastern provinces of the Soviet Union, all in remote and climatically harsh areas, which are yielding an estimated 35,000 metric tons of tin annually, despite extremely costly conditions of production.[3]

Other ore deposits, small and not associated with any extensive belt, have been found and worked in sites scattered around the world, notably in Mexico, Vietnam, Argentina, and Japan. Despite extensive prospecting, no significant tin deposits have ever been found in the United States.

In 1980, 58 percent of the world's tin ore production, measured by tin content, was accounted for by China, Malaysia, Indonesia, Thailand, Burma, and Laos in the Southeast Asian tinbelt; Bolivia added 11 percent; Australia, Nigeria, Zaire, South Africa, Rwanda, Brazil, and the Soviet Union in the secondary tin regions accounted in total for another 26 percent; and in the two European zones, the United Kingdom, Spain, Portugal, East Germany, and Czechoslovakia produced 2 percent. As can be seen in table 1.4, countries in the major and second-rank tin zones have supplied virtually all of the world's output of ore for many years.

Tin is found in various types of deposits, and a number of mining techniques are in use. Virtually all tin is extracted from the mineral cassiterite which, when chemically pure, is 78.6 percent tin in metallic content. Negligible amounts of tin are also obtained from stannite, cylindrite, frankeite, and certain rare sulfide minerals.

Primary or lode tin ore is found only in granite masses or metamorphosed granitic rocks. Geologists think it likely that veins of tin ore were most commonly formed by intrusion into preexisting fissures of either liquid or gaseous magma under pressure, and the process was certainly such that the tin is typically compounded with numerous other substances. Lode mining of primary tin is therefore characterized by hard rock mining in which thin veins are followed for long distances; and the ore extracted from primary deposits is often complex, carrying numerous impurities which make it difficult to smelt and refine.

William Fox has described lode mining in Bolivia in a vivid passage. "The landscape is arid, bleak, cold, sometimes very beautiful and always without humanity. The air is thin; heavy physical effort is difficult. . . . Blasting, often very reckless, is almost continuous. Temperature and humidity levels in the lower sections of a mine may be almost beyond toleration."[4] In the shafts in which Bolivian workers must stoop or crawl, the heat is such that one miner works at the face while another constantly hoses him with water; and every five minutes the two exchange places. Throughout the history of the industry in Bolivia, high accident and fatality rates, a high incidence of silicosis among miners, and short careers in the mines with survivors crippled or debilitated by accident or disease have been common conditions.[5]

Secondary or placer deposits are, to date, yielding much more of the world's supply of tin ore than are primary veins, although geologists believe that vastly more of the earth's ultimately recoverable cassiterite and other tin-bearing ores will come from primary sources than from secondary ones. The secondary deposits consist of alluvial ores eroded from their original setting by action of water and eluvial deposits eroded by wind action. These deposits tend to be more readily accessible from the surface than lode deposits and are generally found in sandy or gravelly soil or in soft rock that can be mined far more easily than hard granite rock masses. Secondary ores are usually of greater purity and easier to smelt than the primary ores.

Secondary tin is mined in several ways. In contrast to the harsh and difficult conditions of Bolivian lode mining, it has been said of the availability of the alluvial ores of Malaysia that "it would be difficult to find any stream having its source in the Main Mountain Range that did not contain in its bed a certain amount of tinstone."[6] Another impression of a part of the Southeast Asian belt can be drawn from the following nineteenth-century description of the Thai island of Phuket:

> The whole island is a gigantic tin mine. The granite of the hills is full of tin, the soil of the valleys is heavy with it. There is tin under the inland forests and tin beneath the sea. In search of tin the indefatigable Chinamen have transformed the scenery. The valleys have been turned inside out, the hills have been cut away, the sea has been undermined and the harbour has disappeared.[7]

Alluvial ores in streambeds can be and still are extracted by simple panning,

Table 1.4. World mine production of tin, by metal content of ore

Country	Long tons									Metric tons	
	1925–29 average	1935	1939	1941–45 average	1950	1955	1960	1965	1970	1975	1980
Malaysia (British Malaya)	56,837	42,374	51,725	26,442	57,537	61,244	51,979	63,670	72,630	64,364	61,404
Bolivia	37,169	25,007	27,211	40,644	31,213	27,921	19,407	23,036	28,944	28,744	27,272
Indonesia (Netherlands Indies)	33,266	20,184	27,755	17,326	32,102	33,368	22,596	14,699	18,761	24,391	32,527
Nigeria	8,319	6,299	9,427	12,275	8,258	8,158	7,675	9,547	7,833	4,652	2,500
Thailand (Siam)	8,204	9,876	17,325	6,914	10,364	11,023	12,080	19,047	21,435	16,406	33,685
China, P. R. (China)	7,085	9,035	10,422	6,282	7,500	18,000	28,000	25,000	20,000	22,000	14,600
Australia	2,830	3,130	3,500	2,778	1,854	2,017	2,202	3,849	8,689	9,678	11,364
United Kingdom	2,658	2,050	1,633	1,334	890	1,034	1,199	1,313	1,695	3,900	2,960
Burma	2,228	4,102	8,536	1,480	1,520	1,130	1,200	677	428	545	1,136
Republic (Union) of South Africa	1,174	622	482	500	643	1,283	1,276	1,671	1,979	2,643	2,800
Zaire (Belgian Congo)	967	5,301	9,663	16,765	13,464[a]	15,028[a]	8,636	6,324	6,356	4,160	3,000
Laos (Indochina)	691	1,309	1,470	675	49	253	397	284	679	522	350
Japan	625	2,197	1,700	1,122	326	896	842	837	780	655	540
Portugal	149	750	1,486	1,152	690	1,445	772	557	428	555	220
Namibia (Southwest Africa)	145	164	156	138	100	357	261	416	1,133	750	1,000
Spain	138	300	106	407	633	822	196	111	436	530	500
Swaziland	98	127	114	96	37	27	6	2	12	12	—
Uganda	98	397	354	275	192	68	32	178	120	117	50
East Germany (Germany)	32	26	300	562	191	669	720	1,000	1,000	1,120	1,600
Argentina	22	700	1,655	990	267	89	238	497	1,153	600	582
Tanzania (Tanganyika)	15	145	224	170	97	41	138	255	104	11	12
Zimbabwe (Southern Rhodesia)	2	12	451	172	69	208	642	510	600	600	960
Mexico	—	621	289	299	440	605	372	503	525	457	20
Cameroon (French Cameroun)	—	217	243	193	67	85	65	40	35	19	8
Italy	—	—	256	133	—	—	—	—	—	—	—
Peru	—	—	47	66	38	—	6	49	101	150	1,000
Canada	—	—	—	308	356	220	278	168	118	283	264
Brazil	—	—	—	90	180	146	1,556	1,810	3,553	5,400	8,000
France	—	—	—	3	81	450	21	447	332	51	—
USSR	—	—	n.a.[b]	n.a.[b]	n.a.[b]	10,300	16,000	23,000	27,000	30,000	36,000

Rwanda (Ruanda-Urundi)	—	—	—	—	—	—	1,277	1,424	1,300	1,250	1,200
Czechoslovakia	—	—	—	—	—	200	200	n.a.	163	176	180
Burundi	—	—	—	—	—	—	—	17	48	100	10
Vietnam (North Vietnam)	—	—	—	—	—	—	—	—	—	250	370
All others	31	92	72	33	146	160	107	177	130	104	133
	163,408	135,037	176,602	139,624	169,304	197,247	180,376	201,115	228,500	225,195	246,247

a. Includes production of Ruanda-Urundi.

b. The Bureau of Mines estimated USSR mine production at an average of 2,500–3,000 long tons for 1934–43; 4,000 LT for 1944; 4,500 LT for 1945; 5,000 LT for 1946; 6,500–7,500 LT for 1947; 7,500–9,000 LT for 1948; 9,000–10,000 LT for 1949; and an average of 8,900 LT for 1950–54.

Source: Minerals Yearbook, various issues.

known as dulang washing throughout Southeast Asia. Other secondary deposits are mined in open cuts and shallow shafts. Ore-bearing ground is dredged from riverbeds, extensive low-lying alluvial deposits, and, in recent years, from offshore seabeds. Mining methods differ widely in scale, capital or labor intensity, and level of technology. Given the various types of terrain in which tin ore is found, no generalization can be made as to the technical or economic superiority of any particular mining technique.

The initial processing of the ore—concentration to extract the cassiterite from the material removed from the deposit—is invariably carried out at the mine site. Concentration is a technologically simple process, although one which may require considerable labor skill and care, involving crushing or grinding the ore and subsequent separation of the cassiterite, which is a much heavier substance than the rock, sand, or gravel in which it was embedded. Some separation is still done by hand washing. The traditional and still most common method is sluicing. Sluicing requires a sloping trench, known in Malaysia as a *palong*, in which ore mixed with water can be run over a series of baffles. Thus, the ideal site for an open pit or shaft mine is on a hillside, where a ground sluice can be constructed, the water can be carried away in valley streams below, and the tailings can be dumped and left to pile up near the bottom of the sluiceway, well below and out of the way of the mine. When mines are not so felicitously situated, it is necessary to construct long trestles for the sluiceways, to convey the ore over considerable distances and heights to the sluice heads, to pump large amounts of water both for sluicing and to remove the excess from the mine site, and to keep the working area cleared of accumulated tailings. These processes can involve widely varying combinations of labor, mechanical pumps and conveyors, and earth-moving equipment.

In Malaysia, gravel-pump mining is the most common method of working open cuts. In a gravel-pump mine, the ore-bearing earth is removed and crushed by pick-and-shovel labor, by mechanical shovels and grinders, or by hydraulicking—high pressure jets of water blasted against the working face—and is then conveyed to the base of a pump which mixes it with the proper amount of water and elevates the mixture to head of the *palong*. The operation requires knowledgeable and conscientious on-site management as well as skilled and careful labor even if only small amounts of simple machinery are used. A major barrier to the entry of European mining companies into the Malayan tin fields before the introduction of dredges in 1914 was the difficulty European managers had in operating labor-intensive mines efficiently with Chinese labor. The problems and frustrations experienced in organizing and supervising the labor force were in part the result of differences in language and culture, but were deliberately exacerbated by local Chinese mine owners and secret societies.[8]

Dredging can be carried out only on certain types of secondary deposits. Such a deposit must be low-lying, so that water can be introduced into and retained in the "paddock" or pool in which the dredge floats; it must be free of boulders and large stumps that could jam or damage the dredge buckets; and it must be extensive enough to justify the capital investment involved in constructing and installing a dredge that at the present time will cost several million dollars. When conditions are suitable, however, dredging has proved to be an extremely efficient form of alluvial mining in terms of volume of ore body handled and saving of labor. Further, dredges can mine deposits that because of depth, water conditions, or soil stability, are inaccessible by any other technique. Dredging is currently limited to Southeast Asia and accounts for less than 25 percent of the world's tonnage. In recent years, very large ocean-going dredges have been in operation off the Thai and Indonesian shores.

In modern dredging operations, concentration is performed on board the dredge by shifting the material on jigs rather than by running it over sluiceways. The tailings from dredges are dumped out the rear, and the dredge mines by digging up the ground in front of it. Thus the dredge moves in its own paddock, and tailings do not pose the disposal problem faced in other methods of tin mining. Jigs and screens are now being used in some open pit mines to supplement concentration by sluicing.

Alluvial concentrates, in the form in which they are shipped from the mine to a smelter, contain from 60 to 76 percent tin metal. As a result, the differences in transportation costs of these tin concentrates and tin metal have been small and unimportant in determining the locations of smelters; smelting is done in producing and consuming countries alike. Lode ore is often of much lower metal content. Some concentrates leave Bolivian mines with as little as 20 percent containing tin.

There are three basic stages to the smelting process, although numerous variations in each stage and combinations of repeat treatments are practiced. The first step, known as roasting, is heating the concentrate to approximately 600° C, at which temperature sulfur, arsenic, and antimony oxidize and can be removed by mechanical means. Other impurities, including metal sulfides, bismuth, zinc, copper, lead, and silver, must be removed by special roasting with certain chemicals added to the charge or by acid leaching.

The second step in the smelting process, making use of the fact that tin metal has a low melting point relative to the other materials that are likely to be left in the roasted ore, is heating the material in a blast or reverberatory furnace to 1,200° C, a temperature at which the tin metal will melt and can be drained out of the furnace. If the slag from the first melting is rich enough, it may be added to the next charge or be remelted by itself.

Tin that has gone through this second step in the process may still contain some other metals or chemicals, most often iron or some compound thereof, which must be removed in a third step, known as refining. There are several refining processes. One of the oldest is "poling," or plunging a bundle of green wood into a vat of molten tin or, in a more modern version, injecting a blast of compressed air or oxygen. Such treatment causes the tin to bubble violently and the iron to oxidize. The oxidized iron rises to the surface as a dross and can be skimmed off. Other refining methods involve magnetic separation of iron and its compounds, further chemical leaching, and electrolysis.

The intensity of the treatment of slag and the appropriate refining process, as well as the initial roasting or leaching, depend on the nature and extent of the impurities in the ore being treated as well as on the level of purity desired in the final product. For this reason, smelters have been constructed to specialize in the treatment of particular ores, and this in turn has contributed to the establishment of channels of trade that have remained fixed for decades in which certain smelters take ores only from certain countries and the countries, in turn, supply ore to only one or a few designated smelters. Differences in smelting have also led to tin metal being stamped and sold by brand names reflecting both smelter of origin and degree of purity, and to certain smelters producing more than one brand and grade of tin.

Modern smelters vary in size from annual capacities of a few hundred tons of metal produced to the 60,000 metric ton per annum capacity of the Straits Trading Company's smelter in Butterworth, Malaysia. The basic unit of most modern smelters is a reverberatory furnace with maximum hearth dimensions of approximately 30 by 12 feet. Larger smelters have banks of such furnaces. Refining equipment varies greatly from one smelter to another, with the most costly and complex being in smelters in Great Britain and the United States that are designed to process Bolivian ores, as well as in Bolivia itself. Although the details of operation of individual smelters are usually not made public, there do not appear to be economies of scale in production beyond the efficient output of a modern reverberatory furnace. Large furnaces typically handle eight- to twelve-ton charges that are smelted for ten to twelve hours. The process is always a batch one, with slag removed after each smelting. Resmelting of slag normally requires sixteen to eighteen hours. Since the charge is a mixture consisting of approximately 80 percent tin ore, which will yield roughly 70 percent tin on smelting and 20 percent on resmelting, it would appear that a smelter with a 1,000 ton annual capacity is one capable of operating a single large reverberatory furnace at an efficient level of intensity.

Recent construction of a number of new smelters, ranging in size from a unit of 120 ton annual capacity established in South Africa in 1967 to Thaisarco's 30,000 ton capacity smelter put on stream in Thailand in 1965, reinforces the impression that technical barriers to entry by plants of widely varying sizes are slight. Table 1.5 shows a steady growth in the number of nations possessing smelters since the 1920s.

There are virtually no end products, other than some pewter and high-quality organ pipes, that make use of tin metal in its pure state. Tin is used as a coating on tinplated steel, and as a component of an alloy in practically all other uses. Bronze is an alloy in which 10 to 30 percent tin is mixed with 70 to 90 percent copper, with small amounts of zinc often added. Tin-bearing metals, such as babbitt metal and white metal, vary widely in composition. The exact mix of many bearing alloys is regarded as a trade secret; but the range runs at least from high-quality aircraft bearings consisting of 93 percent tin, 3.5 percent antimony, and 3.5 percent copper at one extreme down to bearings with 5 to 6 percent tin, 7 to 87 percent lead, and the remainder antimony. Type metal is 5 to 20 percent tin, 25 to 30 percent antimony, and the balance lead. Fine or tinman's solder is two-thirds tin and one-third lead, while plumber's solder is made with exactly reversed proportions. Pewter is traditionally regarded as 80 percent tin and 20 percent lead, but proportions vary. Table 1.6 indicates the various uses of tin by the United States, the major consuming nation.

Table 1.6 shows a pronounced decline in tin consumption in the United States. Although both 1975 and 1980 were years of recession, the long-run drop is largely attributable to the substitution of electrolytic tin plating of steel for an older hot-dip process that consumes as much as three times more tin per unit of area plated than does electrolysis. Increased use of canned foods throughout the world since World War II has just about offset the effect of this replacement on global consumption of tin. Table 1.6 also reflects the widespread substitution of aluminum for tin in foil, a similar substitution of aluminum and plastics in the manufacture of collapsible tubes, development of new solders using less tin, and widespread innovative activity in the bearing field which has led to the displacement of babbitt metal by new antifriction alloys and by improved roller and ball bearings.

Purity of the metal is very important in many uses of tin. Impurities in the tin used for tinplating cause uneven coating of the steel, resulting in both excessive use of tin and incomplete coverage. Very slight contamination by zinc, aluminum, iron, copper, or arsenic, all found in tin ores, has severe effects on the quality of solder, causing it to be brittle, crumbly, coarse, or gritty. Pewterware used for eating or drinking and the coatings of food containers must be free of arsenic and lead poisons. On the London Metal Exchange, the minimum purity of tin acceptable for delivery under a standard contract is 99.75 percent. Tin of less than standard purity can be purchased at a discount. For some uses, high-purity metal of no less than 99.85 percent tin must be specified; and in recognition of growing demands for higher purity, the London Metal Exchange in 1974 authorized trading in a "high grade" contract of no less than 99.85 percent as well as in the standard contract.[9] Some alluvial ores approach the standard after smelting and require little more if any refining, other than "poling" or air injection. Smelted Bolivian tin, on the other hand, must be subjected to extensive and costly refining operations before it can be brought up to the standard grade. An important repercussion of the need for purity is that uses for tin reclaimed from scrap are limited, and reclaimed tin is regarded as unsuitable for use in tinplating, especially electrolytic plating where even minute amounts of foreign substances can lead to tiny "pits" or gaps in the plating that may allow the can wall to corrode or permit leaching into the contents.

Reclaimed or so-called secondary tin (not to be confused with secondary ore

Table 1.5. World smelter production of tin metal

Country	Long tons									Metric tons	
	1925–29 average	1935	1939	1941–45 average	1950	1955	1960	1965	1970	1975	1980
Malaysia (British Malaya)	88,855	60,479	81,536	35,355	68,747	70,632	76,130	72,469	90,049	83,070	72,000
United Kingdom	45,800	29,100	37,400	32,998	28,500	27,241	26,286	16,494	21,687	11,520	6,500
Indonesia (Netherlands Indies)	14,749	11,221	13,941	9,983	405	1,572	1,977	1,189	5,108	17,825	29,100
China, P.R. (China)	7,080	9,700	10,850	6,282	7,000	18,000	28,000	25,000	20,000	22,000	14,600
West Germany (Germany)	3,444	2,042	3,600	717	586	280	782	1,427	1,176	1,306	2,400
Australia	2,952	2,837	3,300	2,809	2,014	2,004	2,254	3,179	5,129	5,254	4,690
Netherlands	1,000	15,600	14,600	—	21,027	26,566	6,393	18,114	5,843	—	1,500
Belgium	720	4,000	3,100	—	9,512	10,432	7,947	4,232	4,190	4,562	3,000
Japan	606	2,036	2,000	1,886	389	1,030	1,260	1,610	1,356	1,212	1,319
Thailand (Siam)	113	—	n.a.	1,508	2	—	—	5,548	21,692	16,630	33,500
Portugal	2	1	30	1,542	209	1,018	601	603	386	409	500
Mexico	n.a.	—	90	263	290	357	365	459	525	1,000	400
Zaire (Belgian Congo)	—	1,588	2,124	11,054	3,238	3,034	2,532	1,815	1,374	575	300
Italy	—	241	146	63	—	—	—	—	—	—	—
Argentina	—	591	1,080	632	253	99	—	—	—	120	120
Norway	n.a.	454	283	—	—	—	—	—	—	—	—
United States	—	—	—	22,172	33,118	22,329	14,026	3,098	n.a.	6,500	3,000
Republic (Union) of South Africa	—	—	—	745	718	779	622	962	603	780	1,100
Spain	—	—	138	386	1,597	608	464	1,787	3,846	5,249	4,500
Canada	—	—	—	308	356	—	—	—	—	—	—
Laos (Indochina)	—	—	—	200	—	—	—	—	621	—	—
Brazil	—	—	—	80	118	1,184	1,311	1,753	3,156	5,400	10,500
Zimbabwe (Southern Rhodesia)	—	—	—	57	80	22	611	494	600	600	934
Peru	—	—	—	55	38	—	—	—	—	—	—
Bolivia	—	—	—	—	392	107	1,002	3,415	301	7,133	18,191

East Germany	—	—	—	—	191	605	600	1,200	1,100	1,120	1,600
Morocco	—	—	—	—	—	8	10	12	12	12	—
USSR	—	—	n.a.[a]	n.a.[a]	n.a.[a]	10,300	16,000	23,000	27,000	30,000	36,000
Nigeria	—	—	—	—	—	—	—	9,321	7,942	4,677	2,000
Czechoslovakia	—	—	—	—	—	—	—	—	—	108	—
Vietnam (North Vietnam)	—	—	—	—	—	—	—	—	—	200	350
	165,321	139,890	174,218	129,095	178,780	198,207	189,173	197,181	223,696	227,262	248,104

a. The Bureau of Mines estimated USSR mine production for 1939–54 at levels shown on table 1.4. Soviet smelting output was assumed to be roughly equal to mining output.

Source: *Minerals Yearbooks*, various issues.

Table 1.6. Consumption of tin in the United States, by finished product (metric tons)

Use	1935	1950	1955	1960	1965	1970	1975	1980
Tinplate	27,290	35,380	33,459	33,238	30,064	25,127	18,869	16,346
Solder	16,644	27,460	22,230	18,278	21,762	20,184	15,075	15,618
Collapsible tubes and foil	5,177	1,666	923	915	1,060	860	538	526
Babbitt metal	5,152	6,409	4,371	3,621	3,666	3,241	2,605	2,380
Bronze and brass	4,830	20,594	19,712	15,336	16,876	13,666	7,555	7,478
Chemicals including tin oxide	4,520	1,716	1,692	1,932	2,833	3,125	3,998	W[a]
Tinning	2,082	2,976	2,613	2,035	2,487	2,143	1,896	2,577
Terne metal	1,064	952	323	469	578	269	352	—
Type metal	1,024	1,980	1,487	1,431	1,300	1,109	265	W[a]
Pipe and tubing	953	440	156	78	66	26	—	—
Galvanizing	620	—	—	—	—	—	—	—
Alloys (misc.)	482	1,707	486	401	476	590	611	134
White metal	397	1,217	1,179	1,542	913	1,274	2,217	914
Bar tin	395	1,434	1,579	1,110	1,832	1,052	637	486
Tin powder	—	—	—	—	—	1,059	850	1,109
Other	543	533	183	174	98	112	322	8,794
	71,173	104,464	90,483	80,560	84,011	73,837	55,800	56,362

a. Withheld to avoid proprietary disclosure. Included in "Other."
Source: Minerals Yearbook, various issues.

deposits) is the final source of the metal that needs to be considered here. Worldwide data on production and use of reclaimed tin are unavailable, in large part because the bulk of such metal never reappears on the market as tin but is worked directly into other materials. *Minerals Yearbook* reported that 88 percent of the secondary tin consumed in the United States in 1975 was used as a constituent in alloys of bronze, brass, solder, bearing metal, and type metal, and as an element in chemical compounds.[10]

In 1971, *Minerals Yearbook* noted that there were five firms in the United States engaged in the recycling of tin, with plants in eleven states.[11] The 1975 edition reported two new entrants.[12] The pattern of recycling in the United States is shown in table 1.7.

The secondary tin recovered at detinning plants is actually a by-product. Because of the shape of tin cans and the nature of other steel-fabricating operations, a large amount of tinplate in the form of clippings and clean scrap is available. The steel in such scrap is far more plentiful and valuable than the tin. However, scrap steel cannot be remelted without first removing any tin coating, since tin alloyed with any ferrous metal has extremely deleterious effects on the latter's working properties.

In 1969, *Minerals Yearbook* cited an estimate by the ITC that 80 percent of the free world's secondary tin was consumed by five countries, with the United

Table 1.7. Tin recovered from scrap processed in the United States, by form of recovery (metric tons)

Form of recovery	1975	1980
Tin metal:		
At detinning plants	1,689	1,677
At other plants	228	26
Bronze and brass:		
From copper-base scrap	6,913	10,352
From lead- and tin-base scrap	42	50
Solder	4,414	4,423
Type metal	877	525
Babbitt metal	569	378
Antimonial lead	442	856
Chemical compounds	670	321
Miscellaneous	12	30
	15,856	18,638

Source: Minerals Yearbook, 1976 and 1980 editions.

States accounting for 35 percent, the United Kingdom for 18 percent, West Germany for 12 percent, Austria for 9 percent, and Australia for 6 percent.[13]

In summary, the following properties of tin and its processing appear to have been highly significant in shaping the development of the worldwide industry. Recoverable tin-bearing ore is known to exist in only a few countries of the world. The natural conditions under which tin ore deposits are found vary greatly, and therefore so do economically efficient methods of tin mining, sizes of mines, and capital and labor requirements. The product shipped from mine to smelter usually has such a high concentration of tin metal that there is no preponderant advantage in locating smelters either near their supplying mines or in the vicinity of consuming centers. Smelters of greatly differing sizes appear to be capable of competing effectively with one another, although some are specially designed to treat ores from particular regions or countries. Tin metal is not an entirely homogeneous commodity, but is sold in varying degrees of purity and under specific brand names as well as by grades. Tin is not in demand for its own sake but rather as an input into other materials and commodities. Tin scrap is an important supplement to the world's supply of the metal, but is limited in its uses.

2. Tin in the Economies of Producing Nations

At the Sixth Special Session of the United Nations' General Assembly in 1974 two resolutions were adopted, one entitled "Declaration on the Establishment of a New International Economic Order," and a second with the title "Programme of Action on the Establishment of a New International Economic Order." Debate on the New International Economic Order (NIEO) has called worldwide attention to a set of issues surrounding the role of primary commodities in the contemporary international economic scene. A 1977 Joint Economic Committee study, excerpted below, encapsulates the strategic influence attributed to commodity production and trade by proponents of the NIEO:

> The NIEO is not solely or even principally concerned with the export of industrial products from the developing world. For many Asian, African and Latin American nations industrialization is still a relatively remote goal. . . .
>
> Just as important in the NIEO scheme of reforms are changes in current trading practices for commodities and raw materials. Many countries are dependent on the export of one or two crops or specific raw materials for the bulk of their foreign exchange earnings.
>
> The developing countries are concerned with both fluctuations in their export earnings from, and the long-term trend in the terms of trade for, their commodity exports. Natural disasters, the entry of new suppliers, and the business cycles of the developed nations can all have a devastating effect on export earnings. Fluctuations in export earnings cause short-run economic and political instability and can seriously disrupt long-run economic plans. . . .
>
> Much of the concern with commodity prices dates to the early work of Raul Prebisch. . . . Prebisch argued that low price and income elasticities for raw materials would severely limit income growth for countries dependent on their export. . . . Prebisch argued that the production of raw materials in developing countries was characterized by relatively competitive markets while the production of industrial goods was concentrated in a number of large oligopolies. The differences in market structure would tend to bias the gains from trade in the direction of the developed countries. Prebisch also felt that highly organized labor in the industrial countries kept at least part of any productivity increases in terms of higher wages. With competitive markets and weak unions, productivity increases in developing countries were more often reflected in lower prices than in higher wages. The combined result was slow growth in the developing world with the gains from trade going almost entirely to the already well off, developed countries. . . .
>
> The concept of commodity agreements or cartels as the new answer to the economic problems of the developing countries seems to rest on two unstated assumptions. The first suggests that raw materials are exclusively or princi-

Table 2.1. Net exports of tin[a] by developing, centrally planned, and industrial market economies,[b] 1980

Nation	Amount (metric tons)	Percentage of total
Developing nations		
Malaysia	60,961	34.1
Thailand	33,955	19.0
Indonesia	25,569	14.3
Bolivia	20,785[c]	11.6
Singapore[d]	13,428	7.5
Nigeria	2,719	1.5
Zaire	2,451	1.4
Brazil	2,158	1.2
Burma	2,087	1.2
Rwanda	1,270	0.7
Argentina	759	0.4
Peru	703	0.4
South Africa	585	0.3
All others	134	0.1
All developing nations	167,564	93.7
Centrally planned economy		
China, P. R.	2,574	1.4
Industrial market economy		
Australia	8,817	4.9
World total	178,955	100.0

a. Net exports are calculated as the sum of exports of tin-in-concentrates, weighed by tin content, plus exports of tin metal, minus the sum of imports of tin-in-concentrates and tin metal. This figure represents the origin of tin metal moving in international trade weighted by tonnage rather than value.

b. Nations are grouped according to classification in World Bank, *World Development Report 1981*.

c. Preliminary estimate.

d. This figure includes re-export of concentrates smuggled into Singapore, most presumably coming from Malaysia, Indonesia, Thailand, and Burma. Singapore reported legal imports of only 2,328 metric tons of tin-in-concentrates for 1980.

Sources: International Tin Council, *Monthly Statistical Bulletin* (March 1982); International Tin Council, *Tin Statistics, 1970–80*.

pally exported by developing countries. The second implies that the production and export of raw materials are at least roughly distributed in the developing world according to the need for further economic assistance. Neither assumption appears to be true.[1]

The above passages have been reproduced at such length because they summarize clearly and perceptively many of the matters that must be discussed in describing the role of tin, as one of these primary commodities, in the economies of both producing and consuming nations.

The distribution of world tin production has been reviewed in chapter 1. Insofar as world trade is concerned, developing countries do predominate in the export of tin, as shown in table 2.1, since Australia consumes rather than exports much of its production, and other industrialized producers such as the United Kingdom and Japan consume far more tin than they produce and are large net importers. The very large amount of tin produced by the Soviet Union does not appear in any export figures.

In terms of other economic characteristics of commodities regarded as crucial in the NIEO debate, the picture as it concerns tin is not so clear. The remainder of this chapter examines the extent of dependence of the economies of producing nations on tin, and the changing role of tin in the economies of these nations. The following chapter discusses the degree to which tin is a necessity for the consuming industrialized nations.

Other features pertinent to the issues raised by the NIEO, including the international tin agreements and the role of the ITC; elasticities of demand for and supply of tin; fluctuations in tin prices, production, and revenues; and the implications for conduct and performance of various features of market structure throughout the world tin industry, are treated in subsequent chapters.

Tin in Prewar Malaysia and Bolivia

It would be difficult to exaggerate the historic importance of tin in shaping both economic and political development in Malaysia and Bolivia. The experiences of these two nations provide archetypal examples of export-dependent economic history shaped by the influence of a single commodity.

Malaysia. Yip Yat Hoong, one of Malaysia's leading economists, has emphasized the importance of tin in his statement that "the early economic development of the Malay States depended *entirely* on the development of the tin mining industry, not in terms of supplying a raw material for domestic industries but in terms of earning foreign exchange."[2] Yip's use of the word *entirely*, which has been italicized here for emphasis, does not seem exaggerated or inappropriate. The early history of tin mining in the Malayan peninsula has been exhaustively researched and described in Wong Lin Ken's detailed study *The Malayan Tin Industry to 1914*;[3] and Wong's interpretation of this history bears out Yip's assertion.

Arab writers of the ninth century referred to the west coast of the Malay peninsula as a source of tin; and at the time of the Portuguese conquest of Malacca in 1511, Malay coins made of tin were in circulation. The Dutch, who captured Malacca from the Portuguese in 1641, established stations on the mainland of the peninsula to engage in trading for tin.[4] Wong cautioned that knowledge of the very early history of Malayan tin mining is "fragmentary and sometimes conflicting," but that a Dutch report of 1790 noted that the largest mine operators in the states of Perak and Selangor at that time were Chinese.[5] Until 1850, most Malayan tin was exported to and consumed in other countries of Asia, with the bulk of the exports going to India and China. Tin had only limited uses in Asia. According to Wong, principal uses were in the manufacture of oil lamps and decorations, for coins, for bronze that in turn was used mainly in musical instruments and weaponry, and, in China, for tin leaf that was burnt as sacrificial offerings.

As the nineteenth century progressed, however, European and American demands for tin rose to levels that the mines of Cornwall could not accommodate, and by 1853 Great Britain had repealed import duties on tin ore and was importing appreciable amounts. Wong has calculated that over the years 1844–48, the Straits Settlements exported 12,228 tons of Malayan tin, of which 4,099 tons went to Europe and 1,072 tons to the United States; in the period 1869–73, total tin exports from the Straits Settlements had risen to 50,433 tons, with 21,884 tons going to Europe and 13,907 tons to the United States.[6] The principal impetus to this expansion was the development of a canning industry which belatedly followed the original use of canned foods by the British army and navy during the Napoleonic Wars.

The expansion of tin mining in Malaya from the 1840s through the remainder of the nineteenth century was primarily in the hands of Chinese miners and laborers. According to Wong, the early Chinese miners came without exception on the invitation of Malay rulers. At the time, he continues, Malay society in the tin-bearing states was in a condition of degeneracy and instability so severe that political survival hinged on pressing all Malay men not needed for subsistence agriculture into the armed forces while, at the same time, exploiting tin resources for immediate revenue. The result was a flood of Chinese immigration that transformed the ethnic composition of the peninsula. Further, by the early 1870s, the Chinese community was large and wealthy enough to exert a strong influence on the British authorities in the Straits Settlements. Wong's account of the period prior to 1874 is a colorful, albeit bloody, one of disputed mining claims; open warfare between rival Chinese triads, or secret societies; massacres of Chinese by Malays; piracy on the rivers and banditry on the trails over which the concentrated tin ores were carried (frequently with the connivance of Malay chieftains); civil wars; and wars between Malay states over disputed tin fields. The final result was British intervention and rule through the device of resident advisors, a step urged by the Chinese mine owners.

Bolivia. H. S. Klein noted that the replacement of silver mining by tin in Bolivia, beginning in the 1890s, led to such a profound change in Bolivia's economic base that the new century has since been labeled as Bolivia's *siglo de estaño*, or the century of tin. Annual production rose from around 1,000 tons in 1890 to 9,000 tons in 1900, and to over 15,000 tons in 1905.[7] Bolivian production peaked at 46,338 long tons in 1929, representing 25 percent of the world's total output for that year.[8]

To put these output figures in proper perspective with respect to their bearing on the economic life of Bolivia, it must be recognized that in terms of population Bolivia is by far the smallest of the major tin producers. Writing in 1966, C. H. Zondag gave the following brief description:

According to current estimates, about 2.3 million people are economically active, of which 1.4 form part of the labor force. A relatively high percentage (40 per cent) of this are women. . . .

The basic structure of the Bolivian economy is relatively simple. Although most of the population is engaged in agriculture, about 25 per cent of the country's exports consists of minerals, with tin accounting in value for about 70 per cent of all mineral exports. To put it differently, less than 2 per cent of the population—the miners—produce almost all of the country's foreign exchange income.[9]

The Bolivian situation conformed to the worst features of the NIEO's dependency scenario in still another important respect in that, prior to 1952, 85 percent of the nation's tin mining was in the hands of three family firms, the Patiño, Aramayo, and Hochschild companies. J. M. Malloy described the view held in the 1940s of these tin barons by the Movimiento Nacionalista Revolucionario (MNR), the political party that subsequently led the revolution of 1952 and then nationalized the mines of the three firms. "The MNR held that, due to the Big Three tin magnates' economic power, they formed a state within the state," Malloy wrote. "This superstate ruled Bolivia through an upper class which the tin barons bought and controlled. This group was known as *La Rosca*, an oligarchy which exploited Bolivia for its benefit and that of its imperialist cronies thereby impoverishing the nation, alienating its patrimony, and submerging its national culture."[10] R. J. Alexander's similar assessment was that "the M.N.R. was undoubtedly motivated in its opposition to the Big Three tin-mining companies by its belief that they had become a State within a State. Controlled from abroad, and with little or no say by Bolivians in the determination of their policies, these companies had a life-and-death grip on the country's economy, and particularly upon its foreign trade.[11]

In another study, written slightly earlier than his previously cited essay, Malloy added his own observation of the overall effect of tin on twentieth-century Bolivia, an evaluation quite consistent with the views of Klein and Zondag, commenting, "The development of the tin industry eventually detracted from and curbed the development of other productive sectors. Finally, it began to dominate all aspects of the country's development, not only as a focal point of economic organization and integration, but also as a primary cause of negative economic effects which led to stagnation in all other spheres of local activity."[12]

Malloy described this as the phenomenon of "Two Bolivias," which he depicted as "based on the extraction and shipment of tin abroad" on the one hand, and on the other as "an agricultural system based either on subsistence farming or on very low-level, regionalized commercial activity." The tin system was "relatively modern" while the agricultural system was "semi-feudal." The result, according to Malloy, was "a hitherto unknown wrenching apart of the country's two economic pivots."[13]

The economic segmentation was heightened by cultural, social, and linguistic

Table 2.2. Value of tin exports of major producing countries in millions of U.S. dollars

Country	1950	1955	1960	1965	1970	1975	1980
Malaysia	$125.2	$123.9	$165.9	$284.4	$305.2	$502.2	$1,151.2
Bolivia	63.3	57.3	42.9	93.0	108.1	182.3	378.2
Thailand	18.0	20.6	25.4	56.2	77.3	106.2	554.2
Indonesia	49.0	59.8	54.4	39.5	48.7	184.6	429.8
Australia	—	—	—	—	4.8	16.7	149.0
Zaire	5.5	6.0	4.7[a]	18.2	20.4	21.2	41.2
Brazil	—	—	—	—	4.1	24.1	64.2
Nigeria	16.9	16.4	16.8	41.3	47.3	28.1	45.7
Total	$278.0	$284.0	$310.1	$532.6	$615.9	$1,065.4	$2,813.5

a. Excludes Katanga and South Kasai.
Sources: International Monetary Fund, *International Financial Statistics*, various issues; International Tin Council, *Statistical Yearbook*, various issues; International Tin Council, *Monthly Statistical Bulletin* (March 1982); United Nations, *Yearbook of International Trade Statistics*, various issues.

barriers that sharply divided the large Indian majority, including the workers in the nation's tin mines, from an ethnically European, Spanish-speaking upper- and middle-class minority, and from the *cholos*, or those of mixed Indian and European blood, who performed the more menial tasks of wage labor and small-scale trade in the urban areas.

Tin in the Postwar Economies of Producing Nations

Whatever may be the relevance of a commodity-based theory of economic determinism to the economic history of Malaysia and Bolivia, it is clear that the conditions assumed in that theory have disappeared in all of the tin-producing nations since the end of the Second World War. The postwar tin industry has been transformed by fundamental changes in the economic and political structures of several of the tin-producing nations. By and large, the tin industries of the various countries have been shaped by events that they did little or nothing to generate or to influence. Malaysia, Indonesia, Nigeria, and Zaire all became independent nations. Bolivia's revolution of 1952 was accompanied by the nationalization of its large tin mines. Zaire and Nigeria were torn by civil wars. Indonesia came under and then overthrew the Guided Democracy of Sukarno. Malaysia found the social fabric of the new nation threatened by racial strife between ethnic Malays and Chinese. At the same time, the importance of tin in the economies of the producing nations tended to decline. Data are shown in several tables reflecting the postwar experiences of the seven present producing members of the ITC, plus Brazil. Similar figures are unavailable for the two other major producers, the Soviet Union and the Peoples' Republic of China.

Table 2.2 indicates the value of tin exports of the eight countries during the postwar period.

Table 2.3. Bolivian output of tin-in-concentrates 1926–67 (long tons)

Year	Amount	Year	Amount
1926–30 (average)	38,351	1957	27,796
1931–35 (average)	22,774	1958	17,729
1936–40 (average)	28,031	1959	23,811
1941–45 (average)	40,375	1960	20,219
1950	31,213	1961	20,664
1951	33,132	1962	21,800
1952	31,959	1963	22,246
1953	34,825	1964	24,199
1954	28,824	1965	23,037
1955	27,921	1966	25,522
1956	26,842	1967	27,283

Source: International Tin Council, *Statistical Yearbook*, 1968.

The uninterrupted increase in the value of the total of tin exports for all eight of these countries masks as much as it reveals. In large part, variations in physical output have been more than offset by a rising trend in price. But even when expressed in terms of monetary value, as in table 2.2, fluctuations are evident for Bolivia, Indonesia, and the African producers.

Bolivia. In the aftermath of the Bolivian revolution of 1952, that country's tin mining industry fell into near chaos. The effects of nationalization and longer-run forces are reflected in annual production figures (see table 2.3).

Bolivian output never again reached its 1929 peak of 46,338 long tons. Long before the revolution, the older and more productive mines were becoming depleted, and the Patiño, Aramayo, and Hochschild companies engaged in only limited exploration for new ore bodies.

The Bolivian tin industry was facing difficulties before 1952. The problems and costs involved in extracting Bolivian primary ore from thin granite veins and in smelting the resulting impure concentrates, as well as the harsh physical conditions of altiplano mining, have been noted previously. Nationalization, however, certainly worsened the situation during the next few years. The MNR was not ideologically committed to socialism and, on economic grounds, did not want to take over and run the mines.[14] However, given the extremely important role played by the tin miners and their unions in the revolution, and the party's own rhetoric that laid much of the blame for Bolivia's ills on the three big mining companies, the new government felt that on political grounds it had no choice but to nationalize the Bolivian properties of the Patiño, Hochschild, and Aramayo companies. Other medium-sized and small mines, accounting for only 15 percent of the country's total output, were left in private hands. From the outset the public mining corporation, Corporacion Minera de Bolivia, known as COMIBOL, failed to obtain effective managerial control over the nationalized mines. Through 1957, output held up fairly well. But employment soared, as did COMIBOL's deficits. At the end of 1951, total employment in mining in Bolivia stood at 24,000 with 55 percent employed in underground work. By 1956, total employment had risen to 35,660, and only 32.1 percent were employed inside the mines.[15] Work discipline was nonexistent or in the hands of the miners' unions. Pilferage of commissary stores was widespread. To add to the financial woes of COMIBOL, the Bolivian government had agreed to insistent demands from the United States that compensation be paid for the seized properties, and the cash drain of meeting these payments was a severe one.

In 1957, as part of a stabilization plan worked out with the United States and the International Monetary Fund, and as a prerequisite for developmental aid from these two, the Bolivian government did begin to reduce employment and tighten managerial control of the mines. The result was a sharp fall in output from 1957 to 1958, and the beginning of a slow recovery over the next several years. Of this stabilization program, R. S. Thorn has written, "The MNR was saved from economic disaster, but whatever economic sovereignty Bolivia had achieved by the nationalization of the mines was now surrendered. The dominant economic role played by the tin barons in Bolivian economic life was now replaced by the more benevolent, but equally foreign, International Cooperation Administration of the United States."[16]

Through the 1960s and mid-1970s Bolivian mine output grew slowly but steadily, to a peak of 33,624 metric tons of tin-in-concentrates in 1977. But production fell to 30,881 tons in 1978, and then to 27,781 tons in 1979 and 27,271 tons in 1980. Over these last years of the decade, Bolivia dropped from being the second largest tin producer in the world to fourth, as Thai and Indonesian outputs continued to grow. In part, the decline can be attributed to a fall in the tin content of the ores mined coupled with a lack of exploration and discovery of new bodies of ore, and to the deferred maintenance and modernization of COMIBOL's facilities. Bolivia's Medium Miners' Association has complained about the special tin tax, levied in 1981 at a rate of 53 percent of the difference between the price received for tin-in-concentrates and a standard or assumed unit cost. In January 1981 the assumed cost was US$3.93 per lb. for the medium mines, although the Medium Miners' Association estimated actual costs at between $4.10 and $5.50 per lb. The medium miners also noted that they are required to deliver their concentrates to a government-owned smelter that levies higher charges than those of foreign smelters treating Bolivian ore.[17]

But these longstanding problems were exacerbated by the political upheavals that have shaken Bolivia since the national election of June 1979 in which neither of the leading candidates received a large enough fraction of the votes to form a government. A president subsequently appointed by the Senate, Walter Guevara Arze, was overthrown by a coup led by Colonel Alberto Busch who, in turn, was replaced by Lidia Gueiler as a result of failure of the coup to obtain adequate domestic support. In July 1980, Señora Gueiler's government was

Table 2.4. Southeast Asian output of tin-in-concentrates, 1936–50 (long tons)

Country	1936–40 average	1945	1950
Malaya and Singapore	63,565	3,125	57,567
Thailand	15,243	1,775	10,364
Indonesia	33,226	1,050	32,102

Source: International Tin Council, *Statistical Yearbook*, 1968.

ousted in a violent takeover by General Luis García Meza. In August 1981 Colonel Busch and General Lucio Anez Rivero led a junta that forced García Meza to resign. Busch and Rivero were themselves forced into retirement in early 1982 by General Celso Torrelio, head of the government the junta had installed.

In the weeks following García Meza's coup, thousands of tin miners struck, destroying roads and arming themselves with dynamite in preparation for a strike-breaking assault by the military. The armed forces, in turn, reportedly bombed mining areas where resistance was strong and arrested a number of trade union leaders, including Juan Lechín, head of the Confederación Obrera Boliviana.[18] As a result of the coup, and the disorder and harsh repression that followed, United States' aid to Bolivia was suspended.

Shortly before García Meza was removed from office, an official of his government described the tin industry as in "a frank state of crisis and decadence."[19]

Indonesia. In Southeast Asia, the mines and smelters of Indonesia, as well as those of Thailand and Malaya, fell into Japanese hands early in the Second World War. By 1945, their output was minimal. This was in part because of deliberate wartime destruction by retreating armies of both sides; but also in at least equal part because the Japanese were able to produce far more tin than they could use even in wartime and were unable to ship any surplus output to their Axis partners in Europe under the Allied blockade. They therefore allowed much of Southeast Asia's mining and smelting capacity to fall into disrepair. Restoration of operations was rapid, as shown in table 2.4.

In both Malaysia and Thailand, tin output and earnings followed reasonably stable paths during the two decades after 1950. Indonesia, in sharp contrast, experienced extreme fluctuations in output, as displayed in table 2.5.

The precipitous fall in Indonesian tin output is only one of the economic reversals to strike that country during the period following the election of President Sukarno in 1955 and the era of Guided Democracy he launched in 1957. It should be noted at the outset that Sukarno came to power at a time when Indonesia faced grave economic and political problems, and it is impos-

Table 2.5. Indonesian output of tin-in-concentrates, 1955–80 (long tons)

Year	Amount	Year	Amount
1955	33,667	1968	16,940
1956	30,055	1969	16,542
1957	27,723	1970	19,092
1958	23,200	1971	19,767
1959	21,613	1972	21,766
1960	22,596	1973	22,648
1961	18,574	1974	25,630
1962	17,310	1975	25,346
1963	12,947	1976	23,418
1964	16,345	1977	25,921
1965	14,699	1978	27,410
1966	12,727	1979	29,440
1967	13,819	1980	32,527

Sources: International Tin Council, *Statistical Yearbook*, 1968; International Tin Council, *Tin Statistics, 1966–76* and *1970–80.*

sible to say what the consequences of some alternative courses of action and policies would have been.

The transition to independence was not an easy one for Indonesia, nor one marked by graciousness on either side. J. F. Cady, in his exhaustive and detailed study, *The History of Post-war Southeast Asia*, wrote, "The economic losses suffered by Indonesia during the course of the Japanese occupation stemmed more from deteriorating facilities and the disintegration of the social fabric than from physical destruction per se; in fact, more overt destruction was suffered during the postwar Dutch police actions than during the Japanese occupation."[20]

The Second World War, Cady continued, had been a time of almost complete erosion of the old social and economic patterns of colonialism and the rise of an enthusiastic and active Muslim nationalist movement among youth and newly formed workers' organizations. In the immediate postwar period, "the colonialist economic and administrative structure disintegrated before any effective alternative could be created."[21]

By 1955, a newly recruited and inexperienced corps of Javanese officials, widely resented and mistrusted on the other islands and among non-Muslims, had replaced the Dutch administrators and managers. In addition, economic problems were looming with the end of the Korean War boom, particularly population pressures and rice shortages in Java and inflation. The disruptive transfer of plantations from Dutch to Indonesian control was accompanied by widespread refusals to plant and by forcible eviction of squatters.

But however intractable the economic problems faced by Sukarno in 1955 may have been, there can be no question that under his regime they were heightened rather than solved. In 1957, Sukarno announced his program of

Guided Democracy; and in 1959, by presidential decree, he banned all political activities and announced abrogation of the existing constitution.

Actions taken by Sukarno's government with regard to the tin industry clearly had more political than economic content. In 1958, all Dutch properties were seized, a move which Cady described as "more of a gesture of nationalist assertion than a move toward socialization."[22] The Billiton Joint Mining Company, five-eighths of which was already owned by the Indonesian government, was dissolved and all tin mining operations were nationalized. Foreign investment in tin mining was forbidden. In 1962, the Indonesian State Mining Enterprises decided not to renew a contract under which Indonesian tin ore had been supplied to the United States' Texas City smelter, and began shipping ore exclusively to Malaysia. The following year, the contract with Malaysia was abruptly cancelled as a result of the confrontation, a bitter Malaysian-Indonesian political rupture; and in an arrangement that must have struck Indonesian nationalists as ironic at best, the State Mining Enterprises arranged to have its ore shipped to the Netherlands for smelting by Billiton at its Arnhem smelter.

The Sukarno era was a decade of general economic retrogression, for which many blame the economic policies pursued as much as the underlying conditions.[23] Inflation rose to the point where numerical indicators of increases in the price level became virtually meaningless (the price index of food in Djakarta, with 1957–58 = 100, stood at 58,851 in January of 1966);[24] urban overcrowding, slums, and unemployment festered in Djakarta; investment fell to minimal levels; and, as calculated by D. S. Paauw, real per capita income stagnated at levels near those of the 1930s.[25] The drop in tin output shown in table 2.5 should thus be seen as a relatively minor aspect of an overall Indonesian economic phenomenon.

In 1965, Sukarno was overthrown and Guided Democracy came to an end. The 1975 financial collapse of the national oil company, Pertamina, and the ensuing scandal indicate that Indonesia still faces major problems in managing its economy. Bureaucratic corruption and inefficiency are still widespread.[26] Nevertheless, tin production recovered well from its 1966 low point as the overall economy revived.

Zaire and Nigeria. Civil wars raged in the 1960s in the two major African tin-producing countries of Zaire and Nigeria, with markedly different effects on the tin industries of the two. Along with Bolivia, these two countries, then colonies, supplied the tin ore for the Allied nations during the Second World War, averaging 15,706 long tons per year for Zaire and 12,173 long tons for Nigeria over the years 1941–45.

Tables 2.6 and 2.7 show Zairoise and Nigerian output under decolonization and subsequent events.

The Belgian Congo (Zaire) obtained its independence in June 1960. Within weeks, the Congolese army had mutinied, the province of Katanga had seceded, and the United Nations had been called upon to restore order by military force.

Table 2.6. Output of tin-in-concentrates by the Belgian Congo and Zaire, 1955–80 (long tons)

Year	Amount	Year	Amount
1955	13,055	1968	6,264
1956	12,832	1969	6,647
1957	12,478	1970	6,458
1958	9,689	1971	6,456
1959	9,190	1972	5,960
1960	9,202	1973	5,442
1961	6,570	1974	4,675
1962	7,197	1975	4,562
1963	7,053	1976	3,723
1964	6,492	1977	3,560
1965	6,211	1978	3,450
1966	7,152	1979	3,300
1967	6,583	1980	3,159

Sources: International Tin Council, *Statistical Yearbook*, 1968; International Tin Council, *Tin Statistics, 1966–76*; International Tin Council, *Monthly Statistical Bulletin* (June 1979 and March 1982).

Table 2.7. Nigerian output of tin-in-concentrates, 1956–80 (long tons)

Year	Amount	Year	Amount
1956–60 average	7,641	1973	5,828
1961–65 average	8,597	1974	5,455
1966	9,687	1975	4,652
1967	9,490	1976	3,710
1968	9,804	1977	3,267
1969	8,741	1978	2,935
1970	7,959	1979	2,750
1971	7,326	1980	2,527
1972	6,731		

Sources: International Tin Council, *Statistical Yearbook*, 1968; International Tin Council, *Tin Statistics, 1966–76*; International Tin Council, *Monthly Statistical Bulletin* (June 1979 and March 1982).

The ensuing warfare and disorder did not end until 1965. In Nigeria, civil war broke out with the Biafran secession in May 1967, and continued until January 1970.

Zairoise tin output fell to half its 1957 level by 1965, over the period of worst internal disorder. In contrast, the steady decline in Nigerian output does not seem to have been triggered by the civil war. Indeed, as table 2.7 indicates, output peaked during the extremely bitter fighting of 1968.

The basic explanation of the differences in the effects of these two civil wars lies in geography: the Nigerian tin mines are located on the Jos plateau in the northeast of the country, far from the area of conflict. The subsequent downward trend in Nigerian output reflects depletion of ore and diversification of

Table 2.8. Malaysian output of tin-in-concentrates, 1965–80 (metric tons)

Year	Amount	Year	Amount
1965	64,689	1973	72,260
1966	69,986	1974	68,122
1967	73,275	1975	64,364
1968	76,270	1976	63,401
1969	73,322	1977	58,703
1970	73,794	1978	62,650
1971	75,445	1979	62,995
1972	76,830	1980	61,404

Source: International Tin Council, *Tin Statistics*, various issues.

the economy away from mining in an area far from any ocean port and with high cost of overland transportation of ore, rather than political or military disruptions.

Malaysia. Malaysia's output of tin-in-concentrates appears to have peaked at 76,830 metric tons in 1972, and then to have fallen steadily over the next five years to a low of 58,703 metric tons in 1977, as shown in table 2.8.

This decline in mine production is commonly attributed to three closely linked causes: increased taxation; the lack of availability of new tin-bearing land under stringent licensing policies; and, underlying both, efforts to promote ownership and management of Malaysian business enterprise by the ethnic Malays, or *bumiputras*, as an integral feature of the New Economic Policy (NEP) that was inaugurated in response to ethnic rioting in May 1969.

These riots, which began with Malays attacking Chinese who were celebrating a political victory with parades and demonstrations, raged through Kuala Lumpur and led to curfews throughout the nation to prevent the spread of disorders and killings. The riots left at least 137 dead and 300 wounded,[27] and put an effective end to the so-called "bargain" of 1957. This bargain, which had been drawn up by the Alliance Party, a coalition of Malay, Chinese, and Indian political organizations that ruled the country following its independence, had been designed to permit the newly independent nation to cope with racial antagonisms so strong that they threatened the viability of the country from its inception. The crux of the bargain was the grant of political predominance to the ethnic Malays in return for assurance that Chinese economic power would be preserved under a free enterprise system.[28]

The NEP has been the cornerstone of the Second (1971–75) and Third (1976–80) Malaysian Plans. It is described in the third plan as follows:

> Comprising two prongs, the NEP seeks to eradicate poverty among all Malaysians and to restructure Malaysian society so that the identification of race with economic function and geographical location is reduced and even-

tually eliminated, both objectives being realized through rapid expansion of the economy over time. . . .

Accordingly, through the second prong of the NEP, the Government aims at providing such assistance as may be necessary for all racial groups in the country to find employment, secure participation and acquire ownership and control in the various sectors of the economy. . . .[29]

The NEP has set as its target the ownership and management by Malays and other indigenous people of at least 30% of the commercial and industrial activities of the economy and an employment structure at all levels of operation and management which reflects the racial composition of the nation by 1990.[30]

The redistributive objectives of the NEP also called for a decline in foreign ownership of Malaysian business enterprises to 30 percent, down from an estimated 62 percent in 1971, and for a moderate increase in ownership by non-Malay (almost entirely Chinese and Indian) Malaysians, from 34 percent in 1971 to 40 percent. The *bumiputra* share of 30 percent was to rise from 4 percent in 1971.

The problems of restructuring the tin industry to achieve the NEP's goal of 30 percent *bumiputra* ownership by 1990 were formidable from the start. Yip Yat Hoong has surveyed changes in patterns of ownership of Malaysian tin mines from 1954 to 1964. He identified thirty-one dredging companies incorporated in Malaysia and operating in that country in 1964. All but one were European companies in the sense of being under European management; but Yip found that local ownership had risen from 22.3 percent in 1954 (prior to the proclamation of Malayan independence in 1957) to 64.1 percent in 1964. Ethnic Chinese were predominant in local individual ownership, accounting for 91.7 percent of private holdings in 1964—up from 89 percent in 1954. Local corporate ownership of dredging companies was principally by Chinese-owned commercial banks, investment companies, and insurance companies.[31]

Ownership of gravel-pump and other small-scale tin mines was almost completely in the hands of Chinese. According to the Third National Plan, the share of equity ownership by Malay individuals and Malay interests in all tin mining was 1 percent in 1970. Malay individuals and Malay interests in that year owned 26 mining units of a total of 1,083 in the nation.[32]

For most sectors of the Malaysian economy, the NEP envisaged that ownership restructuring would occur through growth, rather than through the need for any actual diminution of ownership by non-Malay business interests. The relevant sections of the third plan read as follows:

258. The ownership of equity capital on the part of Malay individuals and Malay interests in the corporate sector is targetted to grow by 25.8% per annum during the next fifteen years. . . .

262. No constraints are expected on the growth of corporate ownership on

> the part of other Malaysians. Their ownership is estimated to have increased by 15.1% per annum under the SMP. It is projected to grow by 15.5% per annum between 1975 and 1990 and to account for 40.0% of the total stock of equity capital by the end of the OPP period [1990]. . . .
>
> 263. The share in the ownership of equity by foreign interests is targetted to decline from 63.3% in 1970 and about 55% in 1975 to 30.0% by 1990. However, the ownership of corporate stock by the foreign sector would still increase by 10.4% per annum in absolute terms over the next fifteen years compared with 10.0% per annum during 1971–75.[33]

In the case of tin, however, the prospect for such a relatively painless transition through growth was bleak. Tin output, which was estimated at 72,600 metric tons in 1970, was projected in the Second Malaysian Plan to be about 70,000 metric tons in 1975.[34] As shown in table 2.8, the output reported for 1975 was considerably lower. Mine output of tin was projected to grow, but only by 0.3 percent per annum, over the five years of the Third National Plan; while in the mid-plan review of 1979 that projection was raised to a much greater but still relatively slow rate of 1.7 percent per annum.[35] The Fourth National Plan notes that tin production must be stimulated in the 1980s if the NEP's redistributive goals are to be attained.

There was agreement among virtually all Malaysians with whom I discussed the matter in 1979, both inside and outside of the government, that due to both the very low level of *bumiputra* participation in the tin industry and the low rate of growth envisaged for that industry in the Second and Third Malaysian Plans, the government has been giving very low priority to dealing with problems of either taxation of tin mining concerns or land allocation for mining. "There are practically no Malays in tin," one government official noted, and therefore the overall distributive objectives of the NEP imply that the government "does not have as great a concern for promoting tin or the interests of the tin miners as is the case with other commodities."

Tin is subject to a number of special taxes in Malaysia, in addition to the normal business income tax of 40 percent. During the Indonesian confrontation a special tin profits tax of an additional 10 percent was levied; initially announced as a temporary emergency expedient, it was never thereafter removed. A particularly sore point with Malaysian tin miners is that each tin mine is treated as a separate unit for profits-tax purposes, so that in the case of single ownership of more than one mine site the losses of one cannot be subtracted from the profits of another, although plantation owners are permitted such offsets. In 1977 the tin profits tax was changed to a progressive tax with a minimum rate of 15 percent on annual profits in excess of M$400,000. In 1978 the maximum rate was reduced to 12.5 percent, but was raised again to 15 percent in 1980. Tin miners are further subject to a development tax of 5 percent of net profits after income taxes.

An export duty is levied on all tin-in-concentrates delivered for smelting. In

1974 an export duty surcharge was added, announced as an anti-inflationary measure in a period of rapidly rising world prices for tin and other commodities. In 1977 the export duty was changed from a flat 16 percent rate to a sliding-scale formula based on current tin prices, leading to an increase in the duty per ton at the prices then prevailing. In 1980, offsetting the rise in the tin profits tax, the scale was revised in such a way as to lower the tax when the price of tin was below M$2,250 per picul. The president of the States of Malaya Chamber of Mines noted that "neither new discovery of reserves nor efficient utilization of existing ores is encouraged by the taxation system," but added that the shift in emphasis away from the export duty and towards a profits tax "should provide stimulus for reworking dredged ground and a more complete exploitation of other low-grade grounds."[36]

State governments receive 10 percent of the tin export duty and, since 1977, a similar percentage of the surcharge. Tin miners stress that they are also affected by heavy import duties imposed on petroleum products and machinery parts.

Some commentators attribute much of the decline in Malaysian tin production after 1972 to the disincentive effects of high taxation. The magazine *Asian Business and Industry*, for example, noted in November 1977, "The reason for the miners' stubbornness is clear: they will not spend an additional cent on their existing operations while the government takes 78 percent of their gross income in taxes. There are at least eight direct and indirect taxes levied on Malaysian tin miners: income tax, development tax, export duty, export duty surcharge, sales tax, tin profits tax, import duties and surcharges on imports."[37] On the other hand, just a few months later, in February 1978, the *Financial Times* commented, "Malaysian tin mining companies are making very good profits, thanks to high prices, but with little investment possibilities, many are distributing most of the profits to shareholders. It is not uncommon for tin companies to pay a 100 per cent dividend."[38] Other commentators, including the States of Malaya Chamber of Mines and the *Far Eastern Economic Review*, have described the total package of tin taxes as taking about 70 percent of mining profits.[39]

A 1981 article by John Thoburn concludes that rates of return on investment in both Malaysia and Thailand are high enough to indicate that the tin mining industry is not too heavily taxed. Thoburn, however, argues as did the spokesman for the States of Malaya Chamber of Mines, that the extent to which the export tax is used raises risk unnecessarily, since the tax is paid even on high-cost and profitless mines if any ore is extracted, and that a further emphasis on a profits tax would be advisable.[40]

In 1977, in addition to converting the tin profits tax into a progressive one for the relief of small miners, revising the export duty schedule, and increasing the share of the export duty plus surcharge rebated to the states in order to increase the states' incentives to make mining land available, the federal government agreed to pay Malaysia's contribution to the ITC's buffer stock in lieu of the former practice of financing this contribution through charges levied on miners. As reported in the *Wall Street Journal*, in an article headlined "Malaysia's Incentives to Spur Tin Output Draw Cool Reception from Major Miners," the revised tax package was designed to provide incentives to the small-scale gravel-pump miners rather than to the dredging companies.[41] The reforms imposed a slightly higher total tax rate on the larger dredging companies at the tin price levels prevailing throughout 1977.

By early 1979 there appeared to be a clear consensus among Malaysian government officials and others outside of the government with whom I talked, with the exception of tin miners, that the overall level of taxation was not a deterrent to investment and to expansion of the output of existing mines. Miners, not surprisingly, complained not only of excessively high taxes but of the special discriminatory taxes levied on the tin industry. However, miners were even more critical of the structure of the tax package than of its overall level, and their views were shared by many others. In particular, the treatment of each mine as a unit, the outdated price schedule on which export duties were based, and the import duties on petroleum products and machinery were most frequently criticized.

There is widespread agreement that land policies and the resulting lack of availability of new mining sites are far more serious hindrances to an expansion of Malaysian tin output than is the tax burden on the miners. Approximately 90 percent of the tin miners operating in Malaysia in the late 1970s were mining marginal ground—a term used to describe land that had already been mined at least once. One miner, whose family has been mining tin in Malaysia for three generations, told me that his company, which had owned fourteen tin mines a decade previously, was at the time down to four mines and was phasing out of tin mining entirely. The basic reason was the inability to obtain prospecting rights and then leases. "There is no future for the Chinese in tin," he stated, and his family was therefore shifting funds into palm oil, rubber, and estate housing as rapidly as the mining investment could be liquidated. None of these, he continued ruefully, is as much "fun and excitement" as tin.

In a recent address to the States of Malaya Chamber of Mines, the vice president of that organization noted the decline in mining activity in 1976 and 1977: "The reason for all this is not hard to find. I can only repeat the call of past MMEA [Malaysian Mining Employers' Association] that more land be made available for mining [*sic*]."[42] This observation merely reinforced that of the chamber's president of the previous year, who had stated: "I regret to say that insecurity of tenure of mining land is today probably the most important factor inhibiting the expansion of the mining industry and certainly is a major contributor to the steady decline in production over recent years. Unless a solution to this problem is quickly found Malaysia's tin production will continue to decrease."[43] The 1978 *Annual Report* of the Bank Negara Malaysia (National Bank of Malaysia) cited the "difficulty and delay in obtaining approval for new land for mining, renewing mining leases and prospecting licenses," as the primary problem of the tin mining industry.[44]

The basic problem of land access has been twofold. First, under the Malaysian Constitution land rights are vested solely in the state governments rather than the federal government; yet the structure of tin taxes has been such that most of the public revenue from mining has gone to the federal government. Despite the tax revision of 1977, under which 10 percent of the tin export duty surcharge as well as 10 percent of the basic duty were remitted to the states, the state governments have had little or no incentive to allow mining on lands that might otherwise be devoted to uses generating greater state revenues.

In 1977, the State of Selangor announced a policy under which it would, through a state corporation, assume all leases to mining areas held by foreign-owned corporations upon their expiration, and then sublease these tracts for a substantial tribute. Later, in 1979, the government of Selangor entered into an arrangement with Berjuntai Tin Dredging, a subsidiary of publicly-owned Malaysia Mining Corporation (MMC), which may set a pattern for the future. The state government, having first refused to renew Berjuntai's long-term lease, assigned the tract to a state-owned corporation, Kumpulan Perangsang Selangor (KPS), for subleasing. KPS, instead of subleasing the tract, entered into a joint venture with Berjuntai and later with a private foreign-owned firm, Pacific Tin Consolidated. Such arrangements may speed up the award of prospecting licenses and leasing of mining tracts, but are hardly stimuli to private investment.[45]

In 1980, following three years of negotiation, KPS entered into an agreement with MMC for the formation of a joint subsidiary firm, Kuala Langat Mining, to mine an extremely rich but difficult site, one in which a very large and promising body of ore lies at depths beyond the capacity of existing dredges to reach. The new firm will be 65 percent owned by KPS and 35 percent by MMC. It is capitalized at M$60 million, but the development cost of the project is estimated to exceed M$200 million. MMC will guarantee 50 percent of the loans made to Kuala Langat Mining.[46]

The second aspect of the land problem is closely related to the NEP's *bumiputra* program. Malaysian government agencies, both federal and state, are delaying release of high-grade land until there is greater *bumiputra* participation in tin. Where land is available for lease, preference is given to firms with *bumiputra* participation, leading to the establishment of so-called Ali-Baba companies that have been described by the *Far Eastern Economic Review* as follows: "These work on the basis that a Malaysian Chinese—'Baba'—will quietly buy up the relevant property and then apply for a prospecting license on the understanding that he has the power of attorney for an unnamed Malay—'Ali.' If successful, he then invites one of the established mining companies to prospect the area, and, if commercial quantities of tin are found, a mine is established and the profit split three ways."[47]

The Ali-Baba device does not, however, work smoothly or solve all problems for the Malaysian Chinese tin miners. In a 1977 address, the president of the Malayan Chamber of Mines noted problems that he believed were retarding prospecting. After a non-Malay firm or individual has obtained prospecting rights, he observed, it is usually impossible to locate a Malay partner willing to risk capital to help finance exploration. The result is that non-Malay but Malaysian firms, after bearing all of the risk of prospecting, will be eligible for only 70 percent of the equity participation in any mine established after successful discovery of a promising deposit—and foreign-owned prospectors will be able to obtain only 30 percent of the equity in a mining firm. Further, mining leases must be obtained after successful prospecting, and there is no assurance that such a lease will be granted by a state government or, if granted, will not contain unexpected and costly conditions. It would be optimistic to assume that five percent of prospecting ventures will be successful. "This background of uncertainty," he concluded, "would naturally inhibit even the relatively cheaper alluvial prospecting were areas to be made available. It would effectively preclude consideration of the much more expensive hard rock prospecting."[48]

Pressures to meet the target of 30 percent *bumiputra* participation in ownership and management of tin mining by 1990 were relieved by the formation of MMC and its takeover of the Malaysian mining interests of the London Tin Corporation (LTC) in 1977. At the time of its formation MMC, initially named New Tradewinds, was 71.35 percent owned by a Malaysian government corporation, Perbadanan Nasional (Pernas),[49] with the remaining ownership interest held by Charter Consolidated Ltd., a British firm which had provided managerial services for Malaysian mines as well as had held ownership interest in three mining companies that were also incorporated into MMC. At the time of acquisition, the mines that were thereby partially acquired by MMC accounted for approximately 24 percent of Malaysia's tin output.

In the 1979 mid-term review of the Third Malaysian Plan, it was reported that Malays owned thirty-eight tin mines in 1975 and forty-two in 1978, with the output share rising from 5.3 percent of mine production to 5.5 percent. The number of mines owned jointly by *bumiputras* and others rose from fifty-one to seventy (presumably including the MMC group). The share of land under prospecting licenses issued to *bumiputras* declined from 20.8 percent in 1975 to 5.7 percent in 1978 as a result of new licenses issued to multinational companies; but ownership of mining land by Malays and other indigenous peoples rose from 18 to 22 percent.[50] The goal of 30 percent *bumiputra* participation in management was moving ahead rapidly by 1979, in large part because of the replacement of retiring expatriates by Malays in MMC and the joint ventures. At the beginning of the year, the number of expatriates employed in Malaysian tin mining was down to 48, as contrasted to a target number of 226.[51] At the time of independence, 1957, there had been roughly six hundred expatriates so employed; by 1981, the number had dropped to forty.[52]

The situation appears to have eased in recent years, following the 1977 tax changes and the establishment of MMC. The mine production of Malaysia evidently rose in 1978 and again in 1979. The dip in 1980 reflected a drop in worldwide consumption with the onset of recession in the industrialized nations. It is commonly accepted that Malaysia has not yet seriously depleted its tin resources; the problem has never been a lack of availability of tin-bearing land, but rather a lack of access to it. Thus, the Ministry of Primary Industries noted in 1977 that tin mining is still concentrated in the areas where it was originally established in the nineteenth century, primarily because of the lack of any systematic prospecting work in other regions of the country.[53] Most tin mining in Malaysia today, including virtually all of the gravel-pump mining, is the reworking of ground already mined at least once before.

At the Mines Research Institute Malaysia, I was told that sooner or later the alluvial deposits would be exhausted and that the nation would have to turn to lode mining; but the ultimately recoverable tin from lode mining is expected to be several times greater than that from the easier and lower-cost alluvial mining. Such a shift need not occur for many years. The currently most promising discovery has been at Kuala Langat, some thirty miles south of Kuala Lumpur, where vast alluvial deposits are believed to exist, estimated by some to rival those of the entire Kinta Valley—the main source of Malaysian tin throughout the present century—although at depths that are extremely difficult to exploit.

The drop and subsequent recovery of Malaysian tin production from 1972 through 1979 may in large part be illusory. There is no doubt that as taxes rose, and as allowable production was limited under ITC export controls during 1973 and 1975–76, the smuggling of tin-in-concentrates out of Malaysia increased sharply. Estimates of the extent of the smuggling vary, but a peak figure of about 7,000 metric tons in 1977 is a conservative estimate.[54] Under the headline "Malaysia's Tin Conundrum," *Tin International* noted that employment in Malaysian tin mines rose from 36,728 at the end of March 1977 to 39,943 one year later, while over the same one-year period the nation's *reported* production of tin-in-concentrates declined from 62,906 metric tons to 58,550 metric tons.[55] Singapore reported legal imports of approximately 2,300 metric tons of tin-in-concentrates per annum in both 1977 and 1978; but in the same two years the two Singaporean smelters exported 7,725 metric tons and 6,122 metric tons of tin metal.[56] British and Dutch smelters have been handling concentrates of unknown origin in recent years. Billiton's Arnhem smelter, closed in 1971, reopened in 1976, and has since been producing about 175 metric tons of tin metal per month. Billiton officials refused to confirm an inquiry whether the source of the tin concentrates used by the Arnhem smelter was Singapore, stating only that "since we are buying mainly from merchants, the origin of the concentrates [is] unknown to us."[57]

An overview of postwar growth. The declines and subsequent recoveries of tin production in individual nations such as Bolivia, Indonesia, Malaysia, and

Table 2.9. Principal exports of major tin producing nations as percentages of total exports

	1955	1960	1965	1970	1975	1980
Bolivia:						
Tin	66.5	81.3	84.9	56.8	41.2	40.1
Petroleum and natural gas	2.0	6.4	0.6	6.9	19.3	25.8
Malaysia:						
Petroleum	—	4.0	2.3	3.9	9.2	23.7
Rubber	53.7	55.1	38.6	33.4	21.9	16.4
Logs and timber	—	4.7	9.5	16.5	12.0	14.1
Palm oil	1.2	1.7	2.8	5.1	14.3	9.3
Tin	12.9	14.0	23.1	19.6	13.1	8.9
Thailand:						
Rice	40.4	29.8	33.5	17.0	13.0	14.7
Tapioca	—	3.3	5.2	8.3	10.2	11.2
Rubber	26.1	29.9	15.4	15.1	7.7	9.3
Tin	6.5	6.2	9.0	11.0	5.0	8.5
Corn	1.2	6.4	7.8	13.3	12.7	5.5
Sugar	—	0.1	0.8	0.6	12.7	2.2
Zaire:						
Copper	n.a.	n.a.	56.0	65.2	54.2	43.2
Cobalt	n.a.	n.a.	18.0	5.9	11.7	21.1
Coffee	n.a.	n.a.	5.5	4.9	6.6	10.1
Diamonds	n.a.	n.a.	7.5	5.1	6.7	4.2
Tin	n.a.	n.a.	5.9	2.6	2.6	2.7
Indonesia:						
Petroleum	22.5	26.3	35.7	38.7	74.4	58.7
Jungle wood	—	—	—	8.8	7.1	8.2
Rubber	45.6	44.9	29.1	18.5	5.1	5.3
Tin	6.3	6.5	5.2	4.2	3.5	2.0
Australia:						
Wheat	8.1	9.0	13.7	9.1	12.2	10.2
Coal	—	—	—	4.2	8.6	8.7
Wool	44.4	38.1	29.0	15.3	9.3	8.5
Beef	—	—	—	6.9	4.2	6.5
Iron ore	—	—	—	7.6	8.2	6.0
Tin	—	—	—	0.1	0.1	0.7
Nigeria:						
Petroleum	—	2.6	25.4	57.5	92.8	95.3
Tin	4.4	3.5	5.5	3.8	0.3	0.6
Cacao	19.3	20.6	15.9	15.0	3.6	n.a.
Palm kernels	14.5	14.7	9.9	2.5	0.4	n.a.
Rubber	4.2	8.5	4.1	2.0	0.3	n.a.
Brazil						
Coffee	55.6	40.1	27.1	35.0	9.5	12.3
Soybeans	—	—	1.2	2.3	15.6	11.2
Iron ore	1.9	6.8	8.6	6.5	10.7	7.7
Sugar	3.7	6.8	4.7	5.0	10.9	4.7
Tin	—	—	—	0.1	0.3	0.3

Sources: International Monetary Fund, *International Financial Statistics*, various issues; International Tin Council, *Monthly Statistical Bulletin* (March 1982)

Table 2.10. Tin exports of major tin producing countries as percentages of gross domestic product

	1950	1955	1960	1965	1970	1975	1980
Bolivia	n.a.	29.6	11.4	15.4	10.6	8.5	6.2
Malaysia	n.a.	7.6	8.6	10.1	7.5	5.4	4.9
Thailand	1.6	1.1	1.0	1.4	1.2	0.7	1.7
Zaire	n.a.	0.5	0.4[a]	1.1	1.1	0.5	0.7
Indonesia	n.a.	n.a.	n.a.	n.a.	0.5	0.5	0.6
Australia	—	—	—	—	0.01	0.02	0.1
Nigeria	1.2	0.7	0.5	0.9	0.6	0.1	0.05[b]
Brazil	—	—	—	—	0.01	0.02	0.02

a. Exclusive Katanga and South Kasai.
b. Based on unofficial estimate of GDP.
Sources: Economist Intelligence Unit, *Quarterly Economic Review of Nigeria*, first quarter, 1982; International Monetary Fund, *International Financial Statistics*, various issues; United Nations, *Statistical Yearbook*, various issues.

the African producers have been short-run fluctuations around a long-run postwar trend of worldwide growth of tin output that has been much more sluggish than the rates of growth of general economic activity in the producing nations. In consequence, tin has tended to become a less important part of the producers' economies.

Table 2.9 shows the relative importance of tin as a percentage of exports, as well as the 1980 ranking of tin among the nations' principal exports. Table 2.10 indicates the contribution of tin exports to total economic activity as measured by gross domestic product.

The general trend for tin percentages is downward in both tables 2.9 and 2.10. Tin continues to dominate Bolivia's reported exports and to account for a substantial fraction of its gross domestic product, despite Bolivian efforts to diversify the economy and reduce the nation's dependence on tin. Part of the explanation lies in Bolivia's success in developing a local tin smelting capacity. The value of tin exports from Bolivia has gone up much more rapidly in recent years than has its rate of output of tin-in-concentrates, reflecting the increased export of tin metal and the related decline in shipments of low-grade concentrates sold to foreign smelters at substantial discounts for impurities. But the statistics also reflect Bolivia's failure to develop other legitimate sectors of its economy. A recent *Wall Street Journal* article noted that "Bolivian government sources estimate that cocaine may be a $1.6 billion-a-year business, making its export value three times that of tin, the country's leading legal export."[58]

Comparing Bolivia and Malaysia, table 2.9 shows that in 1980 the percentage of total exports accounted for by tin was little more than one-fifth as high for Malaysia as for Bolivia; and with the expansion of palm oil and lumber exports, tin had fallen to fifth rank in Malaysia. Nevertheless, when one looks at the contributions made by tin to the total economies of the two countries in table 2.10, the difference is a rather small one. The reason is simply that Malaysia, a peninsula with a central mountain range and only small coastal areas of flat

Table 2.11. Per capita gross domestic product of major tin producing countries, in 1975 U.S. dollars

	1955	1960	1965	1970	1975	1980
Australia	$3,616	$4,203	$4,785	$5,658	$6,338	$7,027
Brazil	n.a.	n.a.	627	795	1,158	1,366
Malaysia	356	353	466	532	780	1,248
Nigeria	202	224	241	271	399	850[a]
Bolivia	293	258	295	329	383	771
Thailand	177	188	226	294	347	501
Indonesia	n.a.	151	149	169	224	311
Zaire	n.a.	n.a.	132	156	155	140

a. Based on an unofficial estimate of GDP.
Sources: Economist Intelligence Unit, *Quarterly Economic Reivew of Nigeria*, first quarter 1982; International Monetary Fund, *International Financial Statistics*, various issues.

arable land, has a more trade-oriented and less self-sufficient economy than Bolivia, albeit a far more prosperous one. There is a sharp break in the series in table 2.10 between Bolivia and Malaysia at the top and the other six countries listed below them, with tin accounting for less than 1 percent of the gross domestic products of all of the lower nations except Thailand. Despite the recent rise of Australia and Brazil to the ranks of the world's major tin producers, and the likelihood that their relative importance within this group will grow in the coming years, the contribution of tin to their overall economies is and undoubtedly will remain negligible.

The decline of tin does not indicate any general decline in the importance of commodities to the eight countries as a group, nor does it reflect diversification of their economies and lessened dependence on one or a small group of commodities. One of the most glaringly obvious features of table 2.9 is the explosive rise in the importance of petroleum to Bolivia, Nigeria, Indonesia, and Malaysia. Copper and cobalt still predominate in Zaire. The diversification of Brazilian exports, notably the rise of soybeans and the decline of coffee in percentage terms, has reduced that country's dependence on a single export but not on primary agricultural commodities.

While it was shown in table 2.1 that virtually all tin exports are accounted for by developing countries, it is certainly not true that tin is a significant source of income to those countries that have recently been singled out as the "poorest of the poor." The World Bank divides developing countries into "low income countries," with 1979 per capita GNP of $370 or less, and "middle-income countries" with per capita GNP for that year of above $370. Of the major tin producers, only Indonesia and Zaire are among the world's thirty-five low-income countries. Brazil, Malaysia, Nigeria, Bolivia, and Thailand are all categorized as middle-income countries.[59] The growth of per capita gross domestic product for the tin producing countries is shown in table 2.11.

The implications of per capita GDP comparisons are far from obvious. It would be fatuous to assert that international commodity agreements or other devices designed to transfer income from developed countries to, say, Zaire and Indonesia, are wrong merely because Bangladesh and Ethiopia are still poorer. It is at least arguable that one or both of the former two can be helped to the transition to sustained growth with less massive stimulation than the latter two would require. And assisting some nations through commodity arrangements does not preclude other forms of assistance to other countries. On the other hand, it would be equally fatuous to maintain that price-enhancing commodity agreements are in the best interests of all Third World nations. Such agreements work to the detriment of those that must import the products in question, and it is an obvious and inescapable fact that many nations are poor simply because their commodity and other resource endowments are poor.

3. Tin in the Economies of Consuming Nations

General Patterns of World Consumption: Industrialized and Developing Nations

As noted in chapter 1, there are virtually no end uses for tin metal in and of itself. Indeed, the products containing tin, such as tinplate, solder, and various alloys, are themselves almost always intermediate products. Because tin is an indirect input into numerous manufactured products, it is only to be expected that the great bulk of the metal is nominally "consumed" in the advanced industrialized countries. It has become standard jargon, which will be followed here, to refer to the users of tin metal as "consumers," although they are in fact the producers of products containing tin metal, usually in very small proportions in terms of both quantity and cost. Table 3.1 shows patterns of consumption of primary tin metal by various nations from 1955 to 1980.

In 1980, the five largest consumers, the United States, Japan, the United Kingdom, West Germany, and France, accounted for 60.4 percent of the reported total. This represented a substantial decline from 1955, a year in which the five largest—the same five countries as in 1980—had accounted for 72.0 percent of the reported total. Within the top five, the United States declined from consumption of 40.4 percent of the reported usage in 1955 to 25.3 percent in 1980, while over the same twenty-five years Japan rose from 4.4 to 17.6 percent. In 1980, nations classified by the World Bank as "industrial economies" accounted for 80.2 percent of the total consumption shown in table 2.12.

The industrialized or developed nations account for an even higher percentage of imports than of consumption, as shown in table 3.2.

Overall, in terms of the distribution of global production and consumption of tin-in-concentrates and tin metal, the world tin market closely approximates the conditions assumed for a commodity-based strategy of income transfer; most important—virtually all world trade is accounted for by exports from developing countries and imports by the wealthy industrialized nations.

It would, nevertheless, be quite misleading to rely solely on consumption data for tin, either in concentrates or as metal, in order to estimate the extent to which income could be transferred to producing nations through an international agreement or other mechanism affecting the commodity's price. To some extent, any increase in the cost of tin will be passed on and thus will be reflected first in the prices of solder, tinplate, and bearings containing tin, and ultimately in the prices of the final goods in which the tin-bearing intermediate products are incorporated. The initial increase in the price of tin may be passed on in part, in full, or possibly in greater than the full amount to final consumers.[1]

The basic point at issue in the matter of international income transfer is that the industrialized nations export many tin-bearing manufactured products, and thus would shift some part of the burden of an increase in the cost of tin to final

Table 3.I. World consumption of primary tin metal[a]

	Long tons					Metric tons
	1955	1960	1965	1970	1975	1980
United States	59,828	51,530	58,550	53,878	43,620	44,342
Japan	6,500	12,933	17,151	24,710	28,115	30,879
West Germany	8,165	27,745	11,662	14,062	11,958	14,271
France	9,700	11,165	10,136	10,500	9,968	10,052
United Kingdom	22,436	21,790	19,256	16,950	12,165	6,445
Italy	3,000	4,550	5,700	7,200	6,300	5,800
Netherlands	2,515	3,052	3,393	5,467	3,583	5,188
Brazil	1,250	1,600	1,980	2,500	4,300	5,012
Czechoslovakia	1,700	n.a.	2,953	3,420	3,400	4,900
Canada	4,018	3,880	4,892	4,640	4,250	4,766
Spain	675	765	1,800	3,040	4,700	4,250
Poland	1,700	n.a.	2,500	3,532	4,300	3,309
Romania	240	250	300	2,567	3,050	3,000
Australia	2,412	3,841	4,425	3,837	3,258	2,845
Belgium/Luxembourg	2,022	2,756	2,378	3,000	4,352	2,601
India	4,200	4,000	3,900	4,800	2,850	2,282
South Africa	1,400	2,007	1,289	2,062	2,322	2,101
Mexico	705	1,181	1,200	1,640	1,600	1,800
Republic of Korea	—	—	—	—	—	1,761
Taiwan	—	—	—	—	—	1,300
Hong Kong	—	—	—	—	—	1,200
Hungary	600	400	400	1,245	1,300	1,200
Philippines	—	—	—	—	—	1,100
Bolivia	—	—	—	—	—	1,000
Yugoslavia	500	1,250	1,500	1,440	1,200	1,000
Argentina	1,600	1,700	1,680	1,800	1,800	700
Turkey	670	750	840	1,208	1,400	500
Denmark	4,950	2,269	545	720	300	123
Other Africa	890	1,400	1,335	1,080	1,100	1,640
Other America	735	920	1,250	1,420	1,600	2,120
Other Asia	1,816	2,337	3,030	4,002	4,952	3,002
Other Europe	3,484	3,666	3,582	4,099	4,457	4,618
Other Oceania	388	350	330	364	364	254
	148,099	168,087	167,957	185,183	172,564	175,361

a. Excludes figures for Peoples' Republic of China, East Germany, and the Soviet Union.

Sources: International Tin Council, *Statistical Yearbook*, various issues; International Tin Council, *Tin Statistics, 1966–76* and *1970–80*; International Tin Council, *Monthly Statistical Bulletin* (March 1982).

consumers in developing countries. In 1980, for example, the nations classified as industrialized by the World Bank accounted for 81.6 percent of total production of tinplate, but for only 68.2 percent of the consumption of that material.

There are no data available to present even rough estimates on how much tin is re-exported as solder or alloys; or as a component of products such as canned food, motors with tin alloyed bearings, and electric appliances with soldered

Table 3.2. Net imports[a] of tin by industrial market, nonmarket industrial, and developing nations, 1980

Nation	Amount (metric tons)	Percentage of total
Industrial market economies:		
United States	42,527	26.4
Japan	31,024	19.2
West Germany	16,068	9.9
France	10,112	6.3
Italy	6,299	3.9
Netherlands	5,586	3.5
United Kingdom	4,980	3.1
Canada	3,075	1.9
Belgium/Luxembourg	2,059	1.3
Other	2,847	1.8
All industrial market economies	124,577	77.2
Nonmarket industrial economies:		
USSR	15,552	9.6
Czechoslovakia	4,600[b]	2.9
Poland	3,309	2.1
Hungary	1,662	1.0
Other	26	—
All nonmarket industrial economies	25,149	15.6
Developing nations:		
Romania	3,000[b]	1.9
India	2,282[b]	1.4
Mexico	1,857[b]	1.2
Yugoslavia	998	0.6
Taiwan	631	0.4
Republic of Korea	571	0.3
Other	2,204	1.4
All developing nations	11,543	7.2
World total	161,269	100.0

a. Net imports are calculated as the sum of imports of tin-in-concentrates plus imports of tin metal, minus the sum of exports of tin-in-concentrates and tin metal.

b. Import and export figures are not available. The figure shown is for consumption minus production.

Source: International Tin Council, *Monthly Statistical Bulletin* (March 1982).

connections; or as bronze and brass musical instruments. Nor can it be ascertained in which countries the ultimate sales of these and other products containing tin are made. A small fraction of these exports will contain tin initially reported as consumed in developing countries in the manufacture of products shipped to industrialized nations.

Even in the absence of reliable numbers, it seems safe to say that nominal "consumers" of tin metal will absorb some of any increase in the metal's cost, and that much of what they are able to pass on in the form of higher prices for their final products will be borne by the ultimate consumers in the developed nations. The above discussion, therefore, is not intended as a denial of the proposition that the burden of an increase in the price of tin would fall most heavily on the developed countries. But it is equally certain that the industrialized countries' share of the overall burden would be less than their share of world consumption of tin metal.

The Uses of Tin and Prospects for Substitution

In the NIEO scenario, the ideal commodity would not only be produced by developing countries for consumption in the industrialized world, but it would also be an essential and irreplaceable good in the consumers' economies.

Tin has a number of economically attractive properties. It is nontoxic and extremely resistant to corrosion. Thus it is an excellent material for lining or plating containers, especially those for foods. It is malleable as a solid and has quite a low melting point. On the other hand, it can be used to harden alloys and still impart to those alloys the anti-friction qualities needed in bearings or the low melting point desired in type metal. In all, there are numerous uses to which such a versatile metal can be put.

The question of the extent to which tin is a necessity for a modern industrialized nation was put to a sharp test in the Second World War. In the 1943 issue of *Minerals Yearbook*, the authors noted that "although it is considered essential for numerous purposes and desirable for many others, the stress of wartime circumstances has shown that tin may be dispensed with or its use sharply limited in some applications."[2] They went on to note the growing use of tin-saving electrolytic plating, the wartime use of tin-free cans, the growth of frozen and dehydrated foods, the substitution of aluminum for tin in collapsible tubes and foil, the decline in nonmilitary demand for bronze and brass, the replacement of tin-based bearings by roller and ball bearings for some uses and by "virtually tin-free" metals in uses with less demanding performance specifications, and the success of regulations requiring use of tin-free and low-tin solders. Two years later, in 1945, *Minerals Yearbook* summed up the implications of the war years by observing that despite consumption of about ten thousand tons of tin per annum in bronzes for naval construction, "there is at hand no evidence of the imposition of undue hardships, at least, on ultimate users of tin-bearing materials. In other words, the record shows that the inelastic portion of the use is substantially lower than commonly has been assumed in recent years."[3]

Over the years following the Second World War, developments influencing the competitive position of tin and its feasible substitutes have continued, although at a much slower pace. In 1957, *Minerals Yearbook* reported that for the first time aluminum cans were being manufactured as containers for motor oil, grated cheese, and meats, and added that, "The Tin Industry (Research and Development) Board of Malaya has announced the availability of a special fund of M$500,000 (US$166,666) for use to refute publicity in favor of tin-less cans."[4]

Tinplate. Manufacture of tinplate represents the single largest use of tin metal, with most of the plate being used to produce food and beverage containers—the ubiquitous tin can. In 1980, 71,700 metric tons of tin, composing 40.9 percent of that year's world consumption of primary tin metal, were used in the manufacture of tinplate. There are a number of substitutes for tinplate in food and beverage packaging. Corrosion-resistant metal cans may be produced from tin-free steel plate (TFS), or from aluminum. Plastic-lined fiber containers can substitute for metal in many uses. Food may be frozen, dehydrated, or packaged in plastic containers or in glass jars. Rapid and refrigerated transportation permits a great deal of food to be sold fresh, at substantial distances from points of origin.

The feasibility of substitutes for tin as a protective element in containers is complicated because the corrosive elements in various types of foods and the strengths of the corrosive and other chemical actions vary over wide ranges. For example, it has proved quite satisfactory to package frozen orange juice in plastic-lined fiber containers, but no plastics have been found that are suitable for many other foods. Tinplate, although it resists corrosion from almost all foods and has no perceptible effect on their taste, odor, or color, cannot be so used as a container for spinach.[5] Lacquers can be used as can lining for only a limited number of foodstuffs, notably certain seafoods. Aluminum is not only susceptible to corrosion from more chemical sources than tin and resistant to coating by the latter, but is a structurally weaker metal than tinplated steel. Such complexities should be recognized in considering recent shifts in can usage. From 1964 through 1977, *Minerals Yearbook* reported U.S. shipments of metal cans by usage, and after 1968 also distinguished between steel and aluminum cans. Table 3.3 summarizes the main trends in these annual tabulations.

The most significant features of table 3.3 are the very large and rapid increases in the use of metal cans for beer and soft drinks, and the concurrent rise in shipments of aluminum cans. The two are, of course, closely related phenomena, as aluminum cans became common beer and soft-drink containers. Thus, the use of steel-base cans has virtually stabilized, although total expenditures on food in the United States rose from $100 billion in 1964 to $137 billion in 1977, expressed in constant 1967 dollars.[6]

In the long run, tin cannot be regarded as an essential component of food packaging, although opportunities for substitution in the short run are severely limited by the need to make extensive changes in capital equipment in order to change over to another material. On the other hand, tin's current position in markets for containers seems to be a strong one in light of the characteristics and costs of substitutes, and also as a result of recent technological changes affecting tinplate itself. In a paper prepared for the ITC some years ago, W. E. Hoare of the Tin Research Institute (TRI) made the following observation.

> Technological alternatives to tin have, from time to time, gained some ground in such applications as wrapping foils, collapsible tubes, bearing

Table 3.3. U.S. shipments of metal cans, by use and kind of metal (thousands of base boxes)[a]

	1964	1965	1966	1967	1968	1969	1970	1971	1972	1973	1974	1975	1976	1977
Use														
Beer	21,792	22,939	25,965	27,537	30,684	33,416	37,593	36,636	44,949	48,438	53,072	52,759	50,604	56,627
Soft drinks	5,591	7,756	11,253	14,580	20,055	23,509	26,197	28,255	31,485	36,049	35,757	33,429	39,491	47,324
Vegetables and vegetable juices	18,088	19,905	20,164	21,972	24,541	23,367	24,058	21,939	21,755	23,914	25,581	24,490	21,895	22,560
Fruits and fruit juices	16,826	15,484	15,923	14,301	14,251	15,300	13,371	13,258	13,639	14,526	14,547	13,235	12,754	13,809
All other foods including soups	36,354	36,573	37,505	36,811	36,729	36,944	37,392	37,407	37,166	37,491	40,051	36,435	36,467	30,986
All nonfood uses	n.a.	n.a.	n.a.	n.a.	19,602	20,090	20,722	20,190	21,678	19,849	18,750	16,960	18,398	17,142
Total	n.a.	n.a.	n.a.	n.a.	145,862	152,626	159,333	157,685	170,672	180,267	187,758	177,308	179,609	188,448
Metal														
Steel-base	n.a.	n.a.	n.a.	n.a.	136,046	140,297	143,064	139,400	141,228	146,625	148,030	133,855	130,264	130,390[b]
Aluminum	n.a.	n.a.	n.a.	n.a.	9,816	12,375	16,235	18,285	29,444	33,642	39,728	43,453	49,345	50,058[b]
Aluminum as percentage of total	n.a.	n.a.	n.a.	n.a.	6.7	8.1	10.2	11.6	17.3	18.7	21.2	24.5	27.5	30.8

a. base box = 31,360 sq. in. of plate, or 62,720 sq. in. of total surface area.
b. Estimated on basis of reported percentage growth.
Source: *Minerals Yearbook*, various issues.

materials, and alloy bronzes, but it may be said that the principal threat to tin producers has always been in the container and packaging industries. The alternatives that have made progress are the use of aluminium alloy sheet and passivated steel sheet (TFS) in place of tinplate. But one might also say that there has also been a gradual substitution of lacquer for tin in the sense that the thickness of the tin coating of tinplate has progressively decreased and that part of the total protective system for the steel is now undertaken by lacquer. However, the production of thin tin coatings of consistent quality was made possible by the introduction of electrolytic tinplate, and although this change was viewed at one time as something of an alarming development for tin, it can now be judged as a life-saving operation in that the availability of highly-efficient product-matched tinplate has kept at bay more serious encroachment by aluminium and TFS. If the only tinplate available to-day were the hot-dipped variety, we should probably be using very little tin for tinplate because economic competition would have been impossible and alternatives would have swept the board.[7]

Recent technological changes further strengthening tin's competitive position have been the development of a thin, "double-reduced" steel sheet and the growing use of this sheet, electroplated with tin, in manufacture of "drawn and ironed" (D & I) cans. Double-reduced plate is 30 percent thinner than plate produced by older methods, and is both stronger and stiffer. The added strength makes it possible to use the thinner plate in cans, but the stiffness requires new processes and equipment in can manufacture. The D & I process produces a two-piece can with a seamless body and only one seamed-on end. The equipment needed for producing such cans is far more expensive than that used for three-piece cans; but if the volume manufactured is high enough the initial cost of the equipment can be recouped by the production of cans with much thinner and therefore less costly double-reduced tinplate sides and bottoms. The tin coating acts as a lubricant in the D & I process, while the absence of such a lubricant makes TFS much less suitable for this process. Aluminum, on the other hand, can be and is used for the manufacture of D & I cans, especially for beer and soft drink containers.

TFS has also been limited in its growth because it cannot be soldered. Further, the plating material used in TFS is chrome oxide. Since chromium must be obtained from either the Union of South Africa or the Soviet Union, TFS has not been an attractive alternative to tinplate from the point of view of strategic materials planners in the consuming nations. The basic factor protecting tin from further inroads by aluminum is that the latter metal is produced only by a highly energy-intensive process involving electrolytic reduction. As a result, costs of production and the prices of aluminum rose steeply relative to those of tin in the years following OPEC's success in boosting petroleum prices.

Plastic containers are also making inroads into the market for tinplate. Curiously, the market for tin was not hurt by the initial introduction of polyvinyl chloride (PVC) containers, since a diorganotin compound, dioctyltin, was used as a stabilizer for PVC; and approximately the same amount of tin was required for this purpose as would have been used to plate a can of similar capacity.[8] But the use of tin as a stabilizer for PVC peaked in 1972, as technological improvements permitted the tin content of the stabilizing compound to be reduced.[9] More recently, polyethylene terephthalate (PET), which requires no tin, has substituted for PVC. By the end of 1981, PET bottles were being used only for soft drinks in the United States, where they accounted for roughly 15 percent of such containers; but they were being used for many other products in Europe, including cooking oils, mineral water, and toilet products.

Solder, bearings, and chemical compounds. The second largest use of tin is in solder, accounting for roughly 25 percent of world output of tin metal in recent years. To date, no acceptable substitutes for tin-based solders have been found for high-quality, reliable soldering operations; and in a strategic as well as an economic sense tin may be regarded as an essential metal in this usage. Solder preforms, in which the soldering material is shaped, sometimes in intricate forms, before being applied, are in growing industrial use. Often, the cost of the preform is many times that of the material used, and reliability is crucial. In such cases, the high cost of tin is no hindrance to its use. Hoare summarized recent changes in solder uses as follows.

> The established traditional uses of solder, as in plumbing, have decreased and the soldered side-seams of cans may contain one twentieth of the tin content used two decades ago, but the sophisticated electronics industry, embracing sectors such as computers, communications and television has given rise to a whole new field of soldering technology where high tin solders are universally used. Electronics soldered joints have become smaller in size as this new technology has rapidly advanced, but this is compensated by the many more connections that are made. Modern techniques such as the mass-soldering of printed circuits allow manufacturers to make several hundred connections at once, amounting to millions of soldered joints each week and procedures for achieving reliability of joints made in this manner are the special background of this paper. For the electronics industries soldering can be applied universally and there is no alternative assembly procedure.[10]

Tin is also an essential and irreplaceable component of the bearing metals required in high-performance uses. The desired characteristics of a bearing include strength, hardness, smoothness, conformability to imperfections in the shaft or mounting, the ability to embed or absorb dirt, good planing qualities, retention of shape and characteristics under heat, and noncorrosibility. The choice of an alloy involves trade-offs among these qualities, but frequently only alloys containing substantial amounts of tin will meet necessary performance specifications.

In his 1974 book, Fox commented that "in a world expanding in every direction in its industrial consumption of metals tin presents no very encouraging picture. Probably the root of the tin problem is not, as is so often held, in over-production but in persistent under-consumption."[11] The tin industry, he continued, suffered "three very severe blows" to its development after 1938. The first was the shift from hot-dipped to electrolytically plated tinplate, which Fox noted is by now virtually complete. The second loss was due to the substitution of aluminum for tin in foil and collapsible tubes. The third loss was the declining use of tin solder in automobile manufacture. Solder consumption in this industry fell from an estimated peak of 19,500 tons in 1936 to probably somewhat under 5,000 tons in 1971, even though automobile output increased sixfold over the same period.[12]

Major objectives of the Tin Research Institute are to counter the growing inroads of substitution and to develop new uses for the metal. In light of the facts just reviewed, it seems evident that TRI has had only limited success in the first of these two objectives. The institute has, however, noted rapid growth of the use of tin in chemical compounds, an area in which its research has been preeminent. According to the institute, triorganotin compounds, which are antimicrobial agents, are being increasingly used in antifouling paints, timber preservatives, and agricultural chemicals. Diorganotin compounds have become important in the manufacture of PVC and as catalysts in production of polyurethane foams; and the inorganic tin compounds have been introduced into ceramic and glass manufacture.[13] By the 1980s, organotin compounds were consuming 8 to 10 thousand tons of tin per annum, all in applications of tin unknown thirty years previously. At the Kuala Lumpur Conference on Tin in the fall of 1981, TRI researchers described chemical uses of tin as the "most exciting in terms of potential market growth."[14]

Future prospects. Ecologically, tin is an attractive metal. Although tin-based organic compounds are frequently highly toxic and thus can be used as insecticides, germicides, fungicides, and antifouling agents, the organotin biocides are degradable and ultimately break down into nontoxic substances. This degradation is speeded by exposure to sunlight and oxygen. Further, tinplate decomposes as it rusts, unlike aluminum and plastic containers that are equally likely to be discarded at unsuitable sites.

Since tin is an expensive metal, a continuing search for substitutes is inevitable. On the other hand, tin has unique and widely useful qualities, most important, resistance to corrosion, low melting point, nontoxicity, and ability to form numerous alloys. Thus, long-run forecasting of the demand for the metal is hazardous. At present, informed observers expect consumption of tin for tinplate to remain fairly stable for some time into the future, as food processing industries and the resulting use of tinplate grow in the developing countries but as the use of tin-saving electroplating also grows. The consumption of tin in soldering has been forecast to grow approximately as rapidly as overall industrial activity, as traditional heavy solder uses, such as in plumbing and automobile manufacture, decline but as usage in electronics increases. Within the electronics field, the amount of solder used per joint has shrunk to minuscule proportions with developments in microminiaturization and printed circuits; but the number of connections to be soldered has grown at a compensating pace. Chemical uses should increase but remain a small component of total consumption.

Future use of tin in various metal alloys is extremely difficult to project. Tin-free bronzes, substituting silicon, manganese, aluminum, and iron, have been developed; but all of these are inferior to copper-tin bronzes. The use of tin in type metal has virtually ceased as a result of innovations in typesetting. TRI has recently discovered promising new uses for tin in iron compounds, especially in the manufacture of sintered iron products.[15] In the 1950s, research at Pilkington Brothers, Ltd. led to the discovery and development of a new process for manufacture of plate glass in which molten glass is floated on a bed of molten tin—a development of revolutionary significance in glass manufacture but requiring only negligible amounts of tin. Another minor use for tin, the manufacture of pewterware, has experienced a boom in recent years. Overall, the most likely outcome is a modest increase in the total demand for tin over the remainder of this century.[16]

It is evident that, in an emergency, a highly developed and industrialized country could cut back drastically on its tin consumption without severe dislocations or heavy real cost. Tin is not, then, an essential commodity as such commodities have been envisaged in the NIEO debate. Even to the extent that tin is necessary and irreplaceable, countries such as the United States which produce no primary tin could fairly easily step up their recovery of secondary tin, as suggested in table 3.4.

Although secondary tin cannot be used for tinplating, it can substitute for primary tin in manufacture of solder, bronze, and bearing metals. As table 3.4 indicates, current output of secondary tin in the United States has fallen to almost one-half of its wartime level. I have not discovered in the literature on the industry any indication that the rate of annual output could not be restored to 30,000 tons or so for some time, or that the cost of doing so would be excessive, in a situation of shortage.

Strategic Stockpiles

Discussion of the role of tin in the economies of consuming nations cannot be concluded without giving some attention to the matter of strategic stockpiles, especially that of the United States. Whatever the actual availability of substitutes, the fact that the United States is totally dependent on imports for her supply of primary tin, with only one major producing country in the same hemisphere, has been inevitably a matter of serious concern to the nation's strategic planners. Other countries have also held national stockpiles of tin. The United Kingdom did so until 1960. In 1974, the Australian government sold its entire strategic stockpile on the basis of Australia's newly achieved self-

Table 3.4. Tin recovered from scrap processed in the United States (metric tons)

1935	24,900
1941–45 (average)	33,140
1950	31,680
1955	28,340
1960	22,050
1965	25,076
1970	20,001
1975	15,856
1980	18,638

Source: *Minerals Yearbook*, various issues.

sufficiency in tin production and in light of a worldwide shortage of tin and rapidly rising prices in that year. Canada also built up and then eliminated a national stockpile.

The United States' concern over dependence on foreign sources was enormously heightened during the Korean War. The year 1951 was one of bitter charges by the United States of price gouging and cartelization of tin, and of counter-charges by the producers that the United States was engaged in economic imperialism and in efforts to break the market. The details of the price gyrations, the United States' strategy of exclusive purchasing by a public agency and domestic controls, the bitter polemics, and the final purchase agreements need not be recounted here: they are given in a full and excellent account in Fox's 1974 book, as well as in Yip's 1969 work.[17]

By the time that the furor had subsided, the United States found itself in possession of a tin stockpile of 349,000 long tons, up from approximately 60,000 tons at the end of the Second World War. Fox, in his concluding observation on this episode, stated:

> Neither party, when stockpiling ceased in 1955 and the U.S. was left sitting on top of around 350,000 tons of tin (the equivalent of over two years' total world production or of seven years of U.S. annual consumption), had yet seriously considered what would happen if the absurdity of these figures were to be fully realised in the U.S.A., and if any part of that gigantic holding were to be released, however gently, on the world market.[18]

Since 1962, the United States General Services Administration (GSA) has been disposing of this aptly labeled "absurd" stockpile. On June 21 of that year, Congress authorized a reduction of 50,000 tons in the size of the stockpile, and later reduced the size of the long-run strategic stockpile objective to 200,000 tons. Disposals from the strategic stockpile are shown in table 3.5 along with average New York tin prices.

During the stockpile buildup, the United States had repeatedly assured the

Table 3.5. Disposals of tin metal from the United States' strategic stockpile and annual average tin prices

Year	Disposals (long tons)	New York price (cents per lb.)
1960	0	101.40
1961	0	113.27
1962	1,400	114.61
1963	10,626	116.64
1964	31,147	157.72
1965	21,733	178.17
1966	16,276	164.02
1967	6,146	153.41
1968	3,495	148.11
1969	2,048	164.43
1970	3,038	174.14
1971	1,736	167.37
1972	361	177.46
1973	19,949	227.22
1974	23,137	396.26
1975	575	339.57
1976	3,456	374.68
1977	2,635	533.26
1978	326	589.24
1979	0	707.29
1980	25	785.73
1981	5,920	735.90

Sources: International Tin Council, *Statistical Yearbook*, 1968; International Tin Council, *Tin Statistics, 1966-76* and *1970-80; International Tin Council, Monthly Statistical Bulletin* (March 1982).

producing nations that, if a decision were ever made to reduce or to eliminate the stockpile, the United States would not begin sales without appropriate consultations. Further, the Strategic and Critical Stockpile Act of 1946 contained a provision that "due regard" must be given "to the protection of producers, processors, and consumers against disruption of their usual markets."[19]

From the inception of the disposal program, the United States government did remain open for such consultations, in particular with the ITC. Following the 1962 Congressional action authorizing stockpile sales, GSA and the ITC entered into discussions that terminated with the cryptic announcement that "progress was made towards a mutual understanding of problems and objectives."[20] In November 1963 a second meeting was held with representatives of the ITC, followed by a similarly uninformative statement that "the [ITC] delegation was informed that the U. S. Government was preparing a long-range tin disposal plan."[21] At the time, the opposition that the ITC could mount against the United States' disposals was weakened by market conditions in which world consumption was substantially in excess of world production and

Table 3.6. Production, consumption, and prices of primary tin metal, 1961–64

Year	Production (long tons)	Consumption (long tons)	Average LME Price
1961	136,500	157,800	£ 888.6
1962	144,700	160,900	896.5
1963	143,000	161,700	909.7
1964	141,000	171,700	1,239.4

Source: International Tin Council, *Statistical Yearbook*, 1968.

prices were rising, as shown in table 3.6. Further, by the end of 1963 the ITC's buffer stock was exhausted following very heavy sales in the face of rising prices during October and November.

In 1964, with prices continuing to rise and with no tin in the ITC's buffer stock, GSA raised its sales limit to 400 and then 600 tons per week and sold a total of 31,147 tons over the year—the largest disposal to date.

Within a few years, the situation had changed considerably. The price of tin fell from an average level of £1,413 in 1965 to £1,296 in 1966, and to £1,206 (adjusting for devaluation of the pound during the year) in 1967. The ITC began to purchase tin for its buffer stock in an effort to stem the fall in price, and by the end of 1967 held 4,755 metric tons of tin metal. In 1966 another round of consultations took place in Washington among representatives of GSA, other interested United States government agencies, and a delegation from the ITC. These discussions led to establishment of an arrangement under which GSA and the ITC's buffer stock manager were to remain in direct contact in order to mitigate any conflicts between GSA disposals and the operations of the ITC's buffer stock. In 1967, the United States Department of State issued a public announcement that "the United States would, in principle, moderate its tin sales program if it should be inconsistent with the contingent operations authorized under the International Tin Agreement."[22] On July 1, 1968, GSA announced that it was suspending commercial sales indefinitely and would not reinstitute them without "appropriate consultations."[23] Disposals under the foreign aid program continued.

Commercial sales from the stockpile were renewed in June 1973, following a reduction in the strategic stockpile objective from 232,000 long tons to 40,500 long tons. This reduction was viewed with particular concern by the ITC and the tin producers since President Nixon's message in submitting stockpile legislation emphasized the objective of reducing inflation, noting that commodity prices had been rising at "unacceptably high rates" and that "by disposing of unneeded items in the strategic stockpile we can strike a critical blow for the American consumer."[24] There were heavy stockpile sales in 1973 and 1974, during which time tin prices were rising steeply under the influence of the OPEC-generated commodity boom.

In October 1976, the strategic stockpile objective was reduced again, this time to 32,499 long tons, which put 167,000 long tons into surplus. In the same month, however, the United States signed articles of accession to the ITC, thereby becoming obligated under Article 43 of the Fifth International Tin Agreement, which reads as follows:

> (a) A participating country desiring to dispose of tin from non-commercial stockpiles shall, at adequate notice, consult with the Council concerning its disposal plans.
>
> (b) At the time when a participating country gives notice of a plan to dispose of tin from non-commercial stockpiles, the Council shall promptly enter into official consultations on the plan with that country for the purpose of assuring adequate fulfillment of the provisions of paragraph (d) of this article.
>
> (c) The Council shall from time to time review the progress of such disposals and may make recommendations to the disposing participating country. Any participating country so concerned shall give due consideration to the recommendations of the Council.
>
> (d) Disposals from non-commercial stockpiles shall be made with due regard to the protection of tin producers, processors and consumers against avoidable disruption of their usual markets and against adverse consequences of such disposals on the investment of capital in exploration and development of new supplies and the health and growth of tin mining in the producing countries. The disposals shall be in such amounts and over such periods of time as will not interfere unduly with production and employment in the tin industry in the producing countries and as will avoid creating hardships to the economies of the participating producing countries.[25]

Despite the 1973 and 1976 reductions in the strategic stockpile objective, Congress did not act to authorize increased disposals, in large part in response to an early 1975 recommendation from the Joint Chiefs of Staff to postpone further tin sales for an indefinite period. In June 1978, with only 70 long tons remaining authorized for sale, GSA suspended all disposals.

On December 21, 1979, Congress approved a bill to authorize disposal of 30,000 long tons of tin and, in addition, to contribute 5,000 long tons to the ITC's buffer stock. This act, the 1979 Strategic and Critical Materials Stockpiling Revision Act, also explicitly forbids any action by stockpile administrators that would disrupt commodity markets.

GSA announced its intention of beginning auction sales at the rate of 10,000 long tons per year in March 1980; but the agency later, at the request of the United States Department of State, postponed initiation of these sales until July 1, 1980. Purportedly, the State Department did not wish the tin sales to begin until after the Bolivian national election on June 29, 1980.[26] The intensity of Bolivian concern with the stockpile disposal program will be noted below. In the same month (March 1980), the United States advised the ITC of its intention to contribute 1,500 long tons of tin to the buffer stock.[27]

The sales program got off to a slow and awkward start. On July 1, the New York price of tin was $7.84 per lb. GSA opened its sales program by announcing that it would offer approximately 500 metric tons every two weeks, on an open bid basis. All of the bids received in the first offering, ranging from $3.90 to $7.60 per lb., were rejected by GSA as unacceptably low. Similarly, on the second offering, bids from $6.53 to $7.53 were rejected. The basic problem seemed to be a widespread concern among potential buyers that the quality of the stockpiled tin was inferior. One buyer was quoted as saying, "We're planning to inspect the material, but the information we're hearing is that the quality [differs] from depot to depot." Another noted, "It's not in the best condition, but it's certainly usable. Many of the ingots have ruptures and the surface is oxidized. It's gray and discolored in appearance."[28] Nevertheless, GSA was not prepared to sell its tin at a discount from the market price. On July 30 the first sale was made, at a discount of 19¢ below the market price. In November, GSA shifted to daily fixed-price offerings, but without success. As shown in table 3.5 only twenty-five tons of stockpile tin had been sold by the end of the year. By that time, the market price had dropped, standing at $6.92 on January 1, 1981.

Stockpile sales picked up in 1981. Users reported that there were no quality problems. By the end of June 1981 sales totaled 855 tons, of which 540 had been sold in May and June. But prices continued to fall, averaging $6.10 in June; and conflict with the ITC emerged as the council began buying tin in April. Although stockpile sales were still small by historical standards, and GSA continued to refuse to quote prices significantly below the New York market price, producers and the trade press reacted with outrage to the unprecedented situation of simultaneous sales by GSA and purchases by the ITC. These buffer stock purchases ended with the entry of the mystery buyer to the market in July. GSA continued to sell, on a rising market, through the end of 1981, but in modest volume.

In June 1981, Associated Metals and Minerals Corporation (ASOMA), the parent company of the firm operating the only tin smelter in the United States, brought suit to halt the stockpile sales, alleging that the current sales program violated the law in being unduly disruptive of the market. In its successful defense, GSA maintained that it was not undercutting prices, but rather was following the market; that as long as producers were not reducing output in the face of falling prices GSA sales could not be viewed as disruptive; and that the disposal was a "crucial national defense program." In its filing, the government contended that "the stockpile revision program . . . is a long-term defense requirement that must proceed with the acquisition of needed materials and the disposal of excess materials, particularly tin, at an accelerated pace. Failure to generate substantial revenues from the sale of tin during the next few fiscal years will seriously jeopardize national defense interests."[29] The reference was to a provision in the 1979 law that the proceeds from sales of tin and silver were to be used to acquire other metals, particularly copper, for the stockpile—a requirement that critics have contended had greater relevance to the depressed state of the domestic United States' copper industry than to the nation's defense.

At the end of the year, GSA set off a renewed round of protests from producers by announcing that it intended to offer tin for sale on world markets as well as domestically. *Business Times* of Malaysia commented that this change in tactics might have been a response to the mystery buyer.[30] In a February 1982 editorial, *Tin International* labeled the new sales policy as "aggressive" and continued, "It is, of course, no thanks to the US that tin prices have not collapsed under the weight of supply surpluses compounded by stockpile sales."[31] Reason for the heightened concern over foreign sales may be inferred from a passage in the States of Malaya Chamber of Mines *Year Book* for 1980 which noted that "it is the threat to dispose of G.S.A. tin and its ready availability rather than the actual disposals which disrupt the tin market. Although only 25 tonnes of G.S.A. tin were sold in 1980, U.S. consumers know they can now purchase their requirements of tin on a day to day basis from the G.S.A. if necessary. Plant inventories of tin have therefore been cut down to the bone."[32] The problem would be multiplied if tin users around the world, and not only in the United States, reduced their inventories in anticipation of the availability of GSA tin.

In May 1980 the stockpile objective was raised slightly, to 42,000 long tons of tin. If target holdings stay at this level (42,674 metric tons), GSA will remain a factor in the tin market for many years. Following sales of 2,580 long tons in January 1982, cumulative disposals totaled 159,118 long tons from an initial stockpile of approximately 349,000 long tons, leaving around 189,900 long tons in the stockpile, of which 147,900 composed a saleable surplus. It is not certain, however, that the strategic objective will remain as low as it is now. In August 1981 a bill was introduced in the United States Senate to establish a general stockpile objective of three years' imports based on a moving five-year average of net imports. If enacted, this bill would raise the objective for tin holdings to about 140,000 long tons.[33] The Department of Defense has endorsed the concept of a stockpile of strategic materials large enough to support a major war on two fronts, lasting three years.

Opinions regarding the effects of the United States' stockpile disposals vary widely. In a 1976 article, Gordon W. Smith and George R. Schink argued that from the establishment of the ITC in 1956 its buffer stocks have been far too small to have stabilized tin prices in the absence of other kinds of market intervention. The power of the ITC to impose export restrictions on its producing members has given the council the ability to prevent prices from falling below a predetermined floor level in almost all instances. "The outcome," Smith and Schink maintained, "has been a loose, informal 'commodity agreement' in which the GSA defended fairly high, but unknown, ceiling prices and the ITA [International Tin Agreements], fairly low, known floor prices. The effective price band was quite wide (±20 to 25% around long-run trends), but market instability was reduced significantly."[34]

Smith and Schink supported their argument with a set of simulations of hypothetical world tin prices and output run on a model of the world tin economy developed by the Wharton School of Business at the University of Pennsylvania, in which they simulated tin market conditions from 1956 to 1973. First, they assumed no intervention by either GSA or the ITC, and second, use of the ITC's buffer stock only. These simulations were compared with the actual experience in which both GSA and the ITC intervened; and the results indicated both that price fluctuations would have been much greater without the GSA disposals and that the ITC's buffer stock operations by themselves would have made very little difference given the limits imposed on the buffer stock manager by the maximum sizes of his stocks. As Smith and Schink noted, these simulations are deficient in one important respect in that the Wharton model is not designed to take account of export controls. They did, nevertheless, conclude that the United States' strategic stockpile had been the "major factor stabilising the market since 1960."[35]

A number of informed and involved people with whom I spoke would disagree strongly with these views that the United States' stockpile program has had a beneficial effect on the world tin industry. It was pointed out that one of the basic reasons for the establishment of the ITC had been fear of "burdensome surpluses" of tin which were in turn the result of excess capacity developed in response to the rapid and massive buildup of the stockpile. Critics also asserted that the 150,000 metric ton deficit of production below consumption that followed the working off of this excess capacity was enhanced and protracted by the damper GSA's role in filling this deficit put on private investment in tin mining. Still another concern is with the lead time that will be needed by the tin industry to adjust to the inevitable cessation of stockpile disposals: unless there are several years in which suppliers have an incentive to invest in the opening of new mines prior to the end of GSA sales, it is feared the world may experience a protracted period of serious tin shortage.

The common thread in these criticisms and concerns is that while the disposals may have had a stabilizing effect on the tin market from year to year, and may have prevented short-run deficits at existing prices, the long-run effect of the entire United States' strategic stockpiling program—both acquisition and disposal—has been unsettling to the industry.

Not all of those who are critical of the disposal program are willing to concede that it has in reality stabilized prices. To the contrary, many would argue that studies relating price movements to actual sales by GSA fail to observe the crux of the issue. No matter how careful and skillful GSA may be, and no matter how closely it may coordinate its disposals with the ITC, the overall political process by which the strategic stockpile policies have been formulated and changed has injected a major element of uncertainty and consequent price instability into the tin market. Recent market forecasts, based more on the predicted fortunes in Congress of one or more stockpile disposal bills than on estimates of production and demand, have often been inaccurate.[36] Speculative price movements on the LME have responded sensitively to rumors and political events in Washington. Tin users have let stocks run down to very low levels in anticipation of disposal programs that failed to materialize and have then driven prices up in their efforts to restore normal working inventories. In an article written a year and a half before final passage of the 1979 Stockpile Act, the *Wall Street Journal* commented on one of the periods of greatest uncertainty, quoting a tin merchant as complaining, "Whenever Congress breathes, selling enters the market. It's ridiculous."[37]

A 1980 editorial in *Tin International* made much the same point, observing that "as the maximum projected annual sale from the US stockpile of 10,000 tons represents only around 5% of current world supplies, the influence of stockpile sales on volatility of the tin market is out of all proportion to their real weight."[38] These sales are, as the editorial suggests, much smaller than those of some previous years. But their impact, especially when the tin price is falling, ought not be minimized. To the extent that published econometric estimates of supply and demand elasticities for tin are correct, an increase of 5 percent in the world supply of tin (i.e., a shift to the right of the supply curve of that scale) should lead to a short-run price decline in the range of 6 to 10 percent.[39]

In a recent unpublished master's degree thesis, Jindarah Phangmuangdee has called attention to a fundamental distinction that must be made between the disposal program from 1962 through 1968 and the program since sales were reinstituted in 1973. The purpose of the earlier program, she noted, was to meet world supply deficits. The later program was publicly proclaimed as being designed to combat inflation within the United States. The 1973 announcement of resumed stockpile sales for this domestic objective was particularly alarming to the producing countries in light of the fact that tin prices to United States' consumers were rising as much because of a decline in the exchange value of the United States' dollar as because of higher production costs and tin prices expressed in the producing countries' currencies. Further, Jindarah observed, it was impossible to reconcile President Nixon's statement that the disposals would "strike a critical blow for the American consumer" with a Washington press communique of June 6, 1973 stating that "GSA intends to conduct disposals in an orderly way so as to minimize effects on world tin market prices and the orderly conduct of world tin markets."[40]

The 1979 Critical Materials Stockpile Act was particularly offensive to Bolivia, a nation whose tin mining costs are the world's highest and which remains dependent on tin for a great deal of its export earnings. In an address opening the Ninth General Assembly of the Organization of American States in La Paz in October 1979, Bolivian President Walter Guevara Arze called the proposed sales "a grave threat to the economy of a friendly country."[41] Subsequently, following President Carter's approval of the stockpile act, the Bolivian delegate won a 13 to 12 vote at the OAS Assembly on his motion to condemn the disposals as an "aggressive" act. To avert a serious rupture within the OAS, a compromise resolution urging the United States "in accordance with the principles and norms of the OAS, not to make sales of tin which would prejudice the economy of Bolivia" was substituted.[42] Recent debate between the Bolivian delegates to the ITC and those from the United States makes it clear that the Bolivians assume, as a matter of course, that the objective of the United States' stockpile disposal policy remains in 1982 as it was explicitly described by President Nixon in 1973—to hold down the price of tin.

Whatever its effects may have been on short-run stabilization or long-run destabilization of the tin industry, it is evident that the United States' strategic stockpile has been a formidable deterrent to any interest on the part of producing nations in abandoning the ITC and establishing an association of producers alone. As the debates on stockpile policy, as well as actual legislation and executive decisions, have shifted pointedly from emphasis on strategic considerations to overt focus on economic and political issues, the threat that the United States would use its stockpile as a weapon to combat any producers' cartel has become more real.

4. From Cartel to Cooperation: The International Tin Agreements

"One thing an economist studying the tin industry really must understand is that there hasn't been a market-determined price for tin in half a century. What we've had is political pricing." In the form just quoted, this remark was made to me by an advisor to one of the national delegations to the International Tin Council; but it was repeated in one variation or another by so many of those associated with the industry that I take it to be a truism of the trade.

The Prewar Agreements and the International Tin Committee

The Bandoeng Pool. Late in February 1921, representatives of the colonial governments of the Federated Malay States and the Netherlands Indies met in the Javan town of Bandoeng to devise and launch a cooperative effort to halt a plunge in the price of tin from a peak monthly average of £396 per long ton in February 1920 to £166 twelve months later at the time of the meeting. The discussions were held in secrecy, with officers of the major tin companies attending as governmental representatives. For the first time, governments of the tin producing territories agreed to cooperate with each other, and with private interests, to intervene in the tin market.[1]

The meeting was held in the middle of a period of worldwide crisis in the tin industry. In early 1920, postwar demand for tin began to slacken. A combination of sustained high levels of production and the renewed availability of shipping facilities following wartime curtailment led to an increase in supplies of tin metal that continued even as demand was falling off. Writing in the authoritative annual publication *Mineral Industry*, E. Baliol Scott described the following year as "the most disastrous year that the tin miner has ever experienced in recent times."[2] The average cash price of tin fell from £296 per metric ton in 1920—itself regarded as a depressed year—to £166 in 1921.

Scott, writing his report for 1921 in mid-1922, stated that at the time of writing he could confirm the existence of an "Eastern Pool" whose establishment was already widely rumored; but, he added, little was yet known of its operations other than that it held an accumulation of 19,138 long tons of tin, and that the participants were the Federated Malay States government, the Netherlands Indies government, and two smelting concerns, the Straits Trading Company and Billiton Maatschappij.[3]

In April 1923, the partners of the Bandoeng Pool, which was by then publicly known by that name, announced a liquidation program under which each party would be free to release 5 percent of its holdings each month. If all participants disposed of the maximum allowed every month, the schedule would amount to a total of 880 long tons of tin per month removed from the pool for twenty months and would exhaust the pool by the end of 1924.

World demand for tin remained strong through 1924; and by the end of the year the Bandoeng Pool dissolved with its stocks exhausted and with the December average price of tin standing at £262 per long ton. Thus the affair ended on a highly satisfactory note for the pool operators and indeed for the worldwide tin industry as a whole.

Subsequent writers have been cautious in their assessments of the extent to which the pool actually contributed to the recovery and in their evaluations of its overall effects. Elizabeth S. May stated that "though the government support undoubtedly steadied the market, it was the growth in world demand that made the success possible."[4] Puey Ungphakorn noted that the decision to begin disposal of the pool's stocks was made at a time when the price of tin stood at £188 per long ton and that at least some of the participants did lose money from their carrying and sale of tin. He concluded that the most important lesson learned from the experience was that it was unwise for private companies to engage in such schemes since this form of market manipulation is by its very nature highly risky and costly in terms of the capital tied up in stocks withheld from the market.[5]

In contrast, Yip Yat Hoong stated that "as the first attempt at international tin control, the Bandoeng Pool was a great success. Besides alleviating to some extent the instability of the tin price during this period, it enabled the Government of the Federated Malay States to make a profit of over half a million dollars."[6] Yip, however, did note that it was very difficult to say whether the 1922–24 price increase was due to the pool or to a general business recovery. Yip followed Klaus E. Knorr in concluding that the pool was unambiguously useful from the producers' point of view, in that it softened the impact of the postwar crisis on producers and their employees, kept some marginal mines from going out of business, and smoothed out the fluctuations in stocks and supplies put on the market during the crisis and recovery. Knorr asserted that the pool "could not stem the downward trend of the market, and, in the final analysis, it was the subsequent expansion of tin consumption that made this intervention a success."[7] William Fox, after noting that the pool was not the only factor in the price rise, did go on to state, "The price recovery which resulted from or coincided with the later months of the Bandoeng pool had very beneficial effects on the export earnings of all producing countries. . . . In the eyes of its proponents, the pool had proved conclusively that a degree of control over stocks meant a degree of control over price. That control was proved profitable."[8]

The crucial aspect of the Bandoeng Pool experience is undoubtedly captured in Fox's emphasis on "the eyes of its proponents." Whatever its effects may have been in reality, the perceived success of the pool set the stage for a transformation of the tin industry, lasting to the present day and by now thoroughly established and apparently viable, in which tin prices and quantities

Table 4.1 Number of tin dredges in Malaya and dredging output

	Number of dredges			Dredging output (pikuls)	
Year	Operating	Stopped	Total	Total	Average
1913	1	0	1	3,780	3,780
1920	20	0	20	78,220	3,911
1921	30	0	30	108,650	3,622
1922	22	11	33	125,600	5,709
1923	33	7	40	193,240	5,856
1924	38	4	42	210,000	5,526
1925	40	2	42	211,630	5,291
1926	41	11	52	213,330	5,203
1927	48	22	70	391,670	6,072
1928	89	0	89	433,772	4,874
1929	105	0	105	605,496	5,767
1930	69	38	107	550,249	7,975

Source: Yip Yat Hoong, "Malaya Under the Pre-War International Tin Agreement," *Malayan Economic Review* 8,1 (April 1963): 83.

put on the market have been subject to manipulation and control under intergovernmental agreements in all but a few years immediately after the disbanding of the pool and following the Second World War. Indeed, as Knorr pointed out, whatever its short-run effects may have been, the Bandoeng Pool set in motion a chain of events which made later joint governmental intervention far more likely, perhaps even inevitable. The gradual disposal of the pool's stocks in 1923 and 1924 dampened investment in new tin-mining facilities at a time when demand was rising above existing production capacity. Then, in 1925–26, tin prices rose to exceptionally high levels following the exhaustion of the pool's holdings. The response to this sudden—and in Knorr's opinion excessive—rise in price was an investment boom that not only eliminated the deficit but overshot and created an excess of world capacity to produce tin. "The pool operations," he concluded, "certainly did not cause the tin boom of the middle 1920's, but they contributed to its magnitude."[9]

The boom of the 1920s. The impact of the investment boom of the late 1920s was heightened by the fact that it followed shortly after the introduction of a major large-scale capital-intensive technological innovation: the first tin dredge had been introduced in Thailand in 1907, and the new technique spread to Malaya in 1912. Table 4.1, taken from Yip's work, shows the growth in both size and number of dredges in Malaya during this period, and the severity of the collapse in 1930.

Yip called attention to the time lag involved in investment in tin dredging, noting that the boom was basically in response to conditions emerging in 1926 but did not lead to substantially increased output until 1928. Hence, the boom overshot and heightened the extent of overcapacity in 1930.

A more general picture of events in the tin industry from what Scott described

Table 4.2. World production, consumption, and average price of tin, 1920–31

Year	World mine production of tin-in-concentrates (long tons)	World consumption of primary tin metal[a] (long tons)	Consumption minus production[b]	Average London price (£sterling)
1920	122,600	125,000	2,400	296.1
1921	116,300	80,000	−36,300	165.4
1922	123,000	130,000	7,000	159.5
1923	125,900	134,000	8,100	202.3
1924	142,000	134,000	−8,000	248.9
1925	145,900	150,000	4,100	261.1
1926	143,500	146,000	2,500	291.2
1927	159,200	150,000	−9,200	289.1
1928	178,000	169,000	−9,000	227.2
1929	195,800	184,000	−10,800	203.9
1930	178,900	163,000	−15,900	142.0
1931	143,900	141,000	−2,900	118.5

a. These consumption figures are estimates and must be viewed with some skepticism. During the 1920s *Mineral Industry* refused to publish figures on total worldwide consumption on the grounds that data submitted to its editors were incomplete and inaccurate, while at the same time publishing production figures that are quite close to those that appeared later in the ITC's *Statistical Yearbook*.

b. To put these very rough estimates into perspective, normal or desired stocks held at all levels of the industry were thought to be around 60,000 to 70,000 long tons.

Source: International Tin Council, *Statistical Yearbook*, 1964.

as the "disastrous" year of 1921 through the years of the Bandoeng Pool and the investment boom to the collapse at the end of the decade is given in table 4.2.

One significant result of the 1920s expansion was to intensify the role of European capital. In 1920, Chinese miners accounted for approximately two-thirds of Malayan tin production; but by 1930 European interests produced 63 percent of Malaya's output.[10] The influx of European investment capital led, in turn, to a sharp increase in worldwide concentration of the tin industry. The single most important contributor to this growth of concentration was the Anglo-Oriental Mining Corporation, Ltd., formed with the support of a group of British banks in February 1928, at the initiative of John C. E. Howeson who subsequently headed the firm,[11] and Simon Patiño whose Bolivian tin mining interests accounted at the time for approximately 55 percent of that nation's output and who also owned substantial smelting interests in Great Britain.

The Netherlands East Indies was the world's only other large tin producer; and in 1929 the industry of that colony was already far more concentrated than that of Malaya, Bolivia, or Nigeria. The government-owned Banka Company and the Billiton Company in which the government had a five-eighths stock interest accounted between them for over 95 percent of the tin-in-concentrates produced by the Netherlands East Indies. All the concentrates produced in the Netherlands East Indies were either smelted in a government-owned smelter on Banka Island or shipped to the Straits Settlements for smelting.

In brief overview of the extent of concentration of the tin industry at the end of the 1920s, large tin-mining interests in Bolivia, Malaya, and Nigeria were interlocked through Anglo-Oriental and a subsidiary known as the London Tin Syndicate (later to become the London Tin Corporation, which was transformed by an exchange of shares into the parent of Anglo-Oriental). In Bolivia, the Patiño, Aramayo, and Hochschild companies controlled about 85 percent of the country's tin mining. In Malaya, European interests had grown from one-third to two-thirds of the tin industry over the decade: in addition to the Anglo-Oriental or London Tin Syndicate group of mines, a second European interest grouping, the so-called Gopeng Group, had arisen among mining companies not connected with Anglo-Oriental but tied together through their common use of the managing agency house of Osborne & Chappel. The British-owned mines of Malaya were also linked with each other by a number of interlocking directorates.[12] In both Malaya and Thailand, however, a large number of small-scale, independent, ethnically-Chinese miners remained outside the consolidation movement. At the smelting level, Consolidated Tin Smelters (a joint subsidiary of Patiño Mines and Enterprises and Anglo-Oriental), the Straits Trading Company of Singapore and Malaya, and the Netherlands East Indies' Banka smelter smelted an estimated 84 percent of the world output of tin metal in 1929. Three governments, the United Kingdom, the Netherlands, and Bolivia, exercised sovereignty over regions that produced 174,998 long tons of tin-in-concentrates out of the world production of 195,800 long tons in that same year, or 89.9 percent of the total.

Thus, the setting in which formal worldwide cartelization of the tin industry took place included the following: a basic technological shift favoring larger-scale and more capital-intensive methods of production;[13] an increasing degree of participation by European operating companies and financial institutions; a decade's experience of severe demand fluctuations; a situation of substantial excess capacity following an unrestrained investment boom; sharply increased but still moderate concentration at the mining level colored by the emergence of a large and powerful financial group; and extremely high degrees of concentration at the smelting and governmental levels.

The International Tin Committee. On July 11, 1929, at a meeting in London, a Tin Producers' Association (TPA) was formed. After protracted negotiations, and the failure of a voluntary output restriction program, TPA announced in January 1931 that an International Tin Control Scheme had been approved by the governments of the Federated Malay States, Nigeria, Bolivia, and the Netherlands East Indies. The scheme was to take effect on March 1 and was to run for two years. It was later agreed to extend the life of the scheme until August 1934. The basic objective of the scheme was to restrict the tin exports of the member nations through the assignment of quarterly export quotas. It was to be implemented by a newly formed International Tin Committee (ITC) made up of representatives from each of the signatory nations.[14] Somewhat ironically, the United Kingdom was not a party to the 1931 agreement (Malaya and Nigeria having joined under the auspices of the Colonial Office); and in fact the uncontrolled output of the tin mines in Cornwall more than doubled between 1931 and 1932, from 598 long tons to 1,346.[15] In September 1931, Thailand joined the ITC after having expanded its output steadily from 7,527 long tons in 1928 to 12,447 long tons in 1931. As the only important producer outside the original members of the ITC, Thailand demanded, and received, highly favorable terms, with its quota set at a flat rate of 10,000 long tons per year.

The Second International Tin Agreement came into effect at the beginning of 1934, covering that year through 1936, and with the Belgian Congo, French Indo-China, Portugal, and the United Kingdom (Cornwall) joining as new signatories in mid-1934. A third agreement, with Portugal and the United Kingdom withdrawing from the ITC, ran from 1937 through 1941; however, the outbreak of the Second World War shifted the emphasis of the ITC from restrictions on production to allocations of output to consuming nations. A fourth agreement was signed, covering 1942 through 1946, but the only note-worthy decision taken under that agreement was not to attempt to draw up a fifth agreement.

The operations of these agreements, particularly the first three, have been described and analyzed in detail by Knorr, Puey, Yip, and Fox; the narrative account will not be repeated here. For the purposes of this study, however, certain aspects of the experience under what was in effect a producers' cartel do need to be reviewed.

Knorr and Yip were cited previously as observing that the operations of the Bandoeng Pool contributed to the magnitude of the subsequent investment boom and to the collapse of the following five years. This view is widely accepted among those who have studied and written on the matter, although I have not found any commentator who argued that the Bandoeng Pool was solely or even primarily responsible for either the boom or the crash. The turmoil in the tin markets of the late 1920s, and its consequences for the industry, did lead quite directly to the formation of the TPA, the abortive voluntary restriction scheme, and to the international tin agreements. Writing in 1936, J. K. Eastham argued that "the present restriction scheme is primarily designed to save the tin trusts from the consequences of mal-investment of capital and that it has further been used for the furtherance of schemes for the more complete integration of the industry."[16]

Yip has taken a somewhat broader view of the motivations of the British government. He noted that in 1929 the British Commonwealth produced 43.3 percent of the world's tin, but smelted 83.7 percent. Great Britain's territorial interests linked it with both the high-cost producers of Nigeria and the low-cost producers of Malaya. Britain's smelters relied on extremely high-cost Bolivian tin ore; and it was considered highly doubtful that the Bolivian industry could have survived the 1930s without an international agreement. "Under these circumstances," Yip concluded, "free competition which would certainly lead to an economic war for the survival of the fittest within the industry would not be in the best interest of Britain. In the face of so severe a depression as during 1929–33, a better way out of the difficult situation would be to advocate restriction, even if by so doing a low-cost producing country like Malaya was being unfairly penalised."[17]

It would thus appear that whatever the motivations may have been, the active role of the British government as exerted through the Colonial Office was essential to hold together the disparate interests of the tin producers—large and small, national and international, high-cost and low-cost, European and Chinese.

As it was, the survival of the tin agreements through the 1930s was precarious. Disputes as to the appropriate levels of export quotas were endemic, typically pitting low-cost producers pressing for higher quotas against high-cost producers seeking greater degrees of restriction. The output of nonmembers was a source of severe strain.

It has already been noted that Thailand joined the first agreement on extremely generous terms involving virtually no meaningful restriction. The second agreement contained a provision under which any signatory nation was free to withdraw from the agreement if the output of tin metal by nonsignatory nations exceeded 25 percent of estimated world production or 15,000 tons in any six-month period. The agreement came close to falling apart almost immediately after its inception, when nonmember output rose to 24 percent of the world total during the first half of 1934. It was saved only by the admission to the ITC of the Belgian Congo, French Indo-China, Portugal, and Cornwall on July 10 of that year. The terms of admission for these four were similar to those that had been granted to Thailand.

The fragility of the prewar ITC is further illustrated by May's comment that "there can be little doubt" that the first agreement was extended only because of "the remarkable change in the tin market in the year 1933," particularly a large and unanticipated increase in deliveries to the United States."[18]

Although neither Knorr nor Fox makes the point, a reading of their accounts of the second agreement indicates that internal dissensions over production in excess of quotas were mounting to an intolerable extent when, in the middle of 1935, Bolivia began to fail to meet its quota. It would therefore appear that the ITC's survival was bolstered by the Chaco War, a conflict waged from 1932 through 1935 that included among its ravages a heavy toll of casualties from Bolivia's tin miners. There was considerable doubt at the time among tin traders whether the ITC would continue to exist. J. W. F. Rowe noted, "In 1936 the price fell considerably, but this was only due to the long protracted negotiations for the renewal of the agreement, and the expectations by merchants of their ultimate failure."[19]

The ITC's membership was limited to producing nations, and it is most accurately described as a producers' cartel. Yet beause of the participation of governments, it was unavoidably sensitive to international political considerations. In response to growing unease and hostility toward the ITC among U.S. tin buyers, the Foreign Affairs Committee of the U.S. House of Representatives conducted a *Tin Investigation* in 1934–35. In its final recommendations, this committee proposed that the United States take appropriate actions to discourage the import of tin metal and encourage the import of tin concentrates to facilitate the development of a U.S. tin smelting industry; to encourage higher levels of exploration for domestic tin ores and research into substitutes for tin; and to control and from time to time prohibit the export of tin scrap.[20] May viewed this investigation and the committee's report as "a serious warning to the powers that dominated the International Tin Cartel."[21] Knorr described the ITC as "perturbed by these reactions."[22]

The impact of these pressures from the principal consuming country, in terms of real effect on tin price and output levels, appears to have been slight. Knorr commented that "the history of international tin control furnishes ample evidence to the effect that neither the protests arising in consuming countries nor their advisory representation at the meetings of the ITC afforded adequate protection to consumers' interests. Indeed, under the prevailing circumstances, it could hardly have been otherwise."[23]

Puey took a similar point of view, noting that the so-called consumer representatives turned out to be from the tinplate industry, one which was in a position to pass on price increases to its customers and whose interests were therefore probably more in assured supplies and uniform prices than in low prices. In his opinion, the effect of consumer representation was negligible.[24]

The final aspect of the prewar ITC's operations to be discussed is the role of pools and buffer stocks. The primary tool of the committee was export restriction. Each country was assigned a standard tonnage based mainly on 1929 production, and export quotas expressed as a percentage of the standard tonnage were set each quarter. In addition, however, there were three tin pools established during the 1930s.

The first of these pools was a private one, formed in late 1931 with the approval of the four ITC signatory governments and with a commitment for mutual cooperation. Indeed, the first agreement was provisionally extended into 1934 in order to assure the pool operators that export restriction and the resulting enhancement of tin prices would continue throughout the period of disposal of the pool's tin; and the chairman of the ITC was made chairman of the pool. The pool operated in complete secrecy, and neither the identities of those involved nor the financial outcome is known. May stated that Anglo-Oriental, Patiño, and the Dutch government were all participants and that the pool had accumulated 21,000 long tons of tin by January 1932.[25] It has also been suggested that the pool acquired most of its tin by transfer from several large holders rather than by purchases on the market. While acquisitions for the pool from whatever source undoubtedly supported prices in 1931, the imposition of extremely stringent export quotas in 1933, designed to facilitate disposals from the pool, stirred harsh debate within the ITC, with representatives of interests not involved in the pool expressing dismay at the severity of the quotas (33.3 percent throughout the year). Two ITC buffer stocks were subsequently established, the first in 1934 and the second in 1938. In 1935, the chairman of the ITC explained, with a degree of sincerity that cannot be ascertained today, that the buffer stock was designed "to protect consumers against a possible sudden price rise due to possible sudden increase in demand which a production quota alteration might not be sufficiently rapid to meet."[26] The 1934 buffer stock has been described by Puey as a "complete failure,"[27] having been accumulated at a time when price was high, demand was growing, and private stocks were low. The stock was liquidated in 1935 following the drop in Bolivian output.

The 1938 buffer stock was acquired in large part as a response to a recommendation of the previous year by the League of Nations' Committee for the Study of Raw Materials that commodity buffer stocks be created in which both producers and consumers would participate. The tin stock was not supported by consumers, but by ITC member contributions under a special 7.5 percent addition to the export quota. A total of 15,611 long tons were accumulated by mid-1939, but the entire stock was eliminated in open sales during the two weeks following the outbreak of World War II in September of that year. These buffer stocks were consistently opposed by the Malayan Chamber of Mines; and the establishment of the 1934 buffer stock led to the resignation of the chairman of the Tin Producers' Association in protest against a device he thought would serve only to perpetuate export restrictions.[28] The minor but contentious role of buffer stocks in the prewar agreements stands in sharp contrast to the major emphasis and reliance placed on the operations of such stocks under the postwar international tin agreements.

Knorr, Puey, and Yip wrote ex post economic analyses of the tin restriction scheme of the 1930s. None was opposed to some form of public control: to the contrary, each noted shortcomings with the functioning of free markets in the tin industry which would warrant governmental intervention. All three agreed that in the short run the demand for tin was extremely price-inelastic, and that in addition demand was unstable and subject to severe fluctuations in response to changes in general business conditions in the consuming countries.

Further, each cited reasons why supply responses to price changes were likely to be inappropriate. Knorr and Yip emphasized the lagged reaction, or the fact that expansion of capacity through investment in new facilities—especially modern dredges—was such that the additional output would not be forthcoming until two or two and a half years after the price rise that had stimulated it. Puey noted this lag and also called attention to the problem of irreversibility: even in the cases of gravel-pump or other forms of mining which could expand more rapidly than the dredge sector, subsequent contraction without severe losses was not feasible.

None of the three was opposed to market intervention that would smooth out price fluctuations and prevent the recurring crises of overcapacity that they thought would otherwise result. Knorr cited the machinations of the Bandoeng Pool and the Anglo-Oriental Corporation in making the point that the structure of the tin industry had evolved in such a way that in any event there would be private interference with the market mechanism which, in the absence of countervailing governmental actions, would be more likely to exacerbate the industry's instability than to exert an equilibrating influence.

Yet despite their receptivity to public action, all three were harshly critical of the prewar ITC. Knorr wrote that the scheme had the apparent purpose of saving both the high-cost and low-cost producers in the tin mining industry, along with all of the excess capacity that existed in 1931. The result, in his judgment, was an inevitably futile effort to eliminate the effects while preserving the cause of the problem.[29]

Puey concluded that the ITC had been quite successful in maintaining an artificially high price level, but only at the long-run costs of stimulating research on substitutes for tin; engendering resentment of control by both consumers and the governments of consuming countries; increasing excess tin smelting capacity, which led to a subsequent decline in the British Empire's share of smelting; and preserving old, high-cost, inefficient mines while preventing the expansion of low-cost producers. The pools, in his opinion, had contributed very little to price stabilization, while the export quotas had the effect of dampening minor short-term price fluctuations at the expense of heightening price reactions to major demand shifts.[30]

Yip argued that the ITC had played an important constructive role in defending the tin industry in the crisis years of 1931 to 1933, but was wrongly continued thereafter as a device for holding up prices after demand had returned to more normal levels. "The overall effect," he concluded, "was to prolong the existence of excess capacity in the tin industry."[31]

The world after the Second World War turned out to be a very different place for the tin industry from that of the 1930s. It seems most unlikely that a set of agreements similar to the prewar series, which led a precarious existence from one renewal to the next as it was, could have survived for long in the new economic and political environment. Yet there were lessons to be learned from the experiences of the International Tin Committee; and the international tin agreements of the 1950s, 1960s, and 1970s can be thoroughly understood and evaluated only in terms of the extent to which the tin producing and consuming countries alike correctly perceived the import of these lessons to changing circumstances.

The Postwar Agreements and the International Tin Council

Preliminary negotiations and the first agreement. The creation of successors to the prewar international tin agreements and the International Tin Committee was neither a quick nor an easy task. In the spring of 1946, while the committee was still discussing the issue of a fifth agreement and its own continuation, a spokesman for the United Kingdom settled the matter by an announcement that his government would not under any circumstances sign such an agreement. In October, the British government called an international tin conference in London, inviting all members of the ITC, other major producers, and major importing and consuming nations. At this conference, the United States took the position that it would support an international organization designed to deal with the problems of "burdensome surplus"[32] in tin markets, provided such an organization was established along the basic principles then being discussed by the United Nations Conference on an International Trade Organization. On the other hand, the United States' delegate expressed his country's unequivocal opposition to any international agreements that restricted markets or fixed prices. Both the United States and the United Kingdom delegates supported the establishment of a tin study group; and the other participants accepted this proposal, in part because it was generally believed that in a period of postwar scarcity, with the need for rehabilitation of Southeast Asian tin producing facilities, and in light of the United States' stockpiling purchases, no burdensome surplus would emerge for at least two years.[33]

The subsequently convened International Tin Study Group held its first meeting in April 1947, with delegates or observers from twenty nations attending. At this meeting, the study group accepted the principle of establishment of an international tin organization within the framework of the Havana Charter.[34] In April 1950, after preparing and reviewing four draft proposals, the study group requested the Secretary-General of the United Nations to convene a tin conference, which met in October and November of that year only to adjourn after failing to come to any agreement. In August 1953, the study group completed another draft agreement which was submitted to the reconvened tin conference and approved on December 9. The agreement was to come into effect when ratified by the governments of countries holding at least 90 percent of the votes provided for producing nations and also by the governments of at least nine countries holding one-third of the votes provided for consumer members. The 1953 agreement did not come into effect until July 1, 1956, following ratification on May 15 by Indonesia.

Chapter VI of the Havana Charter dealt with intergovernmental commodity agreements. Article 55, the first of the chapter's sixteen articles, noted that international commodity trade "may be affected by special difficulties such as the tendency towards persistent disequilibrium between production and consumption, the accumulation of burdensome stocks and pronounced fluctuations in prices," which might have "serious adverse effects on the interests of producers and consumers, as well as widespread repercussions jeopardizing the general policy of economic expansion." Such difficulties might, "at times," justify international commodity agreements.

Article 57 specified the objectives for which intergovernmental commodity agreements would be deemed appropriate. These included prevention or alleviation of "serious economic difficulties which may arise when adjustments between production and consumption cannot be effected by normal market forces alone as rapidly as the circumstances require"; prevention or moderation of "pronounced fluctuations in the price of a primary commodity with a view to achieving a reasonable degree of stability on a basis of such prices as are fair to consumers and provide a reasonable return to producers, having regard to the desirability of securing long-term equilibrium between the forces of supply and demand"; and assurance of "equitable distribution of a commodity in short supply."

Article 60 provided that the membership in any commodity agreement reached under ITO auspices should be open to all member nations and to invited nonmembers; that there should be equitable treatment of nonparticipants; that there should be full publicity given to the agreement and to the deliberations taking place both at its establishment and during its operations; and, most significantly in regard to the tin negotiations, that provision should be made for adequate participation of consuming countries.

Article 62 provided that such an agreement was to be entered into only after a commodity conference had found, or there was general agreement, that a "burdensome surplus of a primary commodity has developed or is expected to develop," or that "widespread unemployment or underemployment in connection with a primary commodity . . . has developed or is expected to develop."

Article 63 set down additional governing provisions, including the following. Agreements should be "designed to assure the availability of supplies adequate at all times for world demand" at fair and reasonable prices and, "when practicable, shall provide for measures designed to expand world consumption of the commodity." Consuming and producing countries should be given equal voting power.

Article 65 specified that commodity agreements should not be established for more than a five-year period.

Although the Havana Charter was never ratified, its provisions remained important to the international economic activities of the United Nations and, more specifically, to the international tin agreements. The General Agreement on Tariffs and Trade (GATT) contained a protocol under which signatories agreed to apply the charter's commercial policy rules "to the fullest extent not inconsistent with existing legislation."[35]

A review of the provisions of the International Tin Agreement of 1953 indicates its fundamental conformity to the Havana Charter, although not absolute adherence to every provision or even, in a few instances, to the spirit of the charter.[36] The preamble found "reason to expect that widespread unemployment or under-employment in the industries producing and consuming tin may develop out of special difficulties to which international trade in tin is subject, including a tendency toward persistent disequilibrium between production and consumption, the accumulation of burdensome stocks and pronounced fluctuations in price." It went on to note "that a burdensome surplus of tin is expected to develop and is likely to be aggravated by a sharp reduction in purchases of tin for non-commercial stocks." The reduction referred to is, clearly, that resulting from completion of the buildup of the United States' strategic stockpile of tin. The preamble asserted the belief that "in the absence of international action this situation cannot be corrected by normal market forces in time to prevent widespread and undue hardship to workers and the premature abandonment of tin deposits." Finally, the preamble recognized "the need to prevent the occurrence of shortages of tin and to take steps to ensure an equitable division of supplies" during the lifetime of the agreement.

The objectives of the agreement, spelled out in Article I, were

> (a) To prevent or alleviate widespread unemployment or under-employment and other serious difficulties which are likely to result from maladjustments between the supply of and the demand for tin;
>
> (b) To prevent excessive fluctuations in the price of tin and to achieve a reasonable degree of stability of price on a basis which will secure long-term equilibrium between supply and demand;
>
> (c) To ensure adequate supplies of tin at reasonable prices at all times; and
>
> (d) To provide a framework for the consideration and development of measures to promote the progressively more economic production of tin while protecting tin deposits from unnecessary waste or premature abandonment.

The core of the agreement was contained in Articles VI, VII, VIII, and IX, which provided for determination of floor and ceiling prices, export control, and establishment of a buffer stock. Under Article VI, the newly established International Tin Council was authorized to set floor and ceiling prices for tin metal and to revise either or both by a distributed simple majority vote. "In so doing," the article stated, "the Council shall take into account the current trends of tin production and consumption, the existing capacity for production, the adequacy of the current price to maintain sufficient future productive capacity and any other relevant factors." To maintain the price range for tin metal, the ITC was to employ the devices of export control and buffer stock purchases and sales.

In Article VII, it was stated that

> The Council shall from time to time determine the quantities of tin which may be exported from producing countries in accordance with the provisions of this Article. In determining these quantities, it shall be the duty of the Council to adjust supply to demand so as to maintain the price of tin metal between the floor and ceiling prices. The Council shall also aim to maintain available in the buffer stock tin and cash adequate to rectify any discrepancies between supply and demand which may arise through unforeseen circumstances.

The ITC was required, not less than once every three months, to estimate the probable demand for tin and the probable changes in commercial stocks over the coming quarter in order to ascertain whether export control should be imposed, continued, or modified. The total permissible amount of exports was to be determined by a distributed simple majority. The total permissible amount was to be distributed among producing members in proportion to the number of votes assigned to each.

Thus, export control and producer voting strength were inextricably tied together. Producers were assigned 1,000 votes in total, as were consumers. Each member was granted a minimum of five votes. The votes of consuming members, in addition to the minimal five, were to be apportioned on the basis of the mean of net imports and consumption over the preceding three years; but no such specific basis for allocation was stated for the producers. Producers' initial votes were distributed on the basis of previous production records. Article VII provided that each year the votes of each producer would be reduced by five percent, and that these votes should be redistributed by a distributed simple majority "in such a way as will, in its [the Council's] opinion, best give effect to the principle of affording increasing opportunities for satisfying national consumption and world market requirements from sources from which such requirements can be supplied in the most effective and economic manner, due regard being had to the need for preventing serious economic and social dislocation and to the position of producing countries suffering from abnormal disabilities." The percentage assigned to a new producing member would also be determined by a distributed simple majority, with the percentages of the other producing members being reduced proportionately to maintain the total number of producer votes at 1,000.

Despite the acknowledgment made to the Havana Charter's call for seeking to supply the world's needs in the most effective and economic manner, producers' votes and export quota percentages have not been determined on any such basis of promotion of global efficiency. Fox noted, to the contrary:

> These agreed tonnages were the result of compromises and the usual blackmail to be expected at international conferences. . . . The agreements did, however, avoid any acceptance of the idea of production capacity (as distinct from performance) as the basis for assessing importance and at no time were they prepared to revert to the mistaken pre-1939 practice of bribing new members into the agreement with the grant of tonnages exempt from the provisions of export control.[37]

Article VII also dealt at length with the problems of actual exports in excess of or below a nation's quota, thus seeking to solve another problem that had arisen in the prewar agreements. If any producing country fell short of 95 percent of its aggregate quota over any four successive control periods, its quota would be reduced by the amount of the deficit under 95 percent, and the ITC could, by vote of a distributed majority, redistribute the quota reduction among the other producing members. Quotas could also be redistributed by distributed majority vote whenever the ITC was "of the opinion that any producing country is unlikely to be able to export in any control period as much tin as it would be entitled to export in accordance with its permissible export amount." In the event of exports in excess of the quota, the ITC was authorized to require the offending producer to make a contribution to the buffer stock equivalent to the amount of the excess, in either metal or cash. More severe penalties were to be levied for persistent violations.

Each producing nation was made responsible for enforcing its own export quotas, and for distributing these quotas among individual miners and smelters. Exports of either tin metal or "the tin content of any material derived from the mineral production of the country concerned" would count against a quota.

Article VIII provided for establishment of a buffer stock, by compulsory contribution from the producing members. The stock was set at 25,000 long tons, with initial contributions to total 15,000 tons, and provision for subsequent calls if needed. Any participating country, whether producer or consumer, could make a voluntary contribution to the buffer stock in either cash or tin metal.

Article IX set forth the responsibilities of the buffer stock manager and the mode of operation of the buffer stock. If the price of cash tin on the London Metal Exchange (LME) was equal to or greater than the ITC's ceiling price, the manager was required to offer tin for sale at the ceiling price until the buffer stock was depleted of metal. If the price was in the upper third of the council's price range, the manager could, at his discretion, sell tin in order to prevent the price from rising too rapidly. He was not permitted to buy or sell when the price was in the middle third of the range, regardless of how the price might fluctuate from day to day, unless otherwise authorized by a distributed simple majority vote of the council. If the price was in the lower third of the range, he could buy tin at his discretion to prevent an unduly steep fall in price. If the price was equal to or below the floor price, the manager was required to buy tin until the buffer stock's funds were exhausted. The buffer stock manager was free to buy or sell either cash or forward on any established market—that is, he could operate on the Penang and New York markets as well as on the LME.

The operations of the buffer stock and the imposition of export control were linked by a provision that no export controls could be imposed unless the buffer stock held a minimum of 10,000 long tons of tin metal or the ITC found by a distributed majority that 10,000 tons was "likely to be held before the end of the current control period, having regard to the rate at which the buffer stock is accumulating."

Thus the buffer stock was made an integral part of the 1953 agreement, in contrast to the sporadic and clearly subordinate role played by such stocks in the prewar control schemes. Most writers would agree with Yip's assessment, that the objective intended for export controls was a limited one, to "prevent prices from remaining persistently below the set floor."[38]

Fox, however, dissented from this view, labeling the buffer stock as "the lesser weapon" in the first three agreements.[39] In recent years, the role to be played by the buffer stock, and therefore its appropriate size, has become a matter of contention within the ITC, with the argument being advanced that if the buffer stock were made large enough it should be possible to do away entirely with export controls. This controversy will be discussed below in more detail.

Lim Chong Yah, in a harshly critical 1960 article, argued that the provisions for a buffer stock and export controls were quite inconsistent with the objectives set forth in Article I.[40] There was nothing in the agreement, he noted, as to how the employment objective was to be attained. Export control, he pointed out, which might raise the price while reducing employment, highlighted "the obvious contradiction between the price objective and the employment objective."[41] Further, the objective of more economically efficient production is not consistent with the use of export controls which affect high- and low-cost producers alike. Speaking of the Malaysian experience with export controls from late 1957 until the time of writing, he asked, "Can we say that the price objective has been achieved at the expense of the employment and economic production objectives?" He answered his own question: "The positive answer . . . is too obvious to need elaboration: the floor price has been obstinately defended with great domestic casualties."[42]

Turning to the objective of assuring adequate supplies at all times, Lim noted that the only device available to the ITC for achieving this objective, other than mere exhortation of producers to produce more, would be sales from the buffer stock. Yet the buffer stock was far too small to serve such a purpose; and in any event it was to be used to defend the ITC's price range, in which case it would inevitably have exhausted its supply of tin metal before a shortage had any consequences other than forcing the price of tin to the ITC's ceiling.

Lim's article was among the earlier ones in a persistent stream of criticism of the 1953 agreement and its successors on the grounds that the protection purportedly afforded to consumers is illusory, or at best that the agreement is heavily weighted in favor of its producing members.

Puey's assessment of the role of the nonvoting consumer representatives to the prewar International Tin Committee was noted in the first section of this chapter.

In his 1965 book, J. W. F. Rowe has an excellent discussion of this matter.[43] The consuming nation, in formulating its policies, as well as in selecting its delegates, will inevitably be influenced by and responsive to the concerns of those most deeply involved with the industry. These will be the manufacturers, merchants, and distributors of products using the commodity. The final consumer is hardly aware of the cost of a number of commodities, most certainly including tin, incorporated in various products purchased. Those immediately interested are likely to be most concerned that the price be the same for all competitors, home and abroad, and secondarily that the price be stable and supplies assured. The level of the price is of tertiary or even minor importance. Merchants' attitudes will be varied. To the extent that dealers are speculators, they will prefer price instability. Arbitrary power to control price, whether by a private monopolist or an international commodity agency, is anathema to speculators. Rowe also pointed out that producing nations might, in certain circumstances, welcome consuming members because the latter could, if willing, impose sanctions not available to producers on defaulting members.

Consuming nations may also have international political interests that are not fully consistent with those of their consumers. Among the consuming country members of the 1953 Tin Agreement, were the United Kingdom with 380 votes, Belgium with 38 votes, and the Netherlands with 52 votes. At that time, neither Malaya nor Nigeria had obtained their independence from Great Britain, yet both federations were producing members of the ITC. British capital was still an important factor in the tin industries of both.[44] The Belgian Congo and Ruanda-Urundi, treated as a single producing member for purposes of the agreement, were still Belgian colonies and the suppliers of tin-in-concentrates to Belgium's smelting industry.

The interest of the Netherlands in supporting its former colony Indonesia is less obvious given the acrimonious circumstances under which independence was achieved; but the Netherlands was at the time a major smelter and re-exporter of tin. The shift in the United States' position from one of general opposition to the 1950 draft tin agreement to enthusiastic support for the 1953 agreement (at least among the delegates to the conference) has been attributed to a growing conviction that stability in tin and rubber markets in Malaya was essential to curbing the communist insurgency which had reached its peak there in 1951.[45]

The nations and dependencies acceding to the First Tin Agreement, and their votes following Thailand's ratification in 1957, are shown in table 4.3. (The agreement came into force on July 1, 1956, following Indonesia's ratification on May 15.)

The great bulk of the world's mine production of tin was accounted for by ITC members. In addition to the nations listed as producing members, participants included Australia, the United Kingdom, and Spain, all of which were in fact producers of tin, although not exporters. The largest tin producers outside of the first agreement were the Soviet Union and the Peoples' Republic of China. Other nonmember producers, listed in order of the importance attributed to them by Fox, included Brazil, Argentina, the Union of South Africa, Rwanda, Laos, Portugal, and Burma.[46] Fox concluded, however, that "control of the tonnage of tin entering the world market was therefore always fairly complete, and the post-1956 agreements were, in general, free from the constant fear underlying the prewar agreements that the restrictions of output on the participating countries would mean a permanent increase in the loss of output to 'outside' countries."[47]

The world's consumers of tin, however, were much less adequately represented. In its *First Annual Report*, for 1956–57, the ITC estimated that its consuming members accounted for approximately 37 percent of the world's total tin consumption, excluding the Soviet Union.[48] Notable among those missing were two of the defeated Axis nations: Japan and Germany (East and West). Clearly, the single most important nation remaining outside the ITC

Table 4.3. Membership and votes, International Tin Agreement, by April 1, 1957

Member country	Number of votes
Consuming members	
Australia	32
Belgium	38
Canada	77
Denmark	79
Ecuador	5
France	165
India	75
Israel	7
Italy	56
Netherlands	52
Spain	14
Turkey	20
United Kingdom	380
Producing members	
Belgian Congo and Ruanda-Urundi	90
Bolivia	213
Indonesia	213
Federation of Malaya	360
Federation of Nigeria	58
Thailand	66

Source: *First Annual Report of the International Tin Council, 1956-57*, pp. 8-9.

was the United States, both as the world's largest consumer of tin and as the holder of a strategic stockpile equivalent to about two years' world production of tin.

Despite the United States' active role in the International Tin Study Group and the 1953 International Tin Conference, on March 5, 1954 the State Department announced, after discussions had taken place between the Department of Commerce and representatives of the "tin-consuming industry," that the United States had decided not to sign the International Tin Agreement.[49] The ITC subsequently, in its *First Annual Report*, described the United States' attitude as one "that was generally regarded as benevolent neutrality."[50]

Throughout the 1960s spokesmen for the executive branch of the United States government repeatedly expressed the hope that their country would join the ITC. However, business interests were staunchly opposed, most notably the tinplate producers and tin traders, although for different reasons. As Fox noted, the tinplate and can manufacturers welcomed price stability while opposing any other form of international regulation that might raise the price of tin; while the dealers, whose livelihood was speculation were opposed to price stabilization.[51] The Department of Commerce was steadfastly opposed to commodity agreements, contending that as a matter of principle they provided solutions to price and output problems inferior to those of the free market.

Undoubtedly, however, the crucial factor militating against United States' membership in the ITC was the nation's strategic stockpile. By 1954, the stockpile was complete, and indeed some 40,000 long tons had been declared as excess. During the period of stockpile acquisition, the United States had committed itself to disposal of the stockpile, if that ever became necessary or desirable, in such a way as to avoid disruption of the world tin market, despite the Congressional rhetoric about cartelization and price-gouging that had accompanied the buildup. And at the time the announcement was made that the United States would not join the ITC, it was simultaneously announced that there would be no disposals of tin from the stockpile except at the president's direction. The United States would have had to accept yet another constraint on its management of the stockpile had it elected to join the ITC. Article XIV of the 1953 agreement stated:

> Participating Governments . . . [s]hall not dispose of non-commercial stocks of tin except upon six months' public notice, stating reasons for disposal, the quantity to be released, the plan of disposal, and the date of availability of the tin. Such disposal shall protect producers and consumers against avoidable disruption of their usual markets. A participating Government wishing to dispose of such stocks shall, at the request of the Council or of any other participating Government which considers itself substantially interested, consult as to the best means of avoiding substantial injury to the economic interests of producing and consuming countries. The participating Government shall give due consideration to any recommendations of the Council upon the case.

Fox commented on the decisive nature of the stockpile issue, observing, "The State Department, however sympathetic to the Council . . . was none too willing to embark on the task of persuading Congress to hand over to any external international organisation rights in a strategic stockpile which had been acquired by American money for American defence."[52]

The membership question was complicated by the fact that the United States, whose "benevolent neutrality" if not actual membership was crucial to the success of the agreements, was "grimly determined"[53] that the Peoples' Republic of China not be admitted to the ITC; and the Federal Republic of Germany, a major importer of tin, was just as adamant in its resistance to admission of the German Democratic Republic.

The major problem dealt with during the first agreement was adjustment to an inflow of tin from the Soviet Union on what Fox referred to as "a massive scale well beyond the comprehension of Western observers."[54] For the first time, the USSR began offering tin for sale on the LME. Soviet exports rose from 3,248 long tons in 1956 to 18,011 tons in 1957, and to 21,948 tons in 1958. Virtually all this tin was sold on European markets, as the United States refused to license any tin imports from the Soviet Union on the grounds that the USSR was in fact re-exporting Chinese tin, and that then current foreign assets control legislation prohibited the import of tin from the Peoples' Republic of China.

The ITC responded by imposing stringent export controls, which lasted for a total of eleven quarters, from December 1957 through September 1960. In addition, the ITC sought to protect the price range by active purchasing for the buffer stock. The full contributions due from members were called up, and the producers agreed to make additional contributions to a special fund. In all, the regular and special fund buffer stocks rose to approximately thirty thousand long tons before the buffer stock manager's cash resources were exhausted. On September 18, 1958, the ITC announced that it was no longer able to support the floor price of £730. The price fell immediately to £640. By October the price had risen back above the floor. This was the only instance in the history of the ITC in which the council was unable to defend the floor.

Member nations also took action. On August 30, 1958, the United Kingdom announced that it was limiting tin imports from the Soviet Union to 750 long tons per quarter; and the Netherlands, Denmark, Canada, and France followed with import limits of their own. At the same time negotiations with the Russians got under way: India, Indonesia, Malaya, and the United Kingdom made diplomatic approaches. The ITC authorized the chairman of the council to invite the Soviet Union to join the ITC as a producing member as well as to urge the Russians to limit their tin exports to the non-Communist world. The Russians responded to the invitation to join the ITC by requesting observer status. On the grounds that the tin agreement did not provide for observers, the Soviet request was denied; but the ITC made a counter offer to remove the restrictions placed by consuming members on Russian imports provided that the Russians in turn agreed to limit total exports of tin to non-Communist nations to no more than 13,500 long tons per year. By early 1959 such an agreement had been reached and the import limitations removed. Soviet exports fell rapidly thereafter, and within a few years disappeared entirely. Table 4.4, taken from Fox, shows the sudden growth and decline of these exports and the apparent role of the Soviet Union as a re-exporter of Chinese tin.

The second agreement. In May and June 1960, a Second United Nations Tin Conference met to draft a new agreement. By the time of the conference, Russian tin exports were well below their previous peak and falling rapidly. Mine capacity in the producing nations had adjusted to the end of the stockpiling boom, and the world's consumption of tin metal was outrunning world mine production. Thus, the 1960 conference, in contrast to that of 1953, convened when expectations were of an impending shortage of tin rather than a burdensome surplus. The changed conditions were recognized in the preamble to the 1960 agreement, which deleted the observation that a "burdensome surplus" was expected to result, and substituted in its stead a section noting that "the situation of the tin market might be aggravated by uncertainties with respect to the disposal of non-commercial strategic stocks unless provision existed for consultation and for the giving of appropriate notice for their liquidation." In light of subsequent actions by the United States, this clause proved apposite.

The second agreement was little changed from the first. One major modifi-

Table 4.4. Soviet Union imports and exports of tin, 1955–72 (long tons)

Year	Imports from China	Imports from others	Exports
1955	16,663	—	2,067
1956	15,452	—	3,248
1957	21,653	—	18,011
1958	18,995	99	21,948
1959	20,471	35	17,816
1960	17,420	44	11,318
1961	11,023	39	5,610
1962	8,563	1,181	492
1963	4,232	3,445	689
1964	1,083	4,330	11
1965	492	5,216	7
1966	492	4,232	7
1967	98	5,512	6
1968	295	6,693	—
1969	30	6,659	—
1970	197	7,967	—
1971	492	3,815	—
1972	784	3,425	—

Source: William Fox, *Tin: The Working of a Commodity Agreement* (London: Mining Journal Books, 1974), p. 294.

cation in the relative strengths of the members was that most of the important decisions, which had required a distributed simple majority vote for approval in the first agreement, required a distributed two-thirds in the second. This alteration gave the Federation of Malaya, with 378 of the producers' 1,000 votes, a de facto veto power, as it would also have done for the United States had that country decided to join the ITC. If such a voting rule had been in effect at the beginning of the first agreement, the United Kingdom, with 380 votes, would have had such veto power. But by the beginning of the second agreement, trade and consumption patterns had shifted so that the United Kingdom had only 277 votes.

The size of the buffer stock was reduced from 25,000 long tons to 20,000. In light of the exhaustion of the buffer stock manager's cash resources and the ensuing failure to defend the floor price in 1958, this might seem an untoward reduction. The producing members, however, felt strongly that the burden of financing the buffer stock, which fell on them alone, was not a fair one. It was also argued that, despite the brief period during which the price had been below the floor, the use of export control could adequately protect the floor price on most occasions and that the buffer stock was of more benefit to consumers than to producers since its most important role would be to dampen price movements threatening the ceiling. But at the 1960 conference, delegates from the consuming countries would not agree to share that burden. This remained a contentious issue for many years. Further, the practical effect of the reduction in size of the buffer stock was offset by authorization given to the buffer stock manager to borrow on the security of tin warrants held by the stock with the council's approval. Such borrowing had been explicitly prohibited in the first agreement.

The conditions under which export control could be instituted were relaxed by providing that only 5,000 long tons of tin need be in the buffer stock when the first three-month control period was declared. The buffer stock had to hold 10,000 tons of metal, however, before additional control periods could be imposed. The council's role in oversight of the buffer stock was somewhat changed. If, in the chairman's opinion, the obligation of the buffer stock manager to sell when the price reached the ceiling and to buy when the price fell to the floor was inconsistent with the basic objectives of the agreement, he could suspend the obligation and immediately convene a meeting of the council to confirm such a suspension. The need for such flexibility had been demonstrated during the first agreement when speculative buying and selling had reinforced price movements near the floor and the ceiling as a result of the speculators' certain knowledge of the actions being taken by the buffer stock manager.

The second agreement went into effect on July 1, 1961; its most important change in membership was the accession of Japan. The expected shortage did appear to be materializing during the early life of the second agreement, as the price of cash tin metal on the LME rose from an average of £800 for the last quarter of 1960 to £953 for the last quarter of 1961. However, the predicted reaction, which had been anticipated in the second agreement's preamble, occurred in September 1961 when the General Services Administration requested authorization to dispose of 50,000 tons of tin from the United States' strategic stockpile. In June 1962, the Congress granted such approval. As summarized by Fox, "The essence of the second agreement of 1961–66 was, first, the acceptance of the fact of shortage, both by the Council and the United States; secondly, the acceptance by both parties that this shortage, or part of this shortage, could be met by releases from the strategic stockpile."[55] He noted that recognition of these circumstances triggered intense debate and controversy as to the appropriate mechanism for stockpile releases and their effects on the economies of the producing nations.

The third agreement. The third agreement was negotiated under a drastically different set of ground rules from those that had governed the first and second agreements. GATT had been coming under increasing criticism from the developing nations as a "rich countries' club."[56] The stress on free and nondiscriminatory trade, with emphasis on reciprocal tariff reductions and the elimination of governmental barriers to the unimpeded flow of goods in international commerce, was considered to be inimical to the economic needs of the developing countries and contrary to the moral obligation of the wealthy developed nations to assist the less fortunate. Most important for the underlying rationale of the tin agreements, it was argued that international commodity agreements should seek to raise the prices of the commodity exports of developing countries relative to the prices of the goods that they imported, rather than merely to stabilize these prices over time. Industrialized consuming countries should be willing to cooperate in arrangements having such an effect.

In May 1964, the United Nations Conference for Trade and Development (UNCTAD) was convened to review the growing dissatisfaction with GATT and to make appropriate recommendations. The developing nations, composing a substantial majority of those attending the conference, voted as a bloc while the developed countries were split, with the United States acting as the principal defender of GATT principles. Among the recommendations passed and subsequently adopted by the General Assembly was one to establish UNCTAD as a permanent organ of the General Assembly.

In March 1964, the ITC had requested the Secretary General of the United Nations to convene another international tin conference to consider renewal of the International Tin Agreement. In light of the General Assembly's action on UNCTAD the following May, which included a provision that "the UNCTAD initiate action, where appropriate, in cooperation with the competent organs of the United Nations for the negotiation and adoption of multilateral legal instruments in the field of trade," the Third United Nations Tin Conference was held under the aegis of UNCTAD in the spring of 1965.[57]

The preamble to the third agreement was a completely rewritten statement. All references to burdensome surpluses, widespread unemployment, and uncertainties associated with the disposal of noncommercial stocks were eliminated. The new preamble opened with recognition of the premise that "commodity agreements, by helping to secure short-term stabilization of prices and steady long-term development of primary commodity markets, can significantly assist economic growth, especially in developing producing countries." It noted the value of continued international cooperation within the framework of an international commodity agreement, the "exceptional importance" of tin to both producing and consuming countries, the need to foster the health and growth of the tin industry, "so as to ensure adequate supplies" in the interests of consumers and producers alike, and "the importance to tin producing countries of maintaining and expanding their import purchasing power."

Article I, specifying the objectives of the agreement, was also drastically rewritten and expanded. Only the section dealing with the prevention or alleviation of widespread unemployment was carried over from the first and second agreements. Article I of the third agreement reads in part:

> The objectives of this Agreement are:
>
> (a) To provide for adjustment between world production and consumption of tin and to alleviate serious difficulties arising from surplus or shortage of tin;
>
> (b) To prevent excessive fluctuations in the price of tin;
>
> (c) To make arrangements which will help maintain and increase the export earnings from tin, especially those of the developing producing countries, thereby helping to provide such countries with resources for accelerated

economic growth and social development, while at the same time taking into account the interests of consumers in importing countries;

(d) To ensure conditions which will help achieve a dynamic and rising rate of production of tin on the basis of a remunerative return to producers, which will help secure an adequate supply at prices fair to consumers and which will help provide a long-term equilibrium between production and consumption.

Fox referred to paragraphs (c) and (d), caustically, as "every question so begged and every viewpoint admirably squared."[58] Despite the sweeping changes in the preamble and Article I, reflecting a new and different sense of the fundamental functions of the ITC, the remainder of the third agreement was virtually identical with the text of the second agreement. Thus, the earlier ambiguities that had been noted and criticized, such as the absence of any mechanisms for dealing with unemployment or for providing meaningful relief to consumers in time of shortage, remained, in addition to the new contradictions.

Fox describes the five years in which the third agreement was in force, 1966 to 1971, as "pleasanter and easier"[59] than those of the first two agreements. The Soviet Union was importing tin from the tin markets of the ITC members rather than exporting to them, and disposals from the United States' strategic stockpile were proceeding in an orderly and moderate fashion. Consumption was rising, as was production, with only small and manageable gaps that were closed by sales from the United States' stockpile. Nevertheless, the ITC did impose export restrictions for the last quarter of 1968 and all of 1969 in response to simultaneous rapid increases in Bolivian, Indonesian, and Thai output; but these restrictions were mild, averaging 4 percent of the most recent uncontrolled exports.

The replacement of the principles of GATT and the Havana Charter by those of UNCTAD made the tin agreement acceptable for the first time to a number of Eastern-bloc and socialist nations. During the third agreement, Czechoslovakia, Poland, Hungary, and Yugoslavia joined the ITC.

The fourth agreement. The Fourth International Tin Conference was convened, again under UNCTAD auspices, in the spring of 1970.[60] The only significant change in the fourth agreement was the increased discretion of the buffer stock manager. The provision that he sell whatever tin remained in the buffer stock at the ceiling price when the market price reached or exceeded the ceiling was reworded to authorize him to sell at the market price. The requirements that he sell tin when the price was in the upper sector of the ITC's range and buy when the price was in the lower sector were relaxed by substituting the word *operate* for *buy* and *sell*. The requirement that he buy tin at the floor price whenever the market price reached or fell below that level was retained.

The Buffer Stock Manager R. T. Adnan observed that the ITC had "established a floor price which is sacrosanct and a ceiling price which is not sacrosanct, but a more general qualification of intent."[61] In the same article, Adnan noted that the manager's broadened authority would permit him to moderate a precipitous rise in the price of tin when the price was in the lower sector or a sudden fall when it was in the upper sector. He might also buy on the Penang market and sell on the LME simultaneously, or vice versa, if such a maneuver appeared advantageous.

The preamble to the fourth agreement made specific reference to UNCTAD, deleting a reference to cooperation "within the framework of an international commodity agreement" and substituting "within the framework of the basic principles and objectives of the United Nations Conference on Trade and Development by means of an international commodity agreement."

The Soviet Union, which had given strong support to the developing countries in their efforts to establish UNCTAD and in formulation of its principles, signed the fourth agreement. The USSR was, perhaps, as suggested by Fox, "embarrassed at not acting on these principles by joining one of the few UNCTAD successes in commodity organisation."[62] As a consequence of its basic change in trading position, the Soviet Union joined as a consuming member. Another basic change in the position of a member nation was recognized when Australia changed its membership category from consumer to producer. Other new member nations acceding to the fourth agreement were West Germany, Bulgaria, and the Republic of China; the last withdrew within a year.

In 1968, the International Bank for Reconstruction and Development (World Bank) had been granted observer status on the ITC, in a reversal of the council's earlier rejection of the Russian request for that status. With the inception of the fourth agreement, UNCTAD and the IMF were granted this status. Another example of the use of ambiguity to resolve controversy was provided when the European Economic Community (EEC) requested an invitation to participate in the fourth tin conference. The Soviet Union and the Eastern European countries were opposed, and the producing members expressed some concern over whether EEC participation would encourage its member nations to vote as a bloc in the ITC. The issue was settled by insertion of a provision, Article 50, which stated that "an intergovernmental organization having responsibilities in respect of the negotiation of international Agreements may participate in the International Tin Agreement. Such an organization shall not itself have the right to vote. On matters within its competence the voting rights of its member States may be exercised collectively." The ITC's annual report for 1971–72 stated merely, "The European Economic Community also participates in the Fourth Agreement," with its status and the nature of its participation left unmentioned.

During the fourth agreement, consuming members did for the first time make voluntary cash contributions to the buffer stock. In July 1972, the Netherlands contributed £671,000; and in January 1973, France contributed £1,215,000. The relative magnitudes of these contributions is indicated by the fact that the cash value of producers' total contributions under the fourth agreement amounted to £27,000,000.

Major structural changes in the world tin markets, which are to be discussed in detail in following chapters, coupled with monetary exchange rate problems, forced a dramatic change in the ITC's determination of the price range during the fourth agreement. As noted by Adnan in early 1972, it was becoming increasingly questionable "whether the prices quoted by the LME are at all times truly reflective of the values of tin as a whole."[63] Japanese purchases of tin had risen rapidly throughout the postwar years, and by 1970 Japan had become the world's second largest consumer of that metal. Most Japanese purchases were made in Penang, Malaysia. Further, as the smelters of Europe declined in output and closed down, and were replaced by increased output from Thai, Indonesian, and Bolivian smelters, the role of the physical market for tin metal in Europe declined and speculative trading forces became more influential on the LME price. Finally, as Adnan noted, Grade A tin, although comprising two-thirds of world output, was not traded on the LME, where contracts were written in terms of standard tin. At the same time, on the purely physical market in Penang, trade was predominantly in the higher grade of tin metal.

Concern over the appropriateness of the LME cash price as a basis for setting the ITC's floor and ceiling prices, and for conducting price stabilization operations, was heightened by the instability of the pound sterling, particularly following the establishment of free exchange rates in June 1972. In contrast to the pound, the Malaysian ringgit was one of the world's most stable currencies. In July 1972 the ITC dropped the LME price as a reference mark for its own operations and began quoting its price range in terms of the Malaysian ringgit price per picul of tin metal on the Penang market.

In January 1973, the ITC imposed export restrictions, after approximately one year of heavy buying by the buffer stock had succeeded in keeping the Penang price fairly constant. The controls were mild, set at roughly the previous export level, and lasted for only three quarters. Indeed, some producers did not meet their quotas, and the effect of the controls was generally viewed as negligible. In late 1973, the tin market was caught up in the worldwide commodity boom. The price of cash tin on the LME rose from £1,591 in December 1972, to £2,788 in December 1973, and to £3,775 the following June. It dropped back to £3,079 in December 1974 and remained around the £3,000 level until early 1976.

The ITC responded to the price decline in the last half of 1974 and to the reduced consumption associated with the recession of 1975 by imposing export control on April 18, 1975. The initial control was harsh, representing an 18 percent cutback from the previous export level. In the first half of 1976, prices rose again, from £3,074 in January to £4,410 in June. On June 30, export control ended; prices continued to rise, reaching £6,906 by December 1977. World consumption recovered to 195,000 long tons in 1976, exceeding production of 190,600 long tons of tin metal.

It must be noted that a portion of the price rise from £1,591 at the end of

1972 to £6,906 by December 1977 was a result of the deterioration of the currency exchange rate of the pound sterling. Over the same five-year period the Penang price rose from M$620 per picul to M$1,752; and the New York price rose from $1.77 per pound to $6.11. These increases are 434 percent for the LME, 283 percent for Penang, and 345 percent for New York.

The fifth agreement. In 1975, the year after the United Nations General Assembly had adopted its resolution declaring a New International Economic Order, the Fifth International Tin Conference met. The preamble to this agreement took account of the NIEO declaration as well as repeating its adherence to the principles of UNCTAD, asserting that the participating nations recognized

> (b) The community and interrelationship of interests of, and the value of continued cooperation between, producing and consuming countries in order to support the purposes and principles of the United Nations and the United Nations Conference on Trade and Development and to resolve problems relevant to tin by means of an international commodity agreement, taking into account the role which the International Tin Agreement can play in the establishment of a new international economic order.

Producers' mandatory contributions to the buffer stock were set at a total of 20,000 metric tons of tin metal or the cash equivalent. Article 22 stated that consuming countries might make voluntary contributions totaling another 20,000 metric tons, which was described as "an implied overall target" for these members.[64] The continuing controversy over consumers' contributions to the buffer stock led to a provision in Article 22 that 30 months after the new agreement came into force the ITC would review the size and extent of these contributions and "may decide that a negotiating conference is to be convened . . . in order wholly or partly to amend this Agreement."

Article 40 provided that the ITC "may make recommendations to producing countries on appropriate measures, not inconsistent with other international agreements on trade, to ensure that, in the event of shortage, preference as regards the supply of tin available shall be given to consuming countries which participate in this Agreement." With what can most generously be described as naiveté, the introduction to the text of the agreement describes this provision as "designed to be a counterpart to the provisions for export control in times of actual or expected surplus of tin on international markets."[65]

The "production costs of tin" were added to the list of factors which the ITC was to take into account in setting its floor and ceiling prices, a seemingly minor addition to a list which in any event ended with "any other relevant factors." The insertion of this phrase, however, proved to be the opening shot of a battle within the ITC during the lifetime of the fifth agreement.

The event that was to have the greatest impact on the ITC was the accession to the fifth agreement by the United States, on October 28, 1976.

The Fifth International Tin Agreement and Recent Issues in International Control

The United States' decision to join the ITC. The United States' decision to join the ITC came at a time when the presuppositions underlying that nation's international economic policies had been questioned—some would prefer to say repudiated—by a series of events and challenges from the developing world. The ascendancy of the UNCTAD viewpoint over that of the GATT, at least within the United Nations Organization, has already been noted.

In late 1973 and 1974, the Organization of Petroleum Exporting Countries (OPEC) struck a major blow at the balance of international market power by the apparently successful cartelization of the world's most important primary commodity and the subsequent quadrupling of the price of oil. Then, in April 1974, the General Assembly, over the opposition of the United States, passed the Declaration of the Programme of Action on the Establishment of a New International Economic Order.

More than a decade earlier, in 1962, United Nations' General Assembly Resolution 1803 had been enacted, stressing the rights of nations to "permanent sovereignty over their natural wealth and resources"; and since that time all tin producing nations had asserted the title of the state to all minerals in the ground. In December 1974, this earlier resolution was reinforced by the Charter of Economic Rights and Duties of States which affirmed, among other rights and duties,

> All States have the right to associate in organizations of primary commodity producers in order to develop their national economies, to achieve stable financing for their development and, in pursuance of their aims, to assist in the promotion of sustained growth of the world economy, in particular accelerating the development of developing countries. Correspondingly all States have the duty to respect that right by refraining from applying economic and political measures that would limit it.[66]

By 1975, the United States was under pressure to fashion some response to these third-world initiatives. In September of that year, at the Seventh Special Session of the General Assembly of the United Nations, Ambassador Moynihan, as head of the United States' delegation, read an address by Secretary of State Kissinger in which, among other things, Kissinger noted that the prices of minerals are particularly sensitive to fluctuations in the general level of economic activity in industrialized nations. "The result," he observed, "is a cycle of scarcity and glut, of underinvestment and over-capacity." He recommended that an advisory body, consisting of members from producing and consuming nations, be set up for every key commodity, to discuss how stability, growth, and efficiency might be promoted in the markets of such commodities. He continued, "President Ford has authorised me to announce that the United States intends to sign the tin agreement, subject to congressional consultations and ratification.

We welcome its emphasis on buffer stocks, its avoidance of direct price-fixing, and its balanced voting system. We will retain our right to sell from our strategic stockpiles, and we recognise the rights of others to maintain a similar programme."[67]

On September 15, 1976, following hearings held by the Committee on Foreign Relations, the United States Senate, by a vote of 71 to 17, gave its advice and consent to ratification of the treaty authorizing the United States to join the ITC; and on October 28, the United States signed the Fifth International Tin Agreement. In the hearings, representatives from Bethlehem Steel Corporation, United States Steel Corporation, and the American Mining Congress submitted statements in opposition to ratification; but the Foreign Relations Committee reported favorably and "without reservation" on the proposed treaty.[68]

The committee's report noted that one obligation of the United States, as a member of the ITC, would be to consult with the ITC concerning plans for disposal of surplus tin from the strategic stockpile. The stockpile objective was being lowered, effective October 1, 1976, to 32,499 long tons. This action meant that slightly over 170,000 long tons would be in surplus, but each future disposal would have to be approved by the Congress. It was pointed out, however, that the agreement also provided that nothing therein should prevent any member nation from taking any action it considered necessary for its national security.

The Foreign Relations Committee's endorsement of United States' membership in the ITC was at least in part based on two points: the United States would benefit from membership since its representation on the council would reduce the risk that the ITC would seek to follow the example of OPEC; and the United States' delegation was to vote against any efforts made in the spirit of UNCTAD to increase the level of market prices of tin for the purpose of transferring resources from consumers to producers.[69]

Subsequent statements from both administrative and congressional sources reinforced this policy position. In June 1977 C. F. Bergsten, Assistant Secretary of the Treasury for International Affairs, was quoted as saying that while the United States was interested in the creation of a number of international commodity agreements, "our primary purpose . . . is to reduce the risk of inflationary pressures in the United States. Indeed, we will sign no commodity agreement unless we are convinced it will promote that objective." He went on to assert that "we reject any thought of agreements which would raise prices above their market levels."[70]

The Joint Economic Committee of the Congress, at almost the same time, recommended a virtually identical policy for the nation, asserting that "the United States should not agree . . . to any attempt to raise prices above market trends."[71]

Thus, the reaction to UNCTAD, the NIEO, and the threat of the spread of OPEC tactics, at least insofar as the United States' posture toward the International Tin Agreements was concerned, seemed to have been one of determined defense of the older order represented by the Havana Charter and GATT. In light of the endorsement of both UNCTAD and the NIEO in the preamble to the fifth agreement, conflict within the ITC was probably inescapable.

United States' concerns: the buffer stock and export control. In several respects, the time at which the United States joined the ITC was an inauspicious one for moderation of the coming clash. From early 1976 through the end of 1977, tin prices virtually doubled in terms of both Malaysian and British currencies. In January 1976, the LME cash price averaged £3,074 per metric ton, and the Penang price M$960 per picul. In December 1977, the LME average price had risen to £6,906, while that on the Penang market was M$1,752.

During the greater part of 1976, prior to the entry of the United States, the behavior of the ITC could only reinforce the conviction that the council was at least as interested in raising the price of tin as in stabilizing it. At the end of 1975, export control was in effect and the buffer stock held 20,071 metric tons, reflecting a drift downward in the price of tin during the recession year of 1975. The price turned up in January 1976. The buffer stock manager continued to be a net buyer of tin through the end of the month. By the time of the council's quarterly meeting in March, the price was in the upper sector of the ITC's price range. The council, apparently impressed by the producers' arguments that the costs of mining had also risen sharply, voted to raise the price range and to retain export controls. At the same time, Bolivia announced that it would not ratify the fifth agreement by April 30, the final date set for signatures by the parties to the tin conference in order to put the agreement into effect on July 1, because even after the increase the ITC's price range was below Bolivian mining costs. A special meeting was called in May, at which time the price range was again increased and the export quota, although retained, was raised from 35,000 metric tons to 40,000 for the second quarter. At its June meeting, the ITC removed export control. The council also announced in June that the buffer stock was down to 2,820 metric tons and that buffer stock operations were being suspended until the fifth agreement went into effect at the beginning of the following month. The price of tin reacted by promptly rising above the ceiling.

There were, thus, several months in which export controls were enhancing prices, while simultaneously the buffer stock was disposing of over 17,000 metric tons of tin—an action which moderated the price rise for a while but left the council unable to check the continued rise or to defend its ceiling over the last half of the year and into the future. In December, the ITC voted to raise the price range for the third time in less than a year, on this occasion over the opposition of the United States, which had been a member only since October. In all, the price range had risen from M$850–M$1,050 per picul to M$1,075–M$1,325 from January to December 1976. On January 13, 1977, the buffer stock was exhausted. Through 1977 and into 1979, the price remained above the ceiling, although the ITC's price range was raised on three more occasions.

Even strong supporters of the ITC found its 1976 actions hard to justify or defend. *Mining Journal* commented editorially that the imposition of export control had been "premature" and was "an important contributory factor in today's [1978's] very high price levels."[72]

A still more critical assessment is that of C. L. Gilbert, who commented in a 1977 article that Fox had been quite correct three years earlier in pointing out that export controls, rather than the buffer stock, had become the main instrument of the ITC.[73] The emphasis on export conrols, in Gilbert's opinion, reflected a basic change in viewpoint of the ITC, associated with the UNCTAD positions that commodity agreements should be permanent arrangements and that it would be a good thing if such agreements were able to raise the relative prices of primary commodities produced by developing countries and consumed by the developed nations. The idea of "burdensome surplus," Gilbert argued, had undergone fundamental change since the time of the Havana Charter. "If one interprets the term 'burdensome surplus' as implying a substantial excess supply likely to persist for many years, then export quotas are the natural tool to support the price," he noted.[74]

In Gilbert's view, the buffer stock played only two subsidiary roles: it could be used to dampen small price variations when the general price level was satisfactory because of either market forces or export controls; and it could absorb short-run market fluctuations of substantial magnitude since export quotas took time to impose or change. Gilbert noted that a number of commentators, both within and outside of the ITC, argued that the council's basic problem was that the buffer stock was too small to defend the ceiling price adequately. In his opinion, however, the size of the buffer stock reflected a deliberate policy choice made by the ITC, since the council was not particularly concerned with maintaining the price at or below the ceiling. He concluded, as had the U.S. Senate's Foreign Relations Committee, as well as Smith and Schink, that there was an "implicit division of responsibilities between the ITC and the GSA, with the ITC bearing the responsibility of protecting the floor price, and GSA selling sufficiently large volumes in periods of shortage to ensure that prices did not become exceptionally high."[75]

As noted previously, one of the first acts of the United States' delegation was to oppose the increase in the price range voted at the December 1976 meeting of the council. Thereafter, through 1980, the United States, usually but not invariably joined by Japan and West Germany (and often by the Soviet Union), was successful in limiting the next four price range adjustments to levels at which the ceiling was almost always below the market price, as shown in table 4.5.

As a result, there were no purchases for the buffer stock, which had become exhausted on January 13, 1977, until the spring of 1981. This four-year period of inactivity for the ITC was protested vehemently by the producing members, who argued that, in the face of the rising costs of tin production, the ITC was being crippled by an unrealistically low price range which had been made

Table 4.5. ITC floor and ceiling prices, and Penang prices on dates of change, 1977–81

Period	Floor price	Ceiling price	Penang price on date of change (M$ per picul)
July 5, 1977–July 14, 1978	1,200.00	1,500.00	1,551.00
July 14, 1978–July 20, 1979	1,350.00	1,700.00	1,750.00
July 20, 1979–Mar. 13, 1980	1,500.00	1,950.00	1,976.00
Mar. 13, 1980–Jan. 13, 1981	1,650.00	2,145.00	2,377.00
Jan. 13, 1981–Oct. 17, 1981	27.28[a]	35.47[a]	32.35[a]
Oct. 17, 1981–	29.15[a]	37.89[a]	35.56[a]

a. M$ per kilogram

Sources: International Tin Council, *Tin Statistics*, 1970–80; *Tin International*, various issues.

possible only by the requirement that a change in the range be voted by a two-thirds distributed majority.

The question of export control could not arise under the circumstances prevailing through 1980, with the ceiling held below the market, nor later with the mystery buyer pushing up prices through the last half of 1981. But following the climax of the mystery buying episode in February 1982, as the price of tin fell toward the ITC's floor price, the council not only began buying for the buffer stock, but obtained members' approval to borrow to strengthen the buffer stock manager's ability to buy, and called up additional cash contributions from members. In April, with the buffer stock holding around 35,000 metric tons of tin, the council members voted to impose a 10 percent export control for the remainder of the quarter ending June 30, 1982.

In sum, the United States made clear from the moment of its entry its opposition to export control and its advocacy of a much larger buffer stock as an alternative. At the same time, and in the face of criticism for inconsistency, the United States opposed compulsory consumer contributions to the buffer stock and would not commit itself to making any voluntary contribution.

Bolivia's concerns: the economic and price review panel. In January 1977, *Tin International* pointed out that the United States' decision to join the ITC and the reiterated refusal of Bolivia to ratify the fifth agreement were closely related. Bolivia had three major objections to the fifth agreement, all heightened by the entry and subsequent behavior of the United States. First, the Bolivians considered the buffer-stock price range to be unreasonably low; second, they objected to the continuation in the fifth agreement of the provision carried over from the earlier agreements under which producers' contributions to the buffer stock were compulsory while those of consumers were voluntary; and third, they objected to the voting system which required a distributed two-thirds majority vote on changing the price range or imposing export control.[76] This last objection was given particular force by the fact that the United States had entered the agreement with 299 votes. By January 1977, as a result of entry to the fifth agreement by India, Romania, and Spain, the United States' votes were reduced to 275. Japan, with 181 votes, had the second largest number of consuming country votes, followed by the United Kingdom with 81 and West Germany with 79.

The fifth agreement had come into provisional effect as scheduled on July 1, 1976, since Bolivia had agreed to a temporary, one-year accession to the agreement. But the agreement would terminate on July 1, 1977 unless previously ratified by at least six countries holding 950 producer votes. The participation of Bolivia, assigned 179 of these votes, was essential.

Bolivia's bargaining position was a difficult and complex one. The high-altitude lode mining of the country was the most costly of any ITC member, and the nation was more dependent on export earnings from tin than any of the others. In addition, the Indian tin miners of the altiplano were a powerful and potentially revolutionary force. Thus, Bolivia had a greater need for an international commodity agreement in tin than did any other member. On the other hand, the Bolivians asserted that their mining costs were well above the ITC's ceiling price—a contention that was not disputed if all taxes, royalties, and export duties were included in costs—and that therefore the country was remaining in the ITC only at a sacrifice.

The Bolivians suggested, pointedly, that another form of agreement, a producers' cartel, would be an alternative if the International Tin Agreement failed. They noted that the Peoples' Republic of China had refused to join the ITC because of ideological opposition to commodity agreements in which the developed, capitalist consuming countries were represented, but had expressed its willingness to participate in commodity associations of the third world developing nations. And, as the *Far Eastern Economic Review* observed, a decline in Bolivia's dependence on tin increased the nation's bargaining power. "Now, however," the *Review* commented in mid-1976, "with new crude oil potential coming on tap, President Banzer could afford to inject some South American political diplomacy into the hitherto colourless International Tin Council."[77] As noted earlier in this chapter, Bolivia's bargaining position with respect to the guidelines of the ITC had been strengthened, since the fifth agreement specifically cited production costs as one of the factors to be considered by the ITC in setting floor and ceiling prices.

After a series of meetings, it was announced in March 1977 that ASEAN and Bolivian delegations had successfully concluded an agreement labeled the La Paz Charter. The charter stated that "both delegations will present . . . a resolution for the establishment of a periodic review and a systematic approach to the determination of an updated price range with a view to strengthening and ensuring the continued effectiveness of the Agreement." In its conclusion, the statement noted that "with the understanding that effective action will be taken by the Council . . . the Bolivian Government will deposit its instrument of acceptance on or before June 30, 1977. In view of the time factor, action by the Council should be taken as quickly as possible."[78]

On March 29, the ITC announced that the government of Bolivia had decided to ratify the Fifth International Tin Agreement, thereby permitting the agreement to come into definitive force. In return, the announcement continued, the council had agreed to create an Economic and Price Review Panel, consisting of four representatives from consuming member nations and four representatives of the producers. Both the ITC and the Tin Research Institute were to provide staff assistance to the panel.[79]

On June 14, 1977, Bolivia deposited the instrument of ratification. At the next meeting of the ITC, in July, the council received the first report from its newly established Economic and Price Review Panel, and on the basis of that report raised its price range from M$1,075–M$1,325 to M$1,200–M$1,500 per picul. The ITC's *Annual Report* for 1977–78 did, however, note, "The market price at the time stood some M$50 above the Council's new ceiling price."[80]

The functions of the Economic and Price Review Panel are described in the ITC's *Annual Report* for 1979–80.

> The work of the Economic and Price Review Panel is to collate and evaluate, at six-monthly intervals, data relevant to the review by the Council of the floor and ceiling prices in the Agreement and to reach findings and conclusions on the appropriateness of the price range for consideration by the Council for action. The purpose behind the establishment of the Panel was to provide a framework within which a more systematic and objective analysis of the factors in Council's approach to determining floor and ceiling prices might be made.[81]

While the Economic and Price Review Panel has developed a great deal of material on mining costs, by type of mine and by nation, both before and after taxes, it has not resolved—nor perhaps even diminished—controversy over the appropriate levels of the ITC's price range. There are two major issues, neither of which can be resolved by "systematic and objective analysis" of cost data. The first stems from a conviction on the part of at least some consumer members that levels of taxation and royalty charges are excessive. Second, and more fundamental, while some producer spokesmen have argued that higher costs justify higher prices, a countervailing consumer position has been to put more emphasis on demand, contending that when price falls because demand is low the higher-cost mines should be closed.

The ITC's voting system and practices. A continuing source of contention has been the agreements' voting system, with a distributed two-thirds majority required for any important decision. As reapportioned in March 1982, the voting strength of the consuming members was as shown in table 4.6.

The voting situation has been exacerbated by the United States' practice of calling for votes on most issues in which it has an interest, in contrast to the council's previous practice of acting by consensus in almost every instance. Indeed, the council's *Rules of Procedure* for the fifth agreement state, "The Council shall aim to reach its decisions by consensus. However, this shall not

Table 4.6. Votes of consuming members of the International Tin Council, as revised effective March 19, 1982

Country	Votes
Austria	8
Bulgaria	10
Canada	31
Czechoslovakia	22
Belgium/Luxembourg	20
Denmark	6
France	59
Germany, F. R.	80
Ireland, Rep. of	5
Italy	38
Netherlands	31
United Kingdom	61
Hungary	12
India	18
Japan	171
Norway	8
Poland	26
Spain	28
USA	260
USSR	93
Yugoslavia	13
Total	1,000

Source: International Tin Council, Press Communique, March 31, 1982.

prevent any participating country from expressing its reservation over such decisions."[82]

In 1977, William Page noted the "club-like atmosphere of the ITC," which he feared was disappearing at the time. "The ITC comes across as a relatively friendly club," he continued, "and the ethos of international commodity discussions is co-operation rather than conflict."[83] This atmosphere, which Page attributed to the smallness of membership (eighteen countries in the first agreement, up to twenty-eight in the fifth) and to sheer good luck, had evolved during the first three agreements before either Japan or West Germany, to say nothing of the United States, had become members. In Page's view, such an atmosphere could not survive the tactics of confrontation that the United States appeared to be bringing to the council. Curiously, in light of its refusal to join the ITC until agreements were negotiated under UNCTAD auspices, the USSR has taken nearly as hard a line as the United States, Japan, and West Germany.

A survey of current issues. Three major substantive issues continue to divide the council—the height and width of the price range, the size and sources of the buffer stock, and the need for the power to impose export controls.

In a 1972 address, ITC Executive Chairman H. W. Allen stated that "the success of the International Tin Council depended very largely on the skill with which the floor and ceiling prices were set. It was the extremely delicate task of the Tin Council to find a particular range of price which, over a period, encouraged further growth of consumption, encouraged the miner to dig for tin, and also encouraged the investor to put money into mining capacity."[84]

It is generally accepted that the United States' delegation has taken the position that actions of the producing nations have negated the role Allen described for the price range, if it was ever attainable. To the contrary, the United States does not wish the ITC's price range to reflect endorsement of special taxes, royalties, and export duties on tin that it views as discriminatory and excessive, much less provide support for such prices by purchases for the buffer stock.

Producers, on the other hand, have argued that the sovereignty over resources proclaimed by the United Nations is meaningless unless the governments of producing nations are free to capture the full value of minerals in the ground, and that the United States' criticisms of their tax systems are both unwarranted by the inherent value of unmined tin and an inappropriate interference in their internal affairs. One of those I interviewed, a national of a producing country other than Bolivia, commented, "The Bolivians have been most outspoken, but you [presumably, the United States] had better take them seriously because they are speaking for a lot of others."

Members of the United States' delegation have pointed out that the necessary size of the buffer stock will rise as the price range to be defended narrows, and vice versa. The United States, therefore, has advocated a wider price range as well as a larger buffer stock.

In May 1979, in an address before the Tin and Lead Smelters Club at the Hague, the ITC's buffer stock manager, P. A. A. de Koning, publicly chided the majority of the consuming member nations for their failure to make adequate voluntary contributions to the buffer stock. In this unusually harsh and direct criticism, he pointed out that one of the principal shortcomings attributed to the ITC by consumer nation delegates—that the council could always defend its floor but not its ceiling prices—was a direct result of an inadequate buffer stock and thus "in fact came down on the heads of their own governments." Further, the inability of the buffer stock to purchase enough tin when the floor was threatened necessitated export controls. The crowning irony, in de Koning's view, was that the tin that could have been purchased in lieu of imposing export controls, had the buffer stock been large enough, could later have been better used to defend the ceiling.[85]

At the time of de Koning's talk, the Netherlands, France, the United Kingdom, Denmark, Japan, Norway, Belgium, and Canada, but not the United States, had made voluntary cash contributions to the fifth agreement's buffer stock.

The United States' offer to the ITC of 1,500 long tons of tin from the strategic stockpile raised a set of problems for the council. The offer was contingent upon the ITC's agreement to value the tin at the price at which it was sold by the buffer stock, and title was to remain with the United States until those particular ingots were sold. In October 1980 the ITC accepted the 1,500 tons on the conditions set by the United States, but over the protest of Bolivia. Producers' contributions of tin had been valued at the floor price, and consumers' voluntary contributions had been in the form of cash. The Bolivians objected to both the new procedure for valuation of the contribution and the equally unprecedented practice of segregating the actual metal contributed from the rest of the stockpile.

Advocates of export control have argued that a defensible floor price is a necessary guarantee to tin miners to induce an adequate level of investment and increased production in an otherwise unacceptably risky world market, and that the ITC's willingness to impose export controls is in turn the only way of providing credible protection of the floor price. Members of the United States' delegation have argued, to the contrary, that the possibility of export control and the resulting compulsory cutbacks in production that would be imposed on miners by the governments of producing countries represents an additional element of risk to miners, and thus deters investment. The probabilities of the price falling below the floor and of miners being required to reduce their rates of production are, of course, both reduced as the size of the buffer stock increases. In any event, producers must have confidence in the judgment of the ITC if the floor price is to play any role in encouraging investment, since the current floor price is unlikely to be an incentive for the development of new capacity, which will begin producing tin in two or three years. What would be important to a prospective investor is the assurance that the ITC would set reasonable floor prices in the future and would have the ability and determination to defend them.

In summary, it seems evident that the International Tin Agreement, whose operations have been virtually immobilized in recent years, simply cannot carry out its functions if political pricing of tin means that both consuming and producing countries view their stake in the agreement solely or predominantly as reducing or raising the current price of tin, as the case may be, and if all members vote in support of these interests under a system requiring a distributed two-thirds majority vote for any significant action. In the past, the agreements survived and continued to function effectively by suppressing divisive issues and deliberately allowing mutually inconsistent policies to coexist without challenge as long as the contradictions did not become severe enough to threaten to disrupt day-to-day operations. This kind of cooperation could continue in the "club-like" atmosphere, with decisions reached by consensus and with a predominance of consuming members who were basically supportive of the producers because of past colonial ties or current investment interests.

Whether the ITC can survive under the new conditions of confrontation, coupled with formal votes, depends on whether both consuming and producing nations perceive common long-run interests, transcending day-to-day prices, that can be served by cooperation between producers and consumers. One fact seems inescapable: if the present system of political pricing of tin were to be eliminated, it would not be replaced by free-market pricing, any more than such market pricing existed before 1956, but rather by some new system of political pricing. In order to obtain insight into new systems that might arise, and therefore to make some assessment of the usefulness of the current tin agreement, it is necessary to examine the structure of the industry at both the mining and the smelting levels, and the operations of the markets for both tin concentrates and tin metal.

5. Supply and Demand: A Preliminary Sketch

The structure of an industry or a market by definition includes all aspects of the external economic environment that influence the decisions of individual businessmen. From such a definition, it follows that structure is the principal explanation of business behavior or conduct. In order to develop a useful approach for a given industry, it is necessary to identify the particular structural elements that are of greatest importance in that industry and then examine their actual influences on the patterns of conduct of the industry's decision makers.

Tin is first mined, then smelted, and finally distributed for consumption by industrial users throughout the world. In one set of markets, tin concentrates are sold by miners to smelters; and in another set, so closely interlinked as to comprise in reality a single worldwide market, tin metal is sold to those who use it as an input for the production of a variety of end products. Elementary supply and demand analysis, applied to a simplified set of assumptions regarding the production and distribution of tin, can provide initial insight into the structural issues most likely to be linked to crucial aspects of conduct and performance, which therefore require more detailed and realistic study.

Supply and Demand Elasticities for Tin

A priori theorizing leads one to expect that the short-run demand for tin would be price-inelastic. The basic characteristics of the market leading to such an expectation are that the demand for tin is a derived demand, and that tin is a small component of most of the final goods in which it is used. Thus, if the value of the tin in a can makes up 5 percent of the price of the canned goods, a 20 percent rise in the price of tin, fully passed on, would lead to a rise of 1 percent in the price of the final product. The final customers are purchasing canned peaches or beer, as the case may be, and the demand for tin is solely a derivative of the demand for the edible or potable contents of the can. If the elasticity of demand for that final product were −2.0, so that the quantity purchased fell by 2 percent in response to a 1 percent price increase, then this would translate into a short-run elasticity of −0.1 in the demand for tin—that is, a 20 percent increase in the price of tin would lead to a decrease of 2 percent in the quantity of tin used to can the goods in question. A short-run adjustment is customarily defined, not in terms of calendar time, but as the response to a price change on the assumption that industrial plant and equipment are held constant. In the long run, capital adjustments are taken into account. When long-run possibilities of substitution, such as frozen peaches instead of canned ones, or aluminum rather than tin cans for beer, are also introduced into the argument, the a priori grounds for expecting the demand for tin to be price-inelastic no longer exist.

The short-run supply of tin has been viewed as price-inelastic by knowledgeable writers primarily on the grounds that the capital equipment, once in place, cannot be used for any other purpose but to mine that particular ore body, and that the price would have to fall very steeply to reach the point where existing producers were not covering the variable costs of operating their mines. Yip, for example, commented, "On the side of supply, elasticity is low, because unlike the resources in industrial production which can usually be more readily re-employed for other purposes, those engaged in tin mining are generally inflexible. Both dredges and gravel-pump mines in Malaya, for instance, are known to have continued producing more or less the same quantity of tin-ore even when the price of tin has undergone some changes."[1]

A survey of econometric studies. There have been only a few attempts to estimate elasticities in the tin market. The empirical econometric studies that have been made give strong support to the a priori view that short-run elasticities of both supply and demand are low, although the researchers have typically been careful to limit their findings and to note the tentative nature of their conclusions in the light of statistical and computational problems that were candidly acknowledged.

Meghnad Desai, in an econometric model of the world tin economy published in 1966, treated disaggregated demand and supply functions alike as completely derived and hence independent of tin's own price.[2] The United States' demand for tin for tinplate, for example, was treated as a function of the output of tinplate and of a time trend to account for increased electrolytic plating. The output of tinplate, in turn, was regarded as a function of the output of manufactured food, which was itself determined in the model by consumer expenditures on food in constant dollars. The United States' demand for tin for other uses was expressed as a function of a composite index of industrial production of the principal tin-using industries, and the index was a function of defense expenditure and gross private domestic investment. Systems of equations were constructed in a similar fashion, with the price of tin excluded, for Canada, the OEEC nations, and the rest of the world. The world's tin output was expressed in one equation as a function of the previous year's output and the existence or absence of ITC export controls. Price entered Desai's system only as a dependent variable determined by stocks and consumption. Desai was careful to note how fundamentally his model differed from the standard structural model in which the quantities both supplied and demanded are treated as functions of price and the system is closed by setting the two quantities equal to each other. He advised the reader that he had formulated and used this novel approach only because in several alternative formulations, with various lags introduced, he had consistently found price to be statistically nonsignificant in both supply and demand equations.

In a comment on Desai's paper, F. E. Banks reported that in a single-equation formulation in which total United States' consumption from 1953 to 1968 was regressed against the previous year's price as well as certain other variables, he had derived a price elasticity of demand of −0.550 in the short run and −1.262 in the long run.[3] Desai, in response, reported that he had calculated a short-run elasticity of −0.35 using world consumption and an unlagged price variable, but added, "I do not feel that much economic meaning can be attached to such highly aggregated 'demand' curves. A disaggregation by region and end use is more meaningful."[4] One major problem with aggregation stemmed from the fact that secondary tin cannot be used for tinplating.

Neither Desai nor Banks described any calculations or results for supply elasticities. K. A. M. Ariff has reported his calculations of West Malaysian output of tin for 1951 to 1967 regressed against the previous year's price and the number of mines in operation, with estimates made separately for each type of mining.[5] At first glance, Ariff's results are quite surprising. He found small but negative output elasticities (implying that a rise in price would cause output to fall) for dredges, hydraulicking, and underground mining; and similarly small but positive elasticities for gravel-pump and open-cast mining. The overall elasticity of output with respect to price was negative, −0.16, but not statistically significant above the 5 percent level.

Ariff, like the other writers discussed here, was quite explicit in warning the reader of the limitations of his findings, noting that it was not possible to calculate supply elasticities as such, because data on stocks were unavailable and because the number of operating mines was the only other theoretically explanatory variable that he was able to obtain, in addition to the price of tin. Measures of the statistical significance of his elasticities for individual mining methods are generally quite low. Nevertheless, Ariff argued that the results "do indicate that the production of tin is very inelastic with respect to price changes."[6]

Ariff also contended that the negative supply elasticities he had estimated were not perverse or unexpected results, but rather should be regarded as accurate reflections of the behavior of a substantial number of tin miners. According to Ariff, extensive low-grade deposits are not workable by large-scale mining methods such as dredging except in periods in which prices are high relative to operating costs. But the shift of some equipment and labor from high-grade to low-grade ore bodies in response to higher prices can cause total output to fall.[7] Ariff did not take notice of the demonstrable fact that this practice would almost invariably lower the present value of the net revenue stream that could be earned over the full lifetime of a mine. Yet, as discussed a few pages below, Ariff was reconfirming Puey's finding of similar behavior among European miners in both Malaya and Thailand in the prewar period.

Ariff also calculated demand elasticities, for purchases of Malaysian tin only, of the leading importers. His figures are given in table 5.1.[8]

Only the elasticities for France, the United States, and the United Kingdom are statistically significant above the 5 percent level: world elasticity is significant only at the 10 percent level.

The final econometric study to be discussed was done by Jasbir Chhabra,

Table 5.1. Demand elasticities of leading importers of Malaysian tin

United States	−0.52
United Kingdom	−0.37
Canada	−0.29
Italy	−0.12
France	−0.47
Japan	−0.19
World	−0.20

Table 5.2 Price elasticity of supply of leading tin exporters

Country or area	Short-term	Long-term
Malaysia	0.31	0.70
Indonesia	0.21	0.91
Thailand	0.60	1.25
Bolivia	0.24	1.34
Developed countries	0.30	0.70
Rest of the world	1.11	2.09

Enzo Grilli, and Peter Pollak of the World Bank.[9] Chhabra, Grilli, and Pollak estimated separate supply equations for each major producing country, lagging price by one, two, or three years depending on evident time of response. Price was adjusted by dividing it by an index of mining costs which was designed to reflect the proportions of various mining techniques used in the several countries. A dummy variable was used to account for the ITC's export control periods. The World Bank study's results are shown in table 5.2.[10]

All of Chhabra, Grilli, and Pollak's estimated supply elasticities are positive, and all except those for Indonesia are significant at the 5 percent level or above. The response lags which give the stongest results vary. For Malaysia, Thailand, and the small producers included in the "rest of the world" category, the apparent lag is one year. Bolivia, Indonesia, and producers in the developed countries appear to have three-year lags. The weighted average for the short-run elasticities is 0.42; and for the long run it is 1.07.[11]

Since the price lag for Malaysia is one year in both Ariff's and the World Bank's equations, the marked difference in the two studies' short-run supply or output elasticities—−0.16 in the former and 0.31 in the latter—calls for explanation. The time periods are somewhat different, and Ariff's result is not statistically significant at the 5 percent level. Ariff, in introducing a separate variable for the number of mines in operation, hoped that this variable would capture the differences in severity of the output restrictions imposed by the Malaysian government in periods of export control, since evidence indicated that firms owning several mines tended to shut some down and work others at undiminished intensity when limited by quotas, and small mines tended to sell their quotas and shut down rather than operate for part of the year.

But in reality the number of mines in operation cannot be, as Ariff assumed for purposes of his model, "an autonomous variable independent of price."[12] In his equations, this variable has a positive coefficient for all types of mines, and a highly significant one for all but underground mines. Since the number of mines in operation must have been influenced by the price of tin, and is also positively correlated with output, the coefficients of the price variables and the elasticities they indicate are understated. The two-valued dummy variable used in the Chhabra, Grilli, and Pollak study fails to distinguish the intensity or duration of quotas as well as Ariff's continuous proxy variable, but the former does avoid the statistical distortion just noted in Ariff's formulation.

Further, Chhabra, Grilli, and Pollak calculated elasticities of supply with respect to what they called "real price," or the price on the London Metal Exchange deflated by indices of production costs for each country. Ariff, on the other hand, used unadjusted Straits prices expressed in Malaysian ringgits. Changes in costs of inputs will shift supply curves, and therefore the approach used by Chhabra, Grilli, and Pollak is conceptually superior. Ariff, writing in 1972, was well aware of the need to take account of such cost changes, but noted explicitly that reliable statistics on costs of tin production in West Malaysia were unavailable. The World Bank study, done several years later, used cost surveys done by the ITC for 1973–75 to weight the costs of fuel, capital, and labor in order to construct indices that were applied to time-series regressions running from 1955 to 1975. Chhabra, Grilli, and Pollak thus incorporated a refinement that captured an important structural feature for their model, but at the cost of making a very dubious assumption that the mix of productive inputs was fixed at 1973–75 proportions despite changes in relative factor costs and technology over the 1955–75 period.

These two models are not being compared in order to arrive at an opinion as to which one is likely to have provided the more accurate elasticities, but rather to draw attention to the point that the difficulties involved in using incomplete and imperfect data force econometricians to make hard choices among alternatives with recognized deficiencies, and that therefore in the present state of the art differences in the findings reported by competent and scrupulous investigators are to be expected.

Chhabra, Grilli, and Pollak also derived demand elasticities, as shown in table 5.3.[13] These authors recognized that the large number of end uses, the changing technologies of tin-using industries, the changing prices of substitutes and complements, and shifts in basic patterns of taste in consumption of food and beverages all make econometric modeling of demand difficult, and they emphasized that their price elasticities of demand were less significant statistically than their supply elasticities. Since their figures seem reasonably consistent with those of others, they will not be discussed further.

Table 5.3. Price elasticities of demand for tin by leading importers

Country or area	Short-term	Long-term
United States		
Tinplate uses	−0.24	n.a.
Non-tinplate uses	−0.13	n.a.
Western Europe		
Tinplate uses	−0.11	n.a.
Non-tinplate uses	−0.30	−0.41
Japan		
Tinplate uses	−0.18	n.a.
Non-tinplate uses	−0.49	n.a.
Other developed countries	−0.37	−1.60
Developing countries	−0.11	−0.15

Two other econometric studies should be noted, although neither need be discussed at length here.[14] Lim calculated very short-term elasticities of supply for various forms of mining in Malaysia, treating quantity produced as a function of the previous quarter's price. He found very low but statistically significant elasticities, some positive and some negative. Behrman, in the course of constructing a model for simulation of proposed international commodity agreements, found a price elasticity of supply for tin of zero in the short run and 0.2 in the long run. More surprisingly, Behrman's model yielded a long-run per capita price elasticity of demand for tin of −5.0. This latter estimate is so divergent from those of other investigators that it must be regarded as extremely improbable, and presumably an indication of the difficulty of modeling a meaningful long-run elasticity of demand.

Shortcomings of the econometric estimates. There are numerous conceptual difficulties in estimating elasticities of the sort just reviewed. Indeed, the problems seem so formidable that only rough approximations can be gained from statistical and econometric work.[14]

The sporadic export restrictions of the prewar International Tin Committee and the postwar International Tin Council give rise to problems in estimating supply elasticities. First, as already noted, there is the question of whether one can devise a statistical variable that will take account of the degree of restriction at any one time or must be content with merely distinguishing between years in which there were restrictions of varying degrees for at least some of the time and those in which there were none. An even more fundamental problem is the extent to which the industry has been and is cartelized. It is a standard theorem of microeconomics that a monopolist's output with respect to price cannot be determined independently of demand, and therefore the supply curve does not exist in a monopolized market. Explicitly or implicitly, all of the above studies have assumed that the world tin industry is a competitive one. In chapters 7 and 8, discussing the industry's structure, the extent of competitive conditions will be examined; but it must be noted here that to the degree that discretionary market power exists, the concept of price elasticity of supply is spurious.

A second major problem with supply elasticity is highlighted by Ariff's discussion of his finding of negative output elasticities for certain types of mining and for West Malaysian tin output in total. While his inclusion of the number of mines in operation may, as I believe, bias his industrywide supply elasticities downward and probably accounts for the negative coefficients, this variable does have the virtue of calling attention to the negative relationship that undoubtedly does exist for certain individual mines.

Puey, some twenty years earlier, had noted the widespread practice of mining poorer deposits in periods of higher prices and turning to the richer ones when prices fell. Unlike Ariff, Puey also recognized and called attention to the fact that the practice is one that on its face is economically irrational, or a business strategy that ceteris paribus does not maximize the present value of the income stream obtainable over time from a given mine or deposit. Puey offered two hypotheses to explain this apparent perversity.[15] First, tin mining is inescapably a highly speculative business. There may be a trade-off between risk and profit; and thus high prices may be necessary to induce the mining of tracts of uncertain richness, while the deposits that are known to be rich and capable of being mined at low cost can be kept in reserve. Second, in the short run, mining companies appear to have placed perhaps undue emphasis on financial soundness or regular and steady profits and a year-to-year cash flow that permitted regular dividends, even at the cost of diminishing the present value of total profits. Such an emphasis could be perfectly rational for a miner whose personal capital resources were limited and who did not have ready access to a broad, sophisticated, and knowledgeable capital market.

It has further been suggested to me that European mine managers, not under close supervision, may have been more interested in maintaining steady flows of remittances that would avoid bringing their operations to the attention of home offices than they were in long-run profit maximization. The dredging and underground sectors of the Malaysian tin mining industry, for which Ariff and Lim both found negative supply elasticities, have been dominated by European mining interests.

There is a third set of factors recognized in present-day treatments of mining economics that should be added to Puey's explanatory hypotheses. There is a time lag between the decision to mine a particular ore deposit and the sale of the tin concentrates from that area. Further, after a tract or vein is opened, it will yield its ore over a period of time, probably several years. Thus, the miner is making both short- and long-run decisions on the basis of expected future costs and prices. Ignoring risk-aversion and cash-flow considerations, if he expects the price-cost margin in the long run to rise more rapidly than the rate at which he discounts future income, he should mine his poorer grades now; but if he expects this margin to fall over time, remain approximately constant, or rise less rapidly than his discount rate, he should step up production and mine the richer deposits now.

The introduction of considerations such as uncertainty, risk aversion, cash-flow needs, capital availability, expectations of future costs and prices, and the theoretical framework for optimal exploitation of a fixed resource over time (i.e., economics of the individual mine), make it obvious that the functional relationship determining the responsiveness of tin output to a change in its price is far more complex than can be captured in an econometric model of a manageable level of abstraction or than can be tested with existing data.

There is at least one more problem with measuring elasticities of supply curves derived by fitting time-series data. As Puey noted, the International Tin Committee of the 1930s maintained that a major cause of instability in the tin industry was that new mines opened up easily and promptly in periods of high prices but did not shut down when price fell later. In other words, entry was and presumably still is easier than exit. It follows that short-run elasticity of supply as measured statistically is greater for rising prices than for falling ones and therefore cannot be calculated from a single curve or be measured by a single coefficient. In this connection, it is apparent that the distinction between the short run as defined for theoretical purposes and the econometrician's short term is a crucial one. In the short run, plant and equipment are held constant—a convenient theoretical abstraction. Econometric short-term elasticities attempt to measure responses over a specified period of calendar time, such as a year.

Turning to demand, all of the estimates described above are in agreement that the short-run elasticity of demand for tin is negative but quite low, although in every case the level of statistical significance is also low. In other words, the influence of the price of tin on the quantity demanded is weakened by other influences on buyers. This is as expected in a market in which demand is derived from a number of diverse end-use demands. Estimation of long-run demand for producers' goods is extremely complicated, in that one must allow not only for variations in use of other inputs, modification of machinery or production processes, and additional investment or disinvestment of plant and equipment in the numerous industries using the goods, but also for the introduction of substitutes for tin that are known but unavailable in the short run.

Perhaps the most pertinent judgment on the long-run elasticity of demand for tin, and certainly one of the best informed, is that of Fox, based on his experience as secretary of the ITC from 1956 to 1971 and previous service with the British Ministry of Supply and the International Tin Study Group. He found it "difficult to swallow" the argument that tin was "a commodity inelastic to price" since the price of tin had risen more rapidly than that of any other nonferrous metal, while the growth of consumption had been much slower, over the previous seventy years.[16]

Fox went on to note that the argument for inelasticity was predicated on the notion that since tin represents only a fraction of the cost paid by the ultimate consumer for an item using tin, the consumer will be indifferent to the price of tin. "This belief," he stated bluntly, "is not relevant to the subject."[17] The purchasers of tin, he pointed out, are the fabricator and the tinplate maker, to whom the price of tin is very important, even if it can be passed on to customers for a while, since substitutes are available or can be found, and since the state of existing technology and the potential for research are such that the quantity of tin used in many products and processes can be reduced if costs make it attractive to do so. In one sense, Fox was merely confirming the belief that demand elasticity is much greater in the long run than in the short run. But of more significance, he was in effect saying, as a man of experience, that the economist's concept of long-run demand elasticity is of little relevance to the tin industry.

The crux of the criticism leveled by Fox is that the price elasticity of demand —or of supply for that matter—is rigorously defined as a measure of the proportionate responsiveness of the quantity purchased, or offered, to a change in price alone, holding everything else constant. In particular, elasticity abstracts from or assumes no changes in all other influences on quantities offered or desired such as the prices of other goods, availability of substitutes, tastes, technology, and income levels. In econometric estimates based on observed changes in prices and quantities actually sold and bought, it is, of course, not the case that every explanatory variable but price remains unchanged; and multiple regression techniques seek to introduce the most important of these variables into the estimating equations in order to isolate the effects of price. But "noise," or unidentified influences on quantity, is inevitably omitted from these equations.

Consumer and producer responses to hypothetical price changes. Another method for singling out the influence of price is to ask suppliers and customers how they would react under existing circumstances to various hypothetical price changes. Questions somewhat similar to these were asked, not of mine managers and tin buyers, but of government agencies, in a study made for the ITC by W. Robertson in the mid-1960s.[18] In 1961, the ITC established a working party "to consider an enquiry into the long-term tin position."[19] Questionnaires were sent to governments of the ITC producing members and to those of major consuming countries, asking them to estimate national levels of tin production or consumption, as the case might be, in 1965 and in 1970, assuming alternative prices of £700, £900, and £1,100 per long ton. The letter to producers was sent in September 1962, and that to consumers the following month. Thus, the specification of the year 1965 permitted approximately two years of adjustment time to be assumed, while that of 1970 allowed for a seven-year lag. The responses, reproduced in Robertson's study, make it possible to calculate arc elasticities over the ranges £700–£1,100, £700–£900, and £900–£1,100.

There are, unfortunately, certain deficiencies in Robertson's data for the present purpose of making theoretically correct elasticity calculations. First, several producer respondents criticized his selection of £700 as the lower price bound for his study. "It seems reasonable," he noted, "to regard a price of £700 as too low in the light of the average price over the last ten or fifteen years. Nearly all producing countries, large and small, forecast a drastic fall in output at a price of £700. . . . In the years 1961–63, the average annual price has been about £900."[20]

Second, the letters requesting the estimates were unclear as to the assumptions to be made. The letter to consuming nations' governments simply noted that the ITC "appreciates the relationship between consumption and the price of tin,"[21] and requested that the estimates be made. The letters to producers asked that estimates be made "on the basis of estimated working costs at the relevant time," and that allowance be made "for any improvements in techniques which may reasonably be expected to be in operation in your country in the years referred to."[22] If these instructions were followed, the reported numbers are not consistent with the ceteris paribus assumptions of both short- and long-run supply curves, in that neither costs nor technology were held constant.

Third, the reliability of the data is dubious, as a number of governments did not reply, and the replies of others were incomplete or questionable. The United States, for example, gave estimates only for a price of £900. Several countries submitted identical quantities for two or three of the prices. I have, nevertheless, calculated the arc elasticities from the Robertson report.

Overall, the supply elasticities shown in table 5.4 contradict Puey's observation that short-term supply should be more responsive to price increases than to decreases, since the prevailing price at the time the estimates were made was around £900. However, in light of the objections voiced by producers to Robertson's use of a price of £700, the first two columns of supply elasticities, for both years, ought to be given little if any consideration. The four overall supply elasticities, calculated on slightly different bases for countries and for methods, are all reasonably consistent with the short- and long-run estimates of the World Bank study, if the 1965 coefficients are taken as short-term and the ones for 1970 as long-term. But individual country estimates are hard to reconcile, even in rank ordering. For example, Indonesia is shown with the highest 1965 elasticity of supply over the range from £900 to £1,100 in table 5.4 but is reported with the lowest short-term elasticity by Chhabra, Grilli, and Pollak. In contrast to Ariff's findings, the Robertson data do not yield negative supply elasticities either for any individual country or for any method of mining.

The 1965 demand elasticities shown in table 5.4 are of roughly the same size as the short-term elasticities derived from econometric estimates. But if the 1970 figures actually approximate long-term demand conditions, the Robertson report does not bear out the econometric consensus that the long-term demand for tin is elastic (greater than unity). The indication that, in general, demand is more elastic between £900 and £1,100 than between £700 and £900 is plausible, as well as illustrative of the econometric distortion that may result from fitting a single estimate of elasticity to a range of observations. If the price fell substan-

Table 5.4. Arc price elasticities of tin demand and supply, as calculated from data furnished to the ITC working party on world tin position, 1962

	1965			1970		
	700–1,100	700–900	900–1,100	700–1,100	700–900	900–1,100
Demand						
France	−0.07	0	−0.17	−0.11	0	−0.25
F. R. Germany	−0.30	−0.17	−0.44	−0.33	−0.21	−0.47
Italy	−0.20	−0.18	−0.23	−0.16	−0.16	−0.16
Belgium	−0.26	−0.11	−0.45	−0.47	−0.10	−0.93
Netherlands	0	0	0	0	0	0
Total EEC	−0.17	−0.09	−0.26	−0.20	−0.10	−0.32
Supply by country						
Bolivia	n.a.	n.a.	0	n.a.	n.a.	0
Malaysia	1.63	2.48	0.57	2.10	2.67	1.58
Thailand	0.69	1.00	0.30	0.88	1.26	0.39
Indonesia	0.64	0.50	0.81	0.83	0.89	0.74
Nigeria	2.60	4.38	0.43	2.74	3.08	2.94
Zaire	1.74	2.80	0.42	2.95	4.72	1.08
All others	1.41	2.13	0.53	2.08	2.54	1.68
Supply by method						
Dredging	1.17	1.71	0.49	1.53	1.62	1.47
Gravel pumping	2.01	3.20	0.57	2.29	2.96	1.69
Hydraulicking	0.47	0.34	0.63	0.58	0.55	0.60
Open cast	2.72	4.67	0.32	2.64	4.41	0.53
Underground	n.a.	n.a.	0.04	n.a.	n.a.	0.26
Other	1.58	2.67	0.20	1.50	1.90	1.03

Source: W. Robertson, *Report on the World Tin Position, with Projections for 1965 and 1970* (London: International Tin Council, 1965).

tially (in this case from £900 to £700 per long ton), the quantity demanded would not be expected to increase very much, assuming that the only factor operating to increase the outputs of tin-using products was a fall in the price of such a minor input as tin. Since the demand is derived, it would be limited by the demand for tin-using products even if the price of tin fell to zero. On the other hand, if the price rose by about 20 percent above the level that had been the equilibrium for some time, there might be rather widespread resort to substitutes: at least, there is some level beyond which price cannot be raised without evoking such a reaction. In addition, there is an arithmetic property of the measure of demand elasticity that causes demand to become more elastic along a straight-line segment of a demand curve as price increases and quantity decreases, since in this circumstance equal absolute changes in price and quantity yield lower percentage changes in price and greater percentage changes in quantity.

In sum, the preponderance of evidence supports the use of models assuming that both the supply of and demand for tin are quite inelastic with respect to price in the short run. On the other hand, one ought to be skeptical of the results of any theoretical analysis in which the conclusions follow from an a priori assumption made about the nature of long-run elasticity of either supply of or demand for tin.

A Simple Supply and Demand Model of the World Tin Market

Supply and demand curves are of interest not only as indices of the responsiveness of quantities bought and sold to various prices, but also as tools for description and analysis of market performance. A simple supply and demand model cannot, alone, provide an adequate picture of the world tin market, but it can illuminate the intrinsic workings of that market and identify certain features of structure and behavior that deserve more detailed attention.

As a working approximation of various equilibrium levels of price and output in tin markets, assume a large number of independent tin miners selling to one or two smelters in each of several producing nations, with the markets at this level of trade completely separated from each other by barriers such as a ban on the export of tin concentrates, prohibitive duties on export of tin concentrates, interlocking ownership of mines and smelter, custom, or financial obligations. Assume further that all of these monopsonistic smelters compete in selling their metal on a single worldwide market to large numbers of industrial customers, brokers, dealers, and speculative traders. In this preliminary approach, we also allow for the fact that the ITC can raise and lower the world price of tin metal through its buffer-stock sales and purchases on the LME and is also capable of indirect limitation of mine production through imposition of export quotas.

The local market for tin-in-concentrates. In the market situation just described, the smelters' role is a crucial one. Figure 5.1 shows diagrammatically the short-run equilibrium price and output that would result if a competitive group of miners sold to a single smelter or to a small group of fully collusive smelters.

In this diagram, let D_t represent the demand for tin facing the smelter. D_t is drawn as a straight horizontal line on the assumption that the smelter sells its tin metal on a world market which is close enough to purely competitive so that the price the smelter receives, OA, is independent of the quantity of tin it offers for sale. Let AB be equal to the cost of smelting one unit of tin. This cost is assumed to be constant throughout the range of the diagram, so that AB equals both the average and marginal costs of smelting. D_c, under this condition, is a schedule of the marginal value product of tin-in-concentrates to the smelter. (The units in which tin-in-concentrates are measured are conventionally expressed in terms of quantity of tin metal, in which case all of the quantities indicated by various curves on figure 5.1 are consistent with each other.) To maximize profit, the smelter will set its marginal cost of tin-in-concentrates equal to this marginal value product.

The competitive supply curve of tin-in-concentrates is given by MC_m, the horizontal summation of those segments of the marginal cost curves of the individual miners that are above their average variable cost. This schedule is

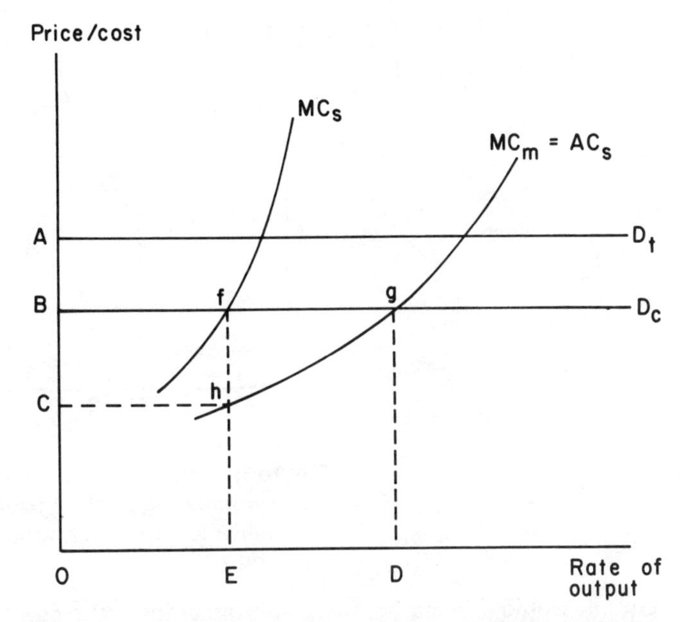

Figure 5.1. Short-run equilibrium prices and rates of output for tin concentrates, assuming a competitive mining industry supplying a single smelter, and a comparison of producers' surplus before and after vertical integration of such a market

also the average cost of tin-in-concentrates to the smelter, AC_s. The profit of the smelter, as a monopsonist, will be maximized when it purchases *OE* units of tin-in-concentrates and pays the miners a price of *OC*, where MC_s is the marginal cost of tin-in-concentrates to the smelter as derived from the average cost curve AC_s. In contrast, if there were a large enough number of independent smelters so that each took the cost of tin-in-concentrates as independent of its purchases, D_c would be the market demand curve and *OD* units of tin-in-concentrates would be sold at a price of *OB*.

The steeper (less elastic) the supply curve, the greater will be the divergence between the price paid to the miners and the marginal value of their product. In other words, the more inflexible the output of the mines in the short run, the more vulnerable are the miners to exploitation by the smelter, where exploitation is defined as the relative difference between the marginal value product of a factor and the payment it receives for its contribution.[23]

If competitive mining and monopsonistic smelting operations were integrated, total producers' surplus could be increased by the area *fgh*, since production would then be carried out to the point where the marginal cost of mining equalled the marginal value product of the tin-in-concentrates, at the level *OD*. (This level coincides with the competitive level only as a result of the assumption that the smelter's tin metal is sold at a world price that it cannot influence.)

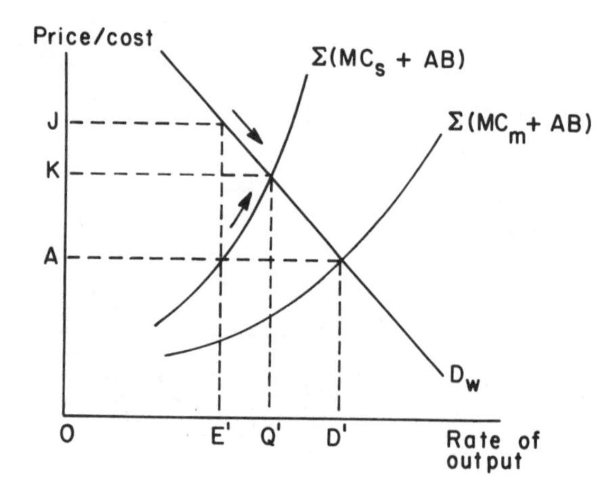

Figure 5.2. Short-run supply of and demand for tin metal, comparing price and output equilibria under alternative local market structures in mining and smelting

Output would expand because, with integration, the new firm would have no incentive to restrict purchases in such a fashion as to transfer the surplus amount *BfhC* from the mining level of the system to the smelting level, at the cost of giving up *fgh*. Upstream integration would also capture, for the smelter, producers' surplus otherwise accruing to the independent inframarginal mines whose average total costs, excluding economic rent, were below *OC*. There would, however, be little or no incentive to the smelter to integrate backwards to obtain this latter surplus, since unlike *fgh* it would be capitalized in the prices that mine owners would ask for their properties. The area *fgh* is distinctive in representing a surplus that could be obtained through vertical integration and that otherwise would be a deadweight social loss to the producing nation in that it would not accrue to either the monopsonistic smelter or to the miners if they were engaged in arm's-length market transactions. The smelter, therefore, would be willing to pay more for the mines than their value to independent owners.[24]

Most of the surplus *fgh* could also be obtained were the miners able to combine and set a firm collusive price very close to *OB*. In that case, the smelter would be unable to reduce its cost of acquiring tin-in-concentrates by restriction of output and would therefore have no incentive to purchase less than the miners offered, provided the price did not quite reach *OB* and thereby transfer every last bit of the surplus *ChgfB* to the miners. The individual miners, in turn, would produce up to the point where their marginal cost equalled the new price, or almost to *OD*. One strategy open to the smelter to prevent this outcome is acquisition of its own mines.

Thus, one of the major tasks of an investigation of the structure of the tin industry would appear to be to ascertain the extent to which independent mining and smelting enterprises exist in the several tin-producing countries and

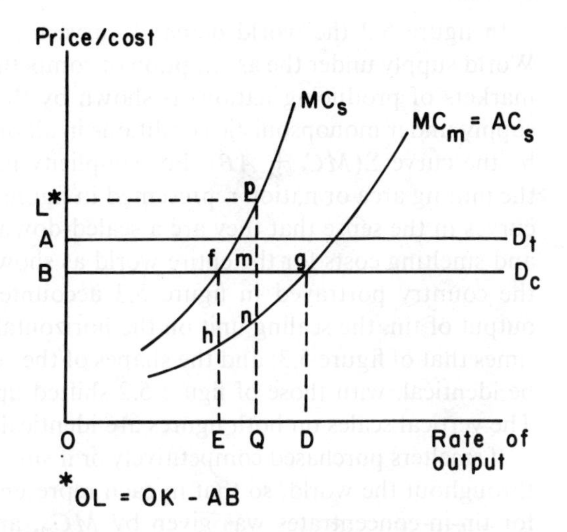

Figure 5.3. Short-run equilibrium prices and rates of output for tin concentrates, comparing costs and benefits to producing nations under alternative market structures

interact on markets for tin concentrates, and to explain why such markets persist.

The world market for tin metal. Turning to the world market for tin metal, we are assuming for the time being that the monopsonistic smelters in various parts of the world are competitive in the sense that each one takes the world price of tin metal as independent of its own output. Under that assumption, each smelter's supply curve of the metal is given by MC_s plus the vertical distance AB in figure 5.1. Since MC_s lies above AC_s and has a steeper slope at every level of output, the individual smelter's supply of tin metal on the world market is less elastic than the supply curve of tin-in-concentrates it faces on a national market. World supply of tin metal is the horizontal summation of the MC_s curves of all the smelters, and thus its elasticity is reduced by monopsonistic elements in the national tin-in-concentrates markets. Under either competitive purchasing of tin-in-concentrates on local markets or vertical integration of mining and smelting, a smelter's supply curve becomes AC_s plus AB; and therefore such integration on the part of a number of smelters would increase elasticity of tin metal supply as well as increase world output.

Figures 5.2 and 5.3 illustrate differences in equilibrium price and output in two such contrasting cases: first, either competitive purchasing of tin concentrates throughout the world or integration of each area's mining and smelting operations; and second, monopsony at the smelting level in all producing areas, still assuming that the smelters in various parts of the world do not collude with each other in their sales of tin metal.

In figure 5.2 the world demand curve for tin metal is designated by D_w. World supply under the assumption of competition in all of the tin concentrates markets of producing nations is shown by the curve $\Sigma(MC_m + AB)$. World supply under monopsonistic conditions in all tin concentrates markets is shown by the curve $\Sigma(MC_s + AB)$. For simplicity in exposition, the cost curves for the mining area or nation represented in figure 5.3 are drawn as representative curves in the sense that they are a scaled-down version of the summed mining and smelting costs for the entire world as shown in figure 5.2. If, for example, the country portrayed in figure 5.3 accounted for 10 percent of the world output of tin, the scaling unit on the horizontal axis of figure 5.2 would be ten times that of figure 5.3; and the shapes of the cost curves in both figures would be identical, with those of figure 5.2 shifted up vertically by the distance AB. The vertical scales on both figures are identical.

If smelters purchased competitively or if smelting and mining were integrated throughout the world, so that in each representative country the supply curve for tin-in-concentrates was given by MC_m, and the world supply curve was therefore $\Sigma(MC_m + AB)$, the equilibrium price of tin metal would be OA, as determined in figure 5.2 by the intersection of the world supply curve with the world demand curve D_w. The price of tin-in-concentrates would then be OB, equal to OA minus AB; and the rate of output would be OD in the representative country and OD' for the the world as a whole.

If, on the other hand, all smelters throughout the world acted as monopsonists, the reaction to a tin-in-concentrates price of OB would be a retrenchment of purchases of tin-in-concentrates from OD to OE in each national market. The immediate impact of such a cutback throughout the world would be a world output of OE', and a world tin price of OJ. In the absence of collusion among the smelters in their sales of tin metal on the world market, the price of OJ would induce each smelter to increase its output to the level of production at which the new price was equal to MC_s. As all smelters expanded along these paths, world output would increase and could be absorbed only at progressively lower prices. As indicated by the arrows on figure 5.2, output would increase and price would fall until a new equilibrium was reached at a world output of OQ' and a price of OK for tin metal. A tin metal price of OK implies a price of OL for tin-in-concentrates, where $OL = OK - AB$.

Figure 5.3 indicates the economic costs and benefits to an individual producing nation of the two alternative equilibria. In the discussion of figure 5.1 it was noted that producers' surplus *fgh* would be lost under monopsonistic pricing of tin-in-concentrates. This conclusion was, however, based on the assumption that the drop in the tin metal output of the single country from OD to OE would not have any effect on the world price. If, to the contrary, all producers were to restrict output of tin metal in order to lower the price of tin-in-concentrates, and if world output therefore first fell to OE' and then recovered to OQ', the representative country's price and output would settle ultimately at OL and OQ, respectively. In this case, the loss of producers' surplus is reduced to *mgn*, and there is an offsetting gain of *BmpL*. Whether or not there is a net increase in producers' surplus (both miners' and smelter's), that is, whether or not $BmpL > mgn$, depends on the elasticities of world demand and supply and the elasticity of the monopsonistic supply curve MC_s. The less elastic the world demand and supply, the greater the price increase associated with any given reduction in output, and hence the greater the area $BmpL$. On the other hand, the greater the elasticity of MC_s, the less the loss in producers' surplus from any initial reduction in output and the greater the recovery of production in response to a higher world tin price. Thus, the ideal situation for a producing country, in contrast to the situation depicted by our device of the representative producer, would be one in which its techniques of production varied from those of other countries in a way such that world supply was highly inelastic while its own supply was highly elastic.[25]

It cannot be determined solely on the basis of this sort of geometric analysis whether an individual producing country would be better off adopting public policy measures designed to set output at OD, such as requiring its smelters to pay miners a fixed price regardless of the quantity of tin concentrates purchased or by encouraging a merger of mining and smelting; or whether it would be to a nation's benefit to permit or even require monopsonistic purchasing by its smelters. From the point of view of an individual producing nation, provided that its output of tin metal is a small fraction of the world's total, the optimal situation would be one in which all other countries restricted output to the point where the world price of tin metal less unit smelting costs equalled MC_s, and it appropriated the full benefit of the ensuing increase in the world tin price by expanding its own output until D_c was equal to MC_m. This is a self-defeating strategy if followed by all producing nations; but were the number of producers large and the share of each in the world market for tin metal small, it would be a highly likely outcome, resulting in a world supply curve of $\Sigma(MC_m + AB)$, a world output of OD', and a world price of OA.

All of the producers might restrict output if world demand and supply were inelastic enough, if their number were small enough so that each one recognized the mutual benefit of restricting output, and if all were aware of the harm that could befall the group if one tried to produce until MC_m equalled D_c and the others followed in retaliation. One large producer might take the lead in reducing output and hope that the others would follow its example. Or, if we drop the assumption that no single producing nation's output of tin metal is large enough for it to be able to take into account the effects that changes in its rate of production would have on the world price, the largest producer might find that world demand was inelastic enough for it to gain from independent cut-backs regardless of whether it was followed by the smaller producers. The policy alternatives suggested by this line of reasoning would appear to be particularly pertinent to Malaysia, a nation mining approximately one-third of the tin entering the world market and smelting an even larger fraction.

The effects of a buffer stock and export controls. Price and output in tin markets are also shaped by the ITC's buffer-stock operations and export con-

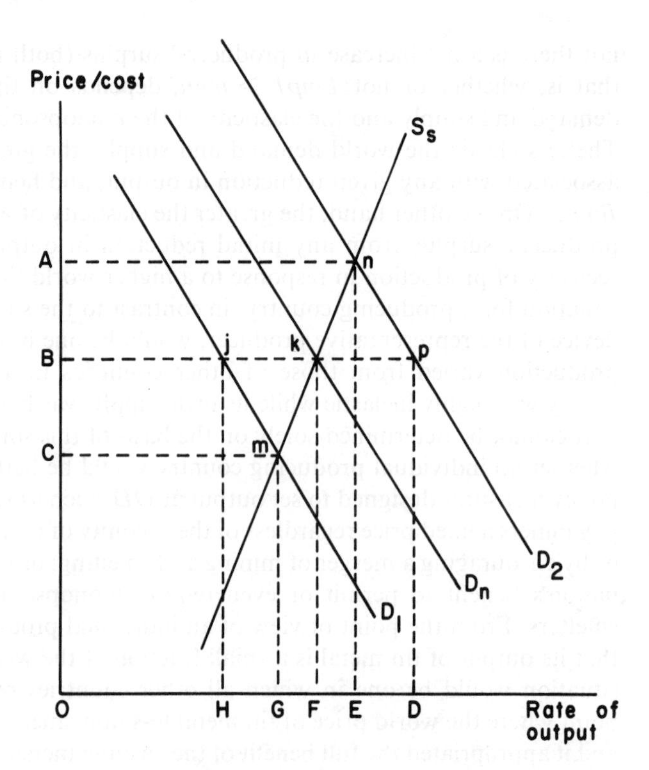

Figure 5.4. The effects on producers' and consumers' surplus of price-stabilizing buffer stock operations, assuming stable supply and fluctuating demand

Source: Adapted from Harry G. Johnson, "The Elementary Geometry of Buffer Stock Price Stabilization," *Malayan Economic Review* 22, 1 (April 1977): 3.

trols. There is a substantial literature dealing with the economic effects on consumers and producers of price stabilization by means of a buffer stock.[26] The basic framework of this literature is illustrated by figure 5.4.

In figure 5.4, depicting the conventional wisdom regarding the world market for tin metal, S_s is a short-run straight-line supply schedule that is held steady, while short-run demand is assumed to vary around a normal linear schedule of D_n, fluctuating between a minimum of D_1 and a maximum of D_2. Demand shifts are assumed to be symmetrical, so that $AB = BC$. Further, we assume that the objective of the buffer stock is to stabilize price at OB.

Suppose, first, that demand falls from D_n to D_1. In the absence of buffer-stock operations, price would fall from OB to OC, and output would fall from OF to OG. If purchases for the buffer stock are made in order to maintain the price at OB, consumer purchases will fall to OH, while HF will be added to the buffer stock. With the buffer stock, producers' revenue is held constant at the area *OBkF*; whereas in the absence of such a stock it would drop to *OCmG*. Producers' surplus is thus increased by the area *CmkB* as a result of the purchases for the buffer stock. Next, suppose that demand rises from D_n to D_2. Without the buffer stock, producers' revenue would rise to *OAnE*, instead of remaining at *OBkF*, for a gain in producers' surplus of *BknA*. Since $AB = BC$, it can be seen that the loss in producers' surplus from stabilizing the price at *OB* when demand rises to D_2 is greater than the gain obtained from such stabilization when demand falls to D_1 (i.e., $BknA > CmkB$). Thus, if demand fluctuations are such that the probabilities of D_1 and D_2 occurring are equal, producers pay for stabilizing revenue at *OBkF* by a loss of total surplus earnings. However, since it is the slope of S_s that determines the magnitude of the difference between *CmkB* and *BknA*, the less elastic the short-run supply curve, the less the net surplus sacrificed by the producers.

A similar analysis can be carried out from the consumers' point of view. When demand falls to D_1, total consumer outlays fall from *OBkF* to *OCmG* without a buffer stock, and to *OBjH* with the buffer stock. If short-run demand is assumed to be inelastic, total consumer spending under demand curve D_1 is higher with the buffer stock than without it. Conversely, if demand rises to D_2, total consumer spending is reduced by the buffer stock's effect of reducing price from *OA* to *OB* while purchases increase from *OE* to *OD*. Thus, the buffer stock stabilizes total consumers' outlay as well as producers' revenue; but it increases the fluctuations in consumers' usage, which would vary from *OG* to *OE* without the buffer stock and from *OH* to *OD* with the stock. The differences, of course, represent quantities purchased for and sold from the buffer stock. Stabilization of price at *OB* lowers consumers' surplus by *CmjB* when demand falls to D_1, but raises it by the larger amount *BpnA* when demand rises to D_2. Thus, while net producers' surplus is reduced by the operations of the buffer stock, net consumers' surplus is increased—continuing to assume equal probabilities of occurrence for D_1 and D_2.

It would therefore appear that if the objective of the buffer stock is stabilization of price at some normal level in the face of fluctuating demand, consumers will gain at the expense of producers. Unless producers put a high value on stability of both price and revenue, their interest in a buffer stock requires explanation. Such an apparently perverse interest within the tin industry is made manifest by the fact that under the postwar tin agreements the cost of operating the buffer stock has been borne almost entirely by the producing-country members of the ITC. One possible explanation is that, contrary to the assumption usually made about nonagricultural primary commodity markets, the supply of tin fluctuates more than its demand. If, in a diagram such as figure 5.4, the supply schedule is allowed to vary while the demand schedule is held constant, it can be shown by a line of reasoning analogous to that used above that producers' surplus is increased and consumers' surplus is reduced by introduction of a price-stabilizing stockpile. Another possibility, discussed by

Behrman and Turnowsky and analyzed for the Malaysian case by Brown, is that the assumptions of parallel shifts of equal magnitude and of linear supply and demand functions are inappropriate or unrealistic. Once the slope of the stable curve is allowed to vary within the presumed range of price and output fluctuations, and the fluctuations of the unstable curve are described by patterns other than equal absolute variations around the mean, neat and unambiguous conclusions regarding producer and consumer gains no longer follow.

Brown simulated price stabilization of tin for the case of fluctuating demand, assuming values for supply and demand elasticities that he regarded as approximately correct for Malaysia. In his simulations, he did not have his buffer stock stabilize the price at the mean but rather, as is the case for the ITC, prevent the price from rising above a ceiling or falling below a floor. He did assume parallel shifts, but allowed for varying elasticities and slopes. For the values that he assumed to be reasonable in his simulations, Brown found that buffer-stock control reduced producers' income, export earnings, and tax revenues but did exert a stabilizing influence, thus reinforcing the conclusions obtained from a simple model with linear schedules and equiprobable divergences from the mean.

In addition to price stabilization through its buffer-stock operations, the ITC may impose export controls on member producing nations. These nations, in turn, are expected to limit the production of individual mines and smelters within their borders. Output restriction raises a classic cartel problem. Different members of the cartel are likely to prefer different degrees of restrictions, if any, and different prices. In general, those with the lower costs of production should prefer lower prices and higher rates of output, although the generality is complicated because at different rates of output different firms may be the lower-cost producers. In the short run, firms with high fixed costs relative to their variable costs are likely to prefer high rates of output over which to spread the fixed costs, even though this entails a lower price than firms with lower fixed and higher variable costs might desire. In a perfect cartel, whatever the level of output, it would be allocated among member firms in such a way as to minimize the total costs of production (i.e., marginal costs would be equalized); and side payments would be used to distribute profits among all members, including those who had agreed to cease production entirely. The ITC does not have the ability to assign rates of production in this manner nor to make side payments. Figure 5.5 illustrates the general problem.

Assume that the world market is in short-run equilibrium at price *OA* and output *OD* as shown in panel A; but that, perhaps because of a drop in demand below normal and inability of the ITC to finance continued price support through buffer-stock purchases, the price has fallen below the costs of production of high-cost producers, one of which has the cost curves drawn in panel B. Panel C depicts the costs of one of the low-cost producers. Only those portions of marginal cost that are above average variable cost (i.e., supply schedules) are drawn. At price *OA*, the high-cost producer would produce an

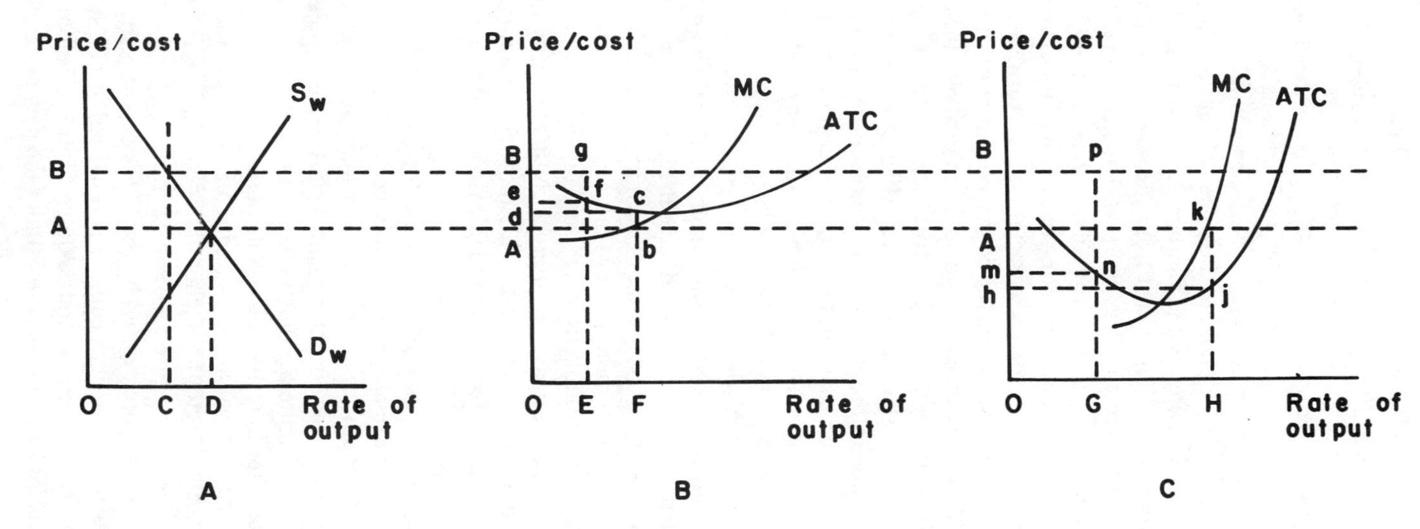

Figure 5.5. A comparison of the effects of output restriction by an imperfect cartel (such as the ITC) on high-cost and low-cost member producers

output of *OF*, and would suffer a loss of *Abcd*. The low-cost producer would produce at *OH*, and enjoy a profit of *hjkA*. Suppose now that the ITC restricts total world exports to *OC* and that, as a result, price rises to *OB* (ignoring domestic consumption of producing nations). Suppose further that the restrictions imposed on our two sample firms by their nations' governments are outputs of *OE* and *OG*. At a price of *OB*, both firms would like to expand output to the point where *MC* is equal to the price *OB*; but on the assumption that cheating is successfully suppressed (another classic cartel problem), it can be seen that the high-cost firm has gained from the restriction, now earning a profit of *efgB* instead of the previous loss of *Abcd*. The situation of the low-cost firm is not so clear. It is now earning a profit of *mnpB* instead of *hjkA* and may be better or worse off than before, depending on which of these two rectangles is the larger. The relative sizes of these rectangles, in turn, depend on the positions and shapes of the cost curves, the degree of restriction imposed on the firm, and the extent to which elasticity of world demand allows the price to rise. A priori, it cannot be said whether the low-cost firm in panel *C* has gained or lost; but there are certainly circumstances under which such firms would, in the short run, suffer a decline in profits from restrictions necessary to allow high-cost firms to break even or earn some profit. If the low-cost firm takes a long-run view, it is even more likely to oppose restrictions that keep its high-cost competitors in business.

It should be noted, though, that if the average total cost curves *ATC* in panels B and C are superimposed on one diagram, there would be rates of output for which the firm in panel B would have the lower costs. *ATC* in panel C has been drawn with a rather long range of declining average total costs and a sharp rise beyond the minimum-cost rate of output. This is a curve typical of a firm with high fixed costs resulting from a capital-intensive mode of operation with plant and equipment designed to produce most efficiently at a specific rate. The *ATC* drawn for the firm in panel B is a more shallow curve, as would be the case for a firm with low fixed costs and fairly constant variable costs over a wide range of output.

The output limitations imposed by producing countries in response to the ITC's export restrictions have almost invariably been set for individual mines rather than smelters. In countries in which the smelters are permitted to act as monopsonists in their local tin concentrates markets, there is a major consideration of equity involved. Figure 5.6 illustrates the profit-maximizing response of a monopsonistic smelter to restrictions on its output.

MC_m and MC_s are as shown above in figure 5.1. D_e is the marginal value product of tin-in-concentrates to the smelter under equilibrium in the world tin metal market (similar to D_c in figure 5.1), and D_r is the marginal value product after the world tin price has been raised by the ITC's export restriction. If the total output of mines in the smelter's country is restricted to *OE*, it need pay a price of only *OD* for its tin-in-concentrates, down from *OC* before the restriction. The smelter may be better or worse off under the new price: its price-cost

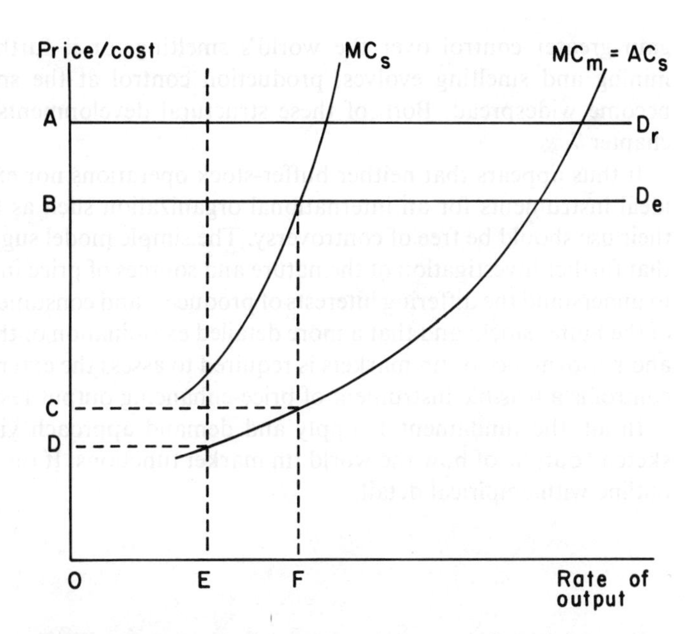

Figure 5.6. The effect on increasing-cost miners of an ouput restriction imposed on a monopsonistic smelter

margin rises from *CB* to *DA*, but the number of units it can sell at the higher margin falls from *OF* to *OE*. Ironically, the rise in the price of tin metal caused by the restriction of output leads to lower unit prices received by miners as well as to reduced levels of production. There is no paradox in this result. The diagram merely illustrates the common-sense notion that if a monopsonist restricts its purchases from a set of competitive increasing-cost suppliers, it can reduce the purchase price it pays them.

On the other hand, if the authorities were more concerned with short-run efficiency than with equitable treatment of the miners, they might well decide to impose output restrictions on the smelter rather than on individual mines during periods of export control. Under the conditions depicted in figure 5.6 the reduced output would be produced by the most efficient mines, each of which would be operating at its most efficient rate of production, given the prescribed level of overall output of tin-in-concentrates. The mines with higher costs at all levels of output would close. Further, in a nation with a large number of relatively small mines, the administrative costs of establishing and enforcing the production limits would be far lower if the control could be set on only one or two enterprises; and the problem of allocating the overall restriction by assigning a quota to each producer would be eliminated or minimized. It might be possible to tax a large portion of the gains that accrued to a smelter from export control and use the proceeds to alleviate the short-term suffering imposed on miners and mine workers. As the governments of producing nations gain greater control over the world's smelting, or if further integration of mining and smelting evolves, production control at the smelting level may become widespread. Both of these structural developments are discussed in chapter 7.

It thus appears that neither buffer-stock operations nor export controls are ideal instruments for an international organization such as the ITC, nor that their use should be free of controversy. The simple model suggests in particular that further investigation of the nature and sources of price instability is needed to understand the differing interests of producers and consumers in maintenance of the buffer stock; and that a more detailed examination of the actual structure and performance of tin markets is required to assess the extent to which export control is a feasible instrument of price-enhancing output restriction.

In all, the fundamental supply and demand approach yields a rough and sketchy outline of how the world tin market functions. It remains to fill in that outline with empirical detail.

6. Price and Output: The Issue of Stability

Output and Price Stability in Tin Markets

One fundamental reason for concern with price elasticities in tin markets is that low elasticities, coupled with sharp and frequent fluctuations in either supply or demand, are associated with severe instability of price as well as of output. For example, a price elasticity of demand of −0.2, indicating that a rise (fall) in price of 10 percent will be associated with a fall (rise) in quantity demanded of 2 percent, also indicates that a shift in the supply curve which causes the equilibrium quantity bought and sold to fall (rise) by 2 percent will lead to a 10 percent price increase (decrease). If either supply or demand is highly volatile and subject to major shifts over time, and if the supply and demand curves are inelastic with respect to price, the commodity in question may be doubly afflicted by large fluctuations in both output (due to the shift) and price (due to the inelasticity). In the case of a supply shift, equilibrium output declines as price rises, so that it is impossible to ascertain whether total revenue will increase or decrease without knowing the elasticity of the demand curve over the range between the two equilibrium positions. When demand shifts, price and output move in the same direction, and therefore total revenue unquestionably moves in the same direction as price.

It has frequently been asserted that models based on unstable as well as inelastic supply and demand curves are particularly applicable to analysis of primary commodity markets. As described in the previous chapter, the hypotheses that both supply of and demand for tin will be highly inelastic in the short run appear to have been borne out by empirical studies.

In the cases of nonagricultural commodities such as tin, which are immune to or only slightly affected by vagaries of rainfall, temperature, insect infestation, and other natural phenomena, instability is attributed primarily to the demand side of the market; it is regarded as largely a consequence of fluctuations in the levels of economic activity of the industrialized consuming countries. Demand for metals and minerals is considered likely to be more unstable than that for many other commodities since metals and minerals are frequently components of producer goods rather than of consumer goods. If demand for a consumer good is growing at a certain rate, the demand for new and replacement machinery and equipment required for production of that good may be constant. In such a situation, a small increase in the rate of growth of demand for the final good may lead to a proportionately much greater increase in investment in new productive equipment, while a decline in the rate of growth of demand for the final good may lead to an absolute reduction in the derived demand for capital goods.

Assume, for example, that the output of a single machine is 100 units of a final good per year and that the useful life of a machine is five years. If the output of the consumer good industry was steady at 10,000 units per year, 100 machines would be required, and replacement demand would be for twenty each year. If output of the final good rose by 20 percent, to 12,000 units per year, the demand for machines would double from twenty to forty—twenty for replacement and twenty to accommodate the expansion. Unless output of the consumer good again rose by 2,000 units in the following year, demand for the machines would fall. A steady rate of production of forty machines per year would require sustained expansion of output of the final product for a number of years.

Further, metals such as tin are often used in the manufacture of durable goods, both producer and consumer; and in many cases the purchase of a durable good can be postponed in a time of economic stringency.

A sharp rise in real interest rates and recession in the industrialized nations have been blamed for the decline in commodity prices and stagnation of commodity markets in late 1980 and 1981. When the interest rates that can be obtained on financial instruments are substantially higher than the expected rate of increase in prices, the income forgone by holding inventories or by investing in commodities, or, in general, the opportunity cost of holding real assets, is also high. Further, the costs of financing long speculation, or of holding claims against commodities to be sold later, rise. A shift of speculative funds out of commodities and into financial markets, coupled with reduced inventory holdings, will depress commodity demand.

Thus, the nations producing commodities such as tin regard themselves as highly vulnerable to shifts in foreign demand due to causes that are beyond their ability to control or even influence, particularly recessions and interest rate movements, in consuming nations. Since demand fluctuations lead to changes in price and output that reinforce each other, the effects of demand fluctuations on export earnings of the producing nations are more severe than those of supply fluctuations of similar magnitude. Industrialized consuming nations, on the other hand, are likely to be concerned about even a moderate curtailment in supply, which for reasons such as political instability in one or more of a few producing countries, could push the price of an essential commodity up an inordinate degree, especially if the demand for that commodity is highly inelastic.

The impact of fluctuating demand on an inelastic supply has been regarded as one of the most serious problems facing the tin industry. Yip, for example, noted, "Since most of the tin produced is imported by the industrial countries of the West, slight changes in the business conditions can cause wide fluctuations in the tin price and hence in the export earnings of the tin producing countries, all of which are economically underdeveloped. In this way all the tin producing countries are tied to the industrial activity of the West. For the tin producing countries this lack of stability is significant because tin export earnings represent a considerable share of their gross national product."[1]

Table 6.1. World output of tin-in-concentrates and consumption of tin metal, 1956–80 (long tons)

Year	Mine production of tin-in-concentrates	Consumption of primary tin metal
1956	166,400	149,800
1957	163,100	143,100
1958	115,800	136,200
1959	119,200	148,200
1960	136,500	162,200
1961	136,500	157,800
1962	141,700	160,900
1963	141,300	161,700
1964	147,200	171,100
1965	151,900	167,300
1966	166,200	176,000
1967	172,900	174,500
1968	183,100	179,900
1969	178,000	186,800
1970	185,600	185,200
1971	187,100	189,000
1972	196,000	191,400
1973	185,900	213,600
1974	183,300	199,600
1975	177,500	172,600
1976	179,800	195,000
1977	188,500	184,000
1978	197,800	184,800
1979	200,700	184,100
1980	199,300	177,400

Sources: International Tin Council, *Statistical Yearbook*, 1968; International Tin Council, *Tin Statistics, 1966–76* and *1970–80*; International Tin Council, *Monthly Statistical Bulletin* (March 1982).

In a 1975 address, Harold W. Allen, the executive chairman of the ITC, made a similar observation, referring to the "inevitable instability" of international markets and noting that "the supply of tin, as of most mined products, is inelastic." In Allen's opinion, consumption of tin is fairly stable; but market demand is nevertheless subject to severe fluctuations because of speculation, changes in inventory positions, use of tin and other commodities as hedges against short-run currency movements, and purchases based on changing price expectations of both consumers and dealers. He concluded, "Because supply does not usually vary significantly in the short-term, the effects of demand movements in terms of price can be very severe, and, if severe enough, could be catastrophic for the producers, or extremely expensive for consumers."[2]

Turning first to quantities of tin produced and consumed, table 6.1 shows output of tin-in-concentrates and consumption of tin metal, excluding the Soviet Union and the Peoples' Republic of China, from 1956, the year of establishment of the ITC, through 1978. Table 6.2 shows annual percentage

Table 6.2. Percentage changes in output of tin-in-concentrates by major producing nations and for the world as a whole, 1956–80

Years	Malaysia	Bolivia	Thailand	Indonesia	Nigeria	Zaire	World
1956–57	−4.8	3.6	8.4	−7.8	4.9	−2.8	−2.0
1957–58	−35.1	−36.2	−42.9	−16.3	−35.2	−22.4	−29.0
1958–59	−2.4	34.3	25.5	−6.8	−11.3	−5.2	2.9
1959–60	38.5	−15.1	24.8	4.5	39.0	0.1	14.5
1960–61	7.8	2.2	9.9	−17.8	1.4	−28.6	—[a]
1961–62	4.6	5.5	10.6	−6.8	5.5	9.5	3.8
1962–63	2.3	2.0	6.2	−25.2	6.3	−2.0	−0.3
1963–64	0.1	8.8	0.1	26.2	−0.1	−8.0	4.2
1964–65	6.1	−4.8	22.1	−10.0	9.5	−4.3	3.2
1965–66	8.2	10.8	18.5	−14.8	−0.1	11.5	8.0
1966–67	4.7	6.9	−0.3	8.6	−2.0	−8.0	4.0
1967–68	4.1	6.7	4.9	22.6	3.3	−4.8	5.9
1968–69	−3.9	1.6	−12.0	−2.3	−10.8	6.1	−2.8
1969–70	1.0	0.2	3.3	15.4	−8.9	−2.8	4.3
1970–71	2.2	0.6	−0.4	3.5	−8.0	—[a]	0.8
1971–72	1.8	7.0	1.8	10.1	−8.1	−7.7	4.8
1972–73	−5.9	−11.8	−5.2	4.1	−13.4	−8.7	−5.2
1973–74	−5.7	2.0	−2.8	13.2	−6.4	−14.1	−1.4
1974–75	−5.5	−2.8	−19.3	−1.1	−14.7	−2.4	−3.2
1975–76	−1.5	−0.7	24.7	−7.6	−21.6	−12.3	0.3
1976–77	−7.4	10.9	18.3	10.7	−11.9	−4.4	5.4
1977–78	6.7	−8.2	24.7	5.7	−15.8	−3.1	4.9
1978–79	0.6	−10.0	12.5	7.4	−6.3	−4.3	2.3
1979–80	−2.5	−1.8	−0.8	10.5	−8.1	−4.3	−0.6

a. Less than 0.05 percent.
Sources: International Tin Council, *Statistical Yearbook*, 1968; International Tin Council, *Tin Statistics, 1966-76* and *1970-80*; International Tin Council, *Monthly Statistical Bulletin* (March 1982).

changes in the output of tin-in-concentrates by the major producing nations and for the world as a whole, once again excluding the Soviet Union and China; and table 6.3 contains a similar display of changes in consumption of tin metal. Since tin concentrates are sold to smelters, and the smelters in turn sell tin metal through dealers to consumers and speculators around the world, there are several levels at which inventories are held between production of tin concentrates and consumption of tin metal. These inventories, as well as sales from the United States' strategic stockpile, serve as buffers, allowing substantial differences between consumption and production to persist for sustained periods of time. Thus, important aspects of supply conditions over the twenty-five-year period are reflected in the figures on production of tin-in-concentrates, as are features of demand conditions in the tin metal consumption figures.

Tables 6.1, 6.2, and 6.3 all show, not surprisingly, that major reductions in total world output of tin occurred during the periods of ITC export controls shown in table 6.4.

The first and most severe set of quarterly control periods, during which

Table 6.3. Percentage changes in consumption of primary tin metal by major consuming nations and for the world as a whole, 1956–80

Years	U.S.A.	Japan	F. R. Germany	U. K.	France	World
1956–57	−10.1	−15.7	9.2	−1.7	7.7	−4.5
1957–58	−11.8	16.3	6.8	−10.8	−1.8	−4.8
1958–59	−4.4	10.7	71.9	7.6	−0.9	8.8
1959–60	12.4	21.4	64.6	5.6	2.4	9.4
1960–61	−2.4	10.7	−7.0	−7.1	−10.0	−2.7
1961–62	8.6	−3.5	−55.0	5.9	11.4	2.0
1962–63	1.1	15.0	−4.0	−3.7	−1.3	0.5
1963–64	6.1	12.9	11.0	−6.7	−0.2	5.8
1964–65	—[a]	−4.4	−5.9	—[a]	−8.1	−2.2
1965–66	2.8	8.6	−7.6	−4.3	1.6	1.4
1966–67	−3.9	8.6	−1.0	−6.0	1.6	0.9
1967–68	1.7	10.2	4.1	−1.0	−10.6	3.1
1968–69	−1.9	14.2	19.1	3.7	10.6	3.8
1969–70	−8.1	−4.5	4.7	−6.1	−1.8	−0.9
1970–71	−2.0	18.6	1.0	−3.1	−0.5	2.1
1971–72	2.9	10.4	1.3	−10.8	4.6	1.3
1972–73	8.7	19.6	10.1	13.3	7.1	11.6
1973–74	−11.2	−12.6	−8.3	−12.9	−3.7	−6.6
1974–75	−16.8	−16.9	−17.8	−15.9	−11.5	−13.5
1975–76	23.6	23.3	24.1	7.8	5.5	13.2
1976–77	−8.0	−14.4	−4.9	−6.1	4.2	−5.7
1977–78	−1.2	−0.8	−4.6	−4.1	−3.9	0.4
1978–79	2.2	5.6	0.9	−8.7	−2.5	0.4
1979–80	−10.4	−1.1	4.5	−41.9[b]	4.1	−5.9

a. Less than 0.05 percent.
b. Reflects thirteen-week strike at British Steel Corporation.
Sources: International Tin Council, *Statistical Yearbook*, 1968; International Tin Council, *Tin Statistics, 1966-76* and *1970-80; International Tin Council, Monthly Statistical Bulletin* (March 1982).

restriction averaged 28 percent, was imposed from December 15, 1957 through September 30, 1960, in response to the sudden and unexpected flow of Russian exports of Chinese tin into the world market (see table 4.4).

The export control of September 19, 1968 through December 31, 1969 was of a much smaller magnitude. Export control was imposed in the last quarter of 1968 because of concern within the ITC over the strong and steady expansion of Bolivian output through the 1960s, the recovery of the Indonesian industry from the era of Sukarno's Guided Democracy, and the rapid growth of Thai output. For reasons that will be discussed below, prices had fallen rather steeply in 1966, 1967, and 1968. The 1968–69 limitation averaged only 4 percent, and the control periods ended under pressure of an 11 percent rise in 1969.

Through 1970, 1971, and 1972, mine output resumed its growth, with Indonesia continuing to provide the largest single impetus, followed closely by Australia which, although not included in table 6.2, increased its production of

Table 6.4. International Tin Council export control periods, through 1981

Period	Total permissible export tonnages	Total actual export tonnages
Dec. 15, 1957–Mar. 31, 1958	27,000	27,470
Apr. 1–June 30, 1958	23,000	22,835
July 1–Sept. 30, 1958	23,000	22,717
Oct. 1–Dec. 31, 1958	20,000	19,931
Jan. 1–Mar. 31, 1959	20,000	20,073
Apr. 1–June 30, 1959	23,000	22,896
July 1–Sept. 30, 1959	25,000	24,924
Oct. 1–Dec. 31, 1959	30,000	29,267
Jan. 1–Mar.31, 1960	36,000	32,578
Apr. 1–June 30, 1960	37,500	35,105
July1–Sept. 30, 1960	37,500	35,448
Sept. 19–Dec. 31, 1968	42,950	42,944
Jan. 1–Mar. 31, 1969	38,000	37,084
Apr. 1–June 30, 1969	38,750	39,270
July 1–Sept. 30, 1969	39,500	38,806
Oct. 1–Dec. 31, 1969	41,500	41,708
Jan. 19–Mar. 31, 1973	35,040	34,655
Apr. 1–June 30, 1973	42,644	41,346
July 1–Sept. 30, 1973	42,644	39,409
Apr. 18–June 30, 1975	26,560	27,379
July 1–Sept. 30, 1975	33,000	33,158
Oct. 1–Dec. 31, 1975	35,000	34,205
Jan. 1–Mar. 31, 1976	32,835	32,979
Apr. 1–June 30, 1976	40,000	n.a.

Sources: International Tin Council, *Annual Report*, various years; International Tin Council, *Tin Statistics, 1970-80*.

tin-in-concentrates from 6,642 metric tons in 1968 to 11,997 metric tons in 1972. The export controls of January 19 to September 30, 1973 were imposed to check this "creeping surplus," as Fox described it,[3] rather than actually to reduce output; and the controls were therefore set to maintain rather than to curtail the existing level of exports. World output nevertheless fell in 1973, 1974, and 1975 in the face of a sharp rise in price. In the third quarter of 1973 the ITC buffer stock sold 5,000 metric tons of tin, and in September the anomaly of buffer-stock disposals coupled with export control was corrected by ending the latter.

The most recent set of control periods, from April 18, 1975 through June 30, 1976, was initiated in response to the sharp drop in world consumption that accompanied the recession of 1975. The limitation, averaging approximately 18 percent, was sharply criticized as excessive and unduly protracted in the face of a strong price recovery in early 1976, as noted in chapter 4.

An interesting feature of table 6.2 is that the rather steady growth in output of tin-in-concentrates between 1960 and 1972 masks quite large year-to-year fluctuations in individual countries. This indicates a substantial degree of flexibility in output, since virtually all change in mine output results from human volition. The principal exceptions to this statement are the dependence of some mines, particularly in Nigeria, on rainfall to provide water for on-site ore concentration, and the need for offshore dredges in Thailand and Indonesia, and suction boats in the former, to cease operations during the monsoon season. Some of the reductions in output were obviously unintended consequences of war or mismanagement, as discussed previously; but some of the large increases were just as obviously the intended results of managerial decisions, often deliberate offsets to unplanned reductions in other countries.

Table 6.1 shows that world consumption of tin fell sharply in 1957 and 1958. A major portion of the decline can be attributed to a sharp drop in U.S. private consumption in 1957 and 1958, which reinforced the impact of the 1955 cessation of government purchases for the strategic stockpile, and to similarly steep declines in Japanese consumption in 1957 and in British consumption the following year. These two years were years of recession throughout the industrialized world.

The next period of severe year-to-year fluctuations in world tin consumption was that of 1973–76. During what ITC Buffer Stock Manager de Koning referred to as the "false boom from mid-1973 to mid-1974," apparent consumption of tin rose, with the increased purchases attributed largely to individuals and to firms who did not customarily trade in tin moving into the tin market as part of a shift out of cash holdings and into commodities in a period of rapid inflation and shifting exchange rates on world currency markets. In late 1974 and 1975, in the opinion of a number of knowledgeable observers, these outsiders' holdings of tin were sold, reducing apparent consumption below the amounts of tin actually acquired and used by industrial consumers. The effect of disposals of these speculative holdings was augmented and prolonged by the worldwide recession of 1975.[4] The drop of nearly 6 percent in world tin consumption from 1979 to 1980 reflects the onset of the 1980–81 recession.

Data from the United States for periods of recession, shown in table 6.5, illustrate the magnified impact of relatively minor changes in overall economic activity on production of producers' goods and thus on the demand for tin.

The entire period from 1956 to 1980 is one in which world consumption of tin metal exceeded mine production of tin-in-concentrates by a margin of approximately 150,000 long tons. From 1956 through the early 1960s, the shortfall was covered by Russian exports of Chinese tin that are not included in the ITC's world production figures. Of much greater magnitude have been disposals of just over 145,000 long tons of tin from the United States' strategic stockpile since 1962.

There are various statistical indices of stability, none of which is without flaws. For purposes of the present analysis of output and price stability in tin markets, I have fitted least-squares regression lines to the natural logarithms of

Table 6.5. Comparative declines in real GNP, real investment in producers' durable equipment, and consumption of primary tin metal in the United States, 1957–58, 1974–75, and 1979–80

Years	Percentage change in real GNP	Percentage change in real investment in producers' durable equipment	Percentage change in consumption of primary tin metal
1957–58	−1.8	−14.0	−11.8
1974–75	−1.1	−11.9	−16.8
1979–80	−0.2	−4.2	−10.4

Sources: United States Council of Economic Advisers, *Annual Report of the Council of Economic Advisers* (Washington D. C.: Government Printing Office, 1982), p. 234. Table 5.3.

a number of output and price series, and calculated the resulting correlation coefficients.[5] If the trend rate of growth is actually as portrayed by the regression line (i.e., if the period is not distorted by initial years above the underlying trend and final years below it or vice versa, and if there is in fact a single trend over the entire period), then the square of the correlation coefficient r^2 can be interpreted as an index of the deviation of individual years from the long-run trend, with stability being greater as r^2 rises toward its maximum value of 1.

The calculation for world production of tin-in-concentrates from 1956 through 1980 yielded the following:

$$\ln Q_p = 11.809 + 0.0176\,Y$$

$$r^2 = 0.662,$$

where Q_p stands for quantity produced and Y represents years from 0 to 24 (1956 to 1980).

The equation for world consumption of tin metal Q_c for the same twenty-five-year period is as follows:

$$\ln Q_c = 11.918 + 0.0120\,Y$$

$$r^2 = 0.657.$$

These equations indicate that the trend rate of growth of mine output of tin-in-concentrates was slightly under 1.8 percent per year, while that of consumption of tin metal was 1.2 percent per year. As a result of the more rapid growth of output over the twenty-five-year period, the gap between consumption and production narrowed, then closed, and finally converted into a surplus of production. The correlation coefficients indicate that fluctuations in production were just about equal to those in consumption, contrary to the common assertion that, in the case of nonagricultural primary commodities, the principal cause of instability is fluctuations in consumption in the advanced industrialized nations.

Table 6.6. Correlation coefficients for the output of tin-in-concentrates and growth rates of production

	r^2	Annual growth rate (%)
Malaysia	0.229	1.18
Bolivia	0.470	1.51
Thailand	0.721	4.20
Indonesia	0.106	1.20
Nigeria	0.563	−4.44
Zaire	0.898	−4.85

Correlation coefficients for the output of tin-in-concentrates by individual producing countries from 1956 through 1980, and trend rates of growth, are shown in table 6.6. With the exception of Zaire, whose production declined quite steadily over the period, and of Thailand, the individual producers' outputs appear much less stable from year to year than that of the world as a whole. It should be noted, however, that the two lowest correlations, those for Indonesia and Malaysia, obscure fairly consistent movements over shorter time periods. As inspection of table 6.2 will reveal, Indonesian output tended to fall throughout the decade of the Sukarno regime and to rise thereafter. Malaysia reported declining output from 1972 through 1977, but it is widely accepted among tin producers and traders that a substantial portion of the reported decline in its output is spurious and actually reflects increased smuggling in a period of rising taxes, as well as a real drop attributed to stringency in the granting of new mining leases. However, the use of a single index for each producer over the entire twenty-five-year period does have the virtue of reinforcing the observation made above, that there is enough flexibility in tin mining to permit planned expansion or contraction from one year to the next in some countries to offset unintended fluctuations in others. World output over the period was also increased by the steadily growing production of two countries not shown on table 6.2, Australia and Brazil.

Movements in tin prices over the period 1956 to 1980, as well as a price index for minerals and metals, are shown in table 6.7. Prices for tin on the Penang market were used for several reasons. The majority of the world's purchases of physical tin metal are made on that market. The Malaysian ringgit was spared the worst of recent currency gyrations, retaining greater exchange value and stability than either the British pound or the United States' dollar. Use of Penang rather than LME prices avoids the adjustment problem posed by the introduction of contracts for high grade tin on the LME in 1974.

Unlike production and consumption, which are flows, a specific price exists at each moment of time. Price stability must therefore be measured in terms of averages and ranges. Table 6.7 depicts annual average prices and price ranges over each year, facilitating comparisons with annual variations in production and consumption. In chapter 9, it will be argued that variations in

Table 6.7. Levels and percentage changes in tin price within and between years, and price index for minerals and metals, 1956–80

Year	Tin Price in M$ per picul high	low	average	Annual range, as percent of average tin price	Percent change in average tin price	Metals and minerals price index (1968–70 = 100)
1956	422.88	363.50	387.03	15.3	—	59.1[a]
1957	390.62	324.50	373.19	17.7	−3.6	56.9
1958	388.00	344.50	369.35	11.8	−1.0	52.9
1959	408.75	381.00	396.94	7.0	7.5	56.1
1960	410.25	385.25	393.68	6.4	−0.8	57.4
1961	500.00	389.00	446.85	24.8	13.5	56.5
1962	487.00	420.00	447.79	15.0	0.2	56.1
1963	515.88	425.00	455.40	20.0	1.7	57.6
1964	835.25	518.38	619.42	51.1	36.0	69.9
1965	796.38	608.25	702.80	26.8	13.5	78.7
1966	720.25	602.00	645.23	18.3	−8.2	94.7
1967	632.50	572.50	600.10	10.0	−7.0	88.4
1968	639.00	546.75	565.54	16.3	−5.8	93.5
1969	710.13	578.25	626.10	21.1	10.7	101.9
1970	716.75	619.88	664.77	14.6	6.2	104.6
1971	659.88	617.00	631.70	6.8	−5.0	96.2
1972	655.50	605.00	626.80	8.1	−0.8	98.1
1973	1026.00	615.50	686.28	59.8	9.5	132.3
1974	1380.00	820.00	1136.63	49.3	65.6	179.9
1975	1050.00	910.00	963.79	14.5	−15.2	164.1
1976	1320.00	957.00	1146.56	31.7	19.0	172.2
1977	1825.00	1314.38	1588.03	32.8	38.5	184.4
1978	2085.00	1476.00	1743.19	34.9	9.8	189.2
1979	2171.00	1750.00	1960.65	21.5	12.5	250.0[b]
1980	2471.00	1881.00	2160.12	27.3	10.2	284.0[b]

a. Last two quarters of year only.
b. New series: metals, ores, and minerals.
Sources. International Tin Council, *Statistical Yearbook*, 1968; International Tin Council, *Tin Statistics, 1966-76* and *1970-80*; National Institute of Economic and Social Research, *National Institute Economic Review*, various issues.

annual averages provide a better measure of the ITC's performance in stabilizing prices than do shorter-term fluctuations. It can be seen from table 6.7 that, in general, the range of prices within a year was greater in those years in which there were the largest annual average changes in price.

The very sharp price increase in the 1963–65 period, according to Fox, resulted from a delayed reaction to the drop in production in Zaire during the chaotic years following independence and to the decline in Indonesian output in the later period of the Sukarno regime. Persistent fears of a shortage in 1962 and 1963 did not lead to heavy speculative buying or large buildups of users' stocks, however, because of an equally persistent and strong fear that GSA would enter the market with massive disposals from the United States' stockpile to check any price rise induced by such a shortage. It will be recalled that in June 1962 the Congress approved a reduction in the size of the United States' strategic stockpile and, for the first time, authorized sales from the stockpile. In late 1963 and 1964 the fear of shortages overcame the fear of GSA, and the subsequent price rise was far more sudden and steep than it might otherwise have been because it had been suppressed for so long. GSA did respond with very much larger stockpile disposals in 1964, 1965, and 1966. World production, especially that of Thailand, Malaysia, and even Nigeria despite its depleting deposits, rose rapidly, as can be seen in table 4.3. As a result, prices declined from their 1965 peak over the following three years.[6] In this instance, it seems indisputable that the cause of the disturbance was the manner in which the United States' stockpile loomed over the commercial market after announcement of the congressional decision to reduce its size, and that the major stabilizing force was the market mechanism of producers' responses to price signals.

The 1973–74 price rise and subsequent fall in 1975 resulted in large part from the speculative purchases and later sales of tin noted above, coupled with the 1975 recession. From 1975 through 1978, tin prices rose at a rate well in excess of the average for all minerals and metals, although over these same years production grew more rapidly than consumption. This price rise set the stage for increased discord within the ITC, although table 6.7 also shows that, in the opening stages of the commodity boom, the minerals and metals price index had risen earlier and even more steeply than the price of tin. A similar period of catch-up occurred between 1978 and 1979, when the minerals and metals index rose much more sharply than the price of tin.

The primary interest of this study with respect to movements of price is the effect, if any, of the ITC on price stability. Since the council ran out of tin metal in January 1977 and remained in enforced inactivity through 1980, price stability is measured by regressing the natural logarithm of price on time only from 1956, the first year of the ITC's existence, through 1976. The regression yielded the following equation:

$$\ln P_a = 5.855 + 0.0506\,Y$$

$$r^2 = 0.840,$$

where P_a represents the annual average price.

According to this regression, price rose over the twenty-one-year period along a trend line of just over 5 percent per year. The price fluctuated less around this trend rate of growth than did either output of tin-in-concentrates ($r^2 = 0.570$) or consumption of tin metal ($r^2 = 0.811$). The difference in the correlation coefficients between price and output is statistically significant at the 95 percent confidence level;[7] but the difference between the coefficients for price and consumption is not statistically significant.

The correlation coefficients also indicate that from 1956 through 1976 production was substantially less stable than consumption. This difference is significant at the 90 percent confidence level but not at the 95 percent level. It was noted earlier that there was no significant difference over the longer period

Table 6.8. Correlation coefficients (r^2) as indices of stability of output, consumption, and price of copper, lead, zinc, and tin, 1956–76

	Copper	Lead	Zinc	Tin
Output	.983	.955	.976	.570
Consumption	.875	.964	.933	.811
Price	.762	.579	.687	.840

Sources: American Bureau of Metal Statistics, *Non-Ferrous Metal Data*, various issues; American Bureau of Metal Statistics, *Yearbook*, various issues; tables 6.1 and 6.2 above.

from 1956 to 1980, which includes four final years marked by the absence of any ITC intervention as well as the beginning of the 1980–81 recession.

For comparison, correlation coefficients similar to those derived for tin were also calculated for the 1956–76 period for the other nonferrous metals traded on the LME—copper, lead, and zinc. Results are shown in table 6.8.

There are strong inferences to be drawn from table 6.8 of generally high statistical reliability. First, over the entire twenty-one-year period the year-to-year variations in both output and consumption of tin were greater than those for any of the other three nonferrous metals. The differences between the correlation coefficients for output of tin and the corresponding coefficients for the other metals are all significant at the 99 percent confidence level. The differences in the coefficients for consumption are much less significant, with only that between tin and lead being significant at the 95 percent confidence level.

The price of tin, on the other hand, has fluctuated less than the prices of any of the other nonferrous metals. Only the difference between tin and lead, however, is significant at the 95 percent level.

Finally, and perhaps of the greatest interest in suggesting the effect of the tin agreements, price has clearly been less stable than either output or consumption for copper, lead, and zinc, as the standard primary nonagricultural commodity model assumes, while the relationship is reversed for tin. All of the differences in the cases of lead and zinc are significant at the 99 percent level. The differences between the coefficients for output and price for both tin and copper are significant at the 95 percent level.

A plausible and intuitively appealing interpretation of these coefficients is that the ITC, with the assistance of disposals from the United States' strategic stockpile, has succeeded in keeping tin prices more stable than those of the other nonferrous metals, but in order to do so the council has had to impose on its members year-to-year fluctuations in production and, probably, consumption that have been more severe than those experienced in the copper, lead, and zinc markets. This conclusion is bolstered by the facts that the principal dips in tin output are those that occurred during periods of export control and that production has varied more than consumption, even though the difference is not a statistically significant one.

This assessment of the ITC must, however, be made with caution. It is impossible to ascertain the fluctuations in either price or output that would have occurred around long-run trends as a result of supply and demand conditions in a free tin market. There has been no such free market for any length of time since the Bandoeng Pool. Further, direct comparisons between tin markets and those of the other nonferrous metals are valid only to the extent that basic physical and economic conditions are reasonably similar in all of these industries. Despite such caveats, it does seem probable that if the ITC has in fact dampened year-to-year fluctuations in price it has, in the process, heightened the instability of output.

Since total revenue is the product of price and quantity sold, price stabilization coupled with heightened instability of output might either stabilize or destabilize the total earnings from production of tin. Stability of total revenue may be an explicit national economic objective, particularly in the cases of developing countries that are dependent on the export earnings from one or a few primary commodities.

A widely used measure of instability is the Coppock Instability Index (I–I), which can be applied to any fluctuating variable but has most often been used to compare the magnitudes of fluctuations in export earnings of various countries or commodities.[8] As the Coppock Index is formulated, higher numerical values represent greater instability. Coppock calculated his index for total value of world exports of various commodities for 1950 through 1958, with selected results as shown in table 6.9.[9] Ariff calculated Coppock Instability Indices for value of exports from Malaysia alone. His figures are listed in table 6.10.[10] The World Bank, using as an index the percentage deviations from a five-year moving average instead of the Coppock Index, reported the following fluctuation indices for total world earnings from 1950 to 1972 (see table 6.11).[11] Coppock Indices are also shown in table 6.12 for total revenue (average price multiplied by reported consumption) of the nonferrous metals for the 1956–76 period, along with the correlation coefficients used throughout this chapter as measures of instability.

By either measure, total revenue from tin has fluctuated less than that for any of the other three nonferrous metals over the 1956–76 period, corroborating Coppock's findings for 1950–58 and those of the World Bank for 1950–72. The differences between the Coppock Index for tin and those for the other three metals appear to be large, but in the absence of any test for statistical confidence in observed differences in the Coppock Index, it is not possible to ascertain how significant these divergences are. The difference in the r^2 for tin and for lead is significant at a 90 percent confidence level. None of the others is significant. Thus, my calculations do not provide reliable independent evidence on the matter, but they do add support to other findings indicating that tin has been among the more stable of the commodities that have been important in world

Table 6.9. Coppock Instability Indices for world exports of selected commodities, 1950–58

Rubber	40.3
Jute bagging	39.6
Zinc	36.9
Iron ore	34.0
Cocoa	23.5
Copper	21.9
Coffee	18.6
Lead	16.6
Tin	16.5
Timber, lumber	16.2
Tobacco	12.6
Petroleum	11.7
Coal	10.6
Fish	6.5

Table 6.10. Coppock Instability Indices for Malaysian exports, 1947–67

Commodity	I-I (1947–67)	I-I (1953–67)
Iron ore	73.5	23.6
Rubber	42.1	22.1
Tin	28.6	22.4
Timber	28.0	22.6
Palm oil	17.7	8.6
All exports	29.1	15.0

trade since the Second World War in terms of total revenue, and more specifically that total tin revenues have fluctuated less than those of copper, lead, and zinc.

In sum, there is statistical evidence from which to conclude, tentatively, that instability on the production side of the tin industry has been at least as severe as instability on the consumption side. The mine production of individual nations has demonstrated a year-to-year flexibility that has contributed to stability of world output by offsetting unanticipated fluctuations in other countries. The price of tin has fluctuated less around its trend than has output, while the reverse has been true for copper, lead, and zinc. The greater price stability of tin, coupled with the ability of inventory holdings to smooth out consumption over time in the face of instability of output, have more than offset the lesser stability of production, so that total sales revenues from tin are more stable (although not significantly so in a statistical sense) than those of the other three nonferrous metals.

It is not possible to say definitively how much of the greater stability of tin prices, or instability of production, can be attributed to the ITC. But there would appear to be a good prima facie case that the tin agreements have had strong effects on these aspects of market performance.

Table 6.11. Percentage deviations from a five-year moving average of total world earnings for selected commodities, 1950–72

Zinc	15.9
Manganese ore	12.9
Rubber	12.4
Copper	11.4
Sisal	11.4
Lead	10.6
Jute	10.0
Cocoa	9.9
Timber	9.1
Tin	8.7
Palm oil	8.2
Iron ore	8.0
Rice	5.2

Table 6.12. Coppock Instability Indices and correlation coefficients for total revenue of nonferrous metals, 1956–76

	I-I	r^2
Copper	42.2	0.844
Lead	28.5	0.795
Zinc	37.0	0.835
Tin	18.6	0.904

Hedging and Speculation on the London Metal Exchange

Discussion of the mechanisms by which world prices and output of tin are determined must include consideration of the role of hedging and speculative trading of tin metal on the LME. Speculation has been defended on the grounds that it broadens markets by bringing in a new group of traders, transfers risk from producers and consumers to speculators by allowing the former to hedge, and stabilizes markets in those cases in which the speculators as a group are correct in their forecasts of the directions of future price changes. Indeed, one early criticism of the prewar International Tin Committee's buffer-stock and export-quota plans was that they would drive private speculative interests out of the tin market.[12] On the other hand, excessive, misinformed, or manipulative speculation may be destabilizing; and private speculation may, as Fox has maintained, heighten the cost and difficulty of implementing stabilization schemes.[13]

The LME operates in a distinctly different fashion from the other leading worldwide market for tin, that in Penang, Malaysia. Penang is a one-way market for physical tin. The smelters located on Penang Island make the tin metal they produce available to various successful bidders. The Penang market is one-way in that the smelters do not buy tin. The sole exception to this rule is made for the ITC, which is permitted both to buy and sell for its buffer stock

Table 6.13. Futures turnover and warehouse deliveries, LME, 1970–76 (1,000 metric tons)

Year	Futures turnover	Deliveries from LME warehouse	Turnover as a multiple of deliveries
1970	152.0	7.8	19.5
1971	144.9	4.2	34.5
1972	170.1	4.0	42.5
1973	169.3	6.0	28.2
1974	242.4	17.7	13.7
1975	205.6	17.8	11.6
1976	334.5	25.5	13.1

Source: Commodities Research Unit, Ltd., *Establishment of a Tin Exchange in Malaysia: Draft Report Submitted to the Ministry of Primary Industries, Federation of Malaysia, August, 1977* (London: Commodities Research Unit, 1977), between pp. 10 and 11.

Table 6.14. Total LME turnover, Penang sales, and world consumption of tin metal (1,000 metric tons)

				LME turnover as a multiple of	
Year	Total turnover LME	Total sales Penang	World consumption	Penang sales	World consumption
1977	397.2	66.3	184.7	5.99	2.15
1978	383.1	66.4	185.0	5.77	2.07
1979	286.0	57.7	184.1	4.96	1.55
1980	285.5	55.6	177.4	5.13	1.61

Source: International Tin Council, *Tin Statistics, 1970–80.*

on the Penang market. The Penang market is a physical market in that the smelters deliver the metal to buyers, with no provision for closing out transactions in any other fashion. The LME, on the other hand, is a two-way market, trading in both spot warrants and forward contracts, on which the majority of sales and purchases are closed out by reverse transactions, with only minor quantities of metal being involved. The predominance of paper transactions in several recent years is shown in table 6.13.

The increase in warehouse deliveries beginning in 1974 may be explained by the introduction that year of trading in warrants for high-grade tin, comparable in purity to the tin produced by the Malaysian smelters and sold on the Penang market. Yet even with high-grade tin available, well under 10 percent of trading on the LME involves actual delivery of metal to a buyer.

There can be no question that trading on the LME does serve the function of broadening the tin market, as indicated by table 6.14. The years 1977–80 are particularly appropriate for assessing the added depth given to the tin market by private speculators and hedgers, since the ITC's buffer stock was exhausted on January 13, 1977, and the council did not engage in any purchases or sales of tin for the buffer stock on either the LME or the Penang market until the spring of 1981.

About the middle of this period, the relative importance of the LME as a forum for market determination of tin prices was heightened. On August 1, 1978, Malaysia Mining Corporation, whose subsidiaries accounted for approximately 24 percent of Malaysia's output of tin-in-concentrates (but less than this fraction of sales on the Penang market since the Malaysian smelters treat substantial volumes of imported ore), withdrew from the Penang market and began having all of its concentrates smelted by the two Malaysian smelters on a toll basis, with the subsequent sales of the metal being handled by a trading firm in London. The implications of this development for the overall functioning of the world tin market will be discussed in the following two chapters; but it should be noted here that the elimination of MMC's tonnage from the Penang market has led to concern whether, as a result, that market has become too thin to reflect underlying market conditions both in terms of day-to-day price and quantity fluctuations and with respect to the appropriateness of quoting floor and ceiling prices for the ITC in Penang prices. The impact of MMC's withdrawal from the Penang market can be seen in the decline of average daily sales on that market from 228 metric tons (July 1, 1977 through July 31, 1978) to 196 metric tons (August 1, 1978 through December 31, 1979). Monthly averages fell from 5,677 metric tons in the first period to 4,866 metric tons in the second.

Far more problematic than market breadth are the related questions of the extent to which speculation on the LME serves the economic function of transferring risk from hedgers to speculators and whether such speculative activities stabilize or destabilize the market. The economic literature on the general subject of speculation gives ambiguous guidance on both matters.

There are two quite different basic views of the interactions between hedgers and speculators. One of these, which may be identified as the Keynes-Hicks view,[14] treats hedgers as those who produce, distribute, and process a real commodity and who wish to shelter themselves from the risks attendant on price changes over the periods during which they hold stocks or have committed themselves to accept future deliveries. Thus, throughout his discussion Keynes refers to the hedger as the producer,[15] while Hicks identifies the hedger as the ordinary businessman.[16] Both Keynes and Hicks regard the speculator as an individual or organization who buys and sells options that he closes out rather than actually taking or making deliveries (i.e., he engages entirely in paper transactions) and whose earnings on speculative trading represent his reward for risk bearing.

The majority of hedgers, in the Keynes-Hicks analysis, will be those holding stocks of the commodity who must buy short on the forward market in order to hedge their inventory positions. Speculators, therefore, will usually be long, or will have bought futures that they expect to sell at a higher spot price on or before the termination date. Thus, the prices of futures should be below the spot prices expected to prevail when the futures come due; and therefore if prices are not expected to rise the futures price should be below the current spot price—a condition Keynes referred to as "normal backwardation."[17]

Holbrook Working has challenged the basic concept of the Keynes-Hicks approach, that of risk transfer from one group of transactors, the hedgers, to another entirely separate group, the speculators. To the contrary, Working asserted, since most hedgers are well informed as to market conditions for the commodity that they are trading in or processing, the practice of "both hedging and speculation" is prevalent. In particular, when prices are depressed "a good many dealers and processors are attracted by the possibilities of profit through speculative holding of the commodity." As a consequence, neither risk nor the responsibility for price formation is shifted in whole, or even in large part, to pure speculators who trade only in futures.[18]

Trading on the LME appears more closely to approximate the Working view of speculation and hedging than it does that of Keynes and Hicks. C. A. J. Herkstroeter, chairman of Billiton Handelgesellschaft, has described the role of the tin trader as follows:

> Because of fluctuating tin prices and currencies the trader is continually at considerable risk in respect of his own open tin position, resulting from unsold purchases, or uncovered forward sales.
>
> Limiting such price risks through hedging transactions on the London Metal exchange is possible, but price risks cannot be cut out completely.
>
> However, as a rule the tin trader does not want to cut this risk out, as its preclusion would at the same time mean the elimination of extra profit opportunities. These extra profit possibilities are indispensable, as margins in the trade are usually very small in relation to the high value of the material, and sometimes even non-existent, due to fierce competition and unforseen circumstances.[19]

An experienced tin trader, D. Lorenz-Meyer, put more emphasis than did Herkstroeter on the dealer's desire to transfer risk but also noted that only the net trading position need be protected. His paper, described by ITC Buffer Stock Manager Jaime Bueno as "one of the most straightforward and clear" expositions of hedging he had heard, is worth quoting at length here.

> Now if I want to buy Straits Tin in the Penang market I must cable my bid in the evening—due to the time difference—to be submitted to the smelters next day in the morning. . . . Now the standard contract stipulates delivery by the smelters within a period of 60 days. The actual delivery time, however, may vary between two to four weeks, and then a suitable shipping opportunity must be found. Considering a shipping time of about 6 weeks I can count on the tin being available in Europe after about 10 weeks. Obviously, I cannot wait to sell the tin for so long a period, as in the meantime anything might have happened. The GSA might have decided to sell another 100,000 long tons of stockpile tin, or the ITC might have decided to revise the floor

and ceiling prices. You name it, I am afraid nearly anything is possible in the tin market.

So under no circumstances can the tin trader in Europe base his sales price for spot material on the actual cost to him of a certain consignment, which he must have bought months ago, until it is readily available in Europe.

. . . .

Now no responsible trader will operate entirely out of the blue, without taking account of all diverse factors affecting market sentiment . . . and here the necessity comes in to adjust the position according to changing circumstances or to balance additional sales by further purchases and vice versa. This is done by "hedging" and the favourite place for hedging is the LME.[20]

The testimonies of such experienced participants in the tin trade as Herkstroeter and Lorenz-Meyer support Working's general view and, more specifically, attest to the importance of paper trading on the LME as an adjunct to the business of trading real tin on the world market. The fact that dealers may be both hedgers and speculators should not detract from the contribution that such trading may make.[21] Working concluded his article by noting, after reiterating that hedging was not primarily a form of insurance, "It is a form of arbitrage, undertaken most commonly in expectation of a favorable change in the relation between spot and future prices. The fact that risks are less with hedging than without is often a secondary consideration." Working then went on, however, to make a highly debatable assertion that this arbitrage performed the "more important service of promoting economically desirable adjustment of commodity stocks, thereby reducing price fluctuations."[22]

Working, writing in 1953, apparently accepted the conventional wisdom of the time that speculation served the economic function of dampening price fluctuations, since "speculators are people of better than average foresight who step in as buyers whenever there is a temporary excess of supply over demand and thereby moderate the price-rise."[23] The most widely cited argument in support of this viewpoint is that of Milton Friedman, who noted, "People who argue that speculation is generally destabilizing seldom realize that this is largely equivalent to saying that speculators lose money, since speculation can be destabilizing in general only if speculators on the average sell when the currency is low in price and buy when it is high."[24] Friedman conceded that amateurs might actually lose enough money to cause speculation as a whole to be unprofitable, but noted that in the absence of any evidence the presumption had to be the contrary.

W. J. Baumol has propounded a counter-argument, illustrating one case in which speculation can be both profitable and destabilizing. Speculators, he observed, might sell after the peak in price and buy after the trough. "By so doing," he conceded, "they give up any chance to skim off the cream but hope in return significantly to reduce their risks."[25] Such a tactic would have both stabilizing and destabilizing elements, since if it were profitable it would involve speculative sales at prices higher than those of speculative purchases; but at the same time it would add selling pressure when prices were falling and buying pressure when they were rising. The major part of Baumol's article was devoted to a proof that it was logically possible for the effect to be on balance destabilizing.

Various writers have noted that speculation may be destabilizing, as well as unprofitable, because speculators are wrong in their opinions. Also, excessive speculation may be destabilizing. "Speculators may do no harm," Keynes wrote, "as bubbles on a steady stream of enterprise. But the position is serious when enterprise becomes the bubble on a whirlpool of speculation."[26] Speculation, in Keynes's view, becomes excessive when the "the professional investor is forced to concern himself with the anticipation of impending changes, in the news or in the atmosphere, of the kind by which experience shows that the mass psychology of the market is most influenced."[27] In such a situation, the knowledgeable speculator is more concerned with what he thinks others expect than with his own expectations regarding fundamental supply and demand forces—or what he thinks others will think the majority will expect, and so on in a possibly infinite chain. Further, as noted by Commodities Research Unit, it is possible that a particular market will be overwhelmed by a movement of speculative activity out of another, larger market, as evidently happened in 1973 and 1974 when there was a speculative flight out of currencies and into commodities, including tin.

J. W. F. Rowe's *Primary Commodities in International Trade* contains an excellent and nontechnical discussion of the role of organized futures markets in commodity trading.[28] Rowe concluded that speculators in commodity markets are likely to foresee the shapes of major events forthcoming, although not their precise magnitudes, will discount their effects, and thus smooth out larger intermediate and long-run price fluctuations. Commodity speculation, however, tends to be oversensitive to such short-run phenomena as rumors, political events, strikes, business failures, and changes in related prices, leading to exacerbation of day-to-day price oscillations around the trend. The net result, Rowe concluded, is a greater frequency of price changes but a reduction in their violence.

Rowe did single out two major problems with organized commodity futures markets. First, they are subject to manipulation and even cornering, particularly by collusive rings of traders. Such episodes will be transient, he maintained, as speculation cannot alter the fundamental determinants of price trends; and they are likely to hurt other speculators rather than producers or dealers. Rowe dismissed Keynes's concern that speculators may be more interested in what they think other, perhaps amateur or poorly informed, speculators are going to do than they are in real market forces. He argued that this phenomenon is less serious on commodity markets than on securities markets, since most of the speculators in the former—for most of the time—are knowledgeable specialists. When, however, the general public is attracted to a commodity market, most often because of some widespread rumor of impending shortage or outside events, price increases can be amplified. Rowe noted that the classical economic doctrine on this issue is that the professional speculators will become bears if the rumors are wrong or exaggerated and will profit by checking the price rise. On typical commodity markets, however, Rowe feared that the professionals would lack the financial resources to perform this function; and, even more important, in such circumstances the Keynes effect would become operative.

At least one experienced tin trader, former Buffer Stock Manager R. T. Adnan, put more weight on the role of the Keynes effect in day-to-day tin trading than Rowe ascribed to it on commodity markets in general. "It is well known," Adnan observed, "that it is sometimes more profitable for speculators to forecast the psychology of another speculator rather than assess the real demand and supply situation. It also seems to pay to devote one's intelligence to anticipating what average opinion expects the average opinion to be."[29]

One of the characteristics favorable to a futures market and the development of lively speculation is the absence of private monopolistic power or discretionary power over price by any other body. Thus, Rowe noted that national strategic stockpile liquidations and the buffer stock operations of the ITC have "made things difficult at times"[30] for the LME. To cite a recent example, in September 1978, at a time when there was great uncertainty as to the intentions of the United States' government regarding disposals of tin from the strategic stockpile, the *Wall Street Journal* cited one dealer as noting that "fabricators, merchants and customers are those primarily making the market currently," and quoted another as saying that "speculative interest in the market at the moment is the smallest for many, many years."[31]

Virtually everyone I spoke to with experience in or knowledge of the LME would concede that the LME tin market is small enough so that manipulation and occasional attempts at a corner exist. A leading commodity analysis service, that of Rayner-Harwill, commented recently, "It is apparent that such a small market (200,000 tonnes a year is no big thing) can be so manipulated by the trade as to create short term problems of some proportion."[32] A viewpoint expressed by several of those with whom I discussed the matter was that profits from speculative trading in tin were normally so small that these "short term problems" have to be expected and tolerated.

The issues of market manipulation and the vulnerability of the LME to a corner on tin were given new prominence in 1981 and early 1982 by the mystery buyer affair. The buying, which began in July 1981, and pushed LME prices up from around £6,500 to a spot price of just over £9,000 in February 1982, is estimated to have required expenditures in the range of US$500 million to $600 million. Having switched from purchases for future delivery to purchases on the spot market in November, the mystery buyer was compelled to make cash payments and acquire tin metal. After the sudden fall in price in late February, the mystery buyer was believed to have been left holding 40,000 to 50,000 tons of tin.

If the mystery buyer's objective was to obtain a short-run profit from a squeeze or corner on the spot market in late February, the attempt was clearly a failure. Yet a pending crisis and extremely heavy losses to many dealers and traders was averted only by the imposition of a special rule by the LME limiting the penalty to £120 per ton for failure to deliver as forward contracts came due. This action has led to criticism of the LME, particularly from Malaysian circles, and casts some question on the continued viability of trading in tin on that exchange. (Technically, the LME's Exchange Committee acted before it could have declared a corner, as tin metal was being made available on the spot market at all times during the squeeze. The mystery buyer had broken no trading rules.) "I didn't think the London market would ever do that," one trader was quoted as saying, continuing that he was concerned about the future repercussions on the credibility of the LME's trading rules.[33] Inaction, however, might have had still worse consequences. Another trader observed that "the exchange has suffered because the manipulation of the market has hurt its reputation as a hedging mechanism."[34]

The magnitude of the financial resources committed by the mystery buyer, and the apparent failure of the venture, may discourage future attempts at prolonged and large-scale manipulation. The question remains whether smaller-scale, day-to-day speculation in tin on the LME serves a useful function.

In August 1977 the LME tin market went into a backwardation that, with only a few interruptions, lasted until July 1979 and then reappeared the following month. Overall, in 1977, 1978, and 1979, the tin market was in a backwardation for 537 trading days and in a contango (futures prices higher than the current spot price) for 180 days. In addition, as noted above, during these three years the ITC held no tin in its buffer stock and did not make any purchases or sales on the LME, thereby eliminating one damper to unconstrained speculative activity. Further, disruption of commercial markets by sales from the United States' strategic stockpile was minimal. All of these features make the 1977–79 period an interesting one for observing the effects of speculation on price stability.

A prolonged period of marked backwardation is significant in examining these effects because of the nature of forward trading on the LME. There is a standard forward contract in tin, calling for settlement ninety days after the date of initial issuance. Longer-term tin contracts, or transferable futures specifying a month of settlement rather than a specific date, are not traded on the LME. As an illustration of a hedging transaction, assuming a contango, consider a dealer who has purchased a ton of tin at £7,000 and anticipates selling that tin later, allowing roughly two to three months for smelting and shipping. At the time of his purchase, he does not know the price at which he will be able to sell the tin; but on the date of sale he expects to quote that day's settlement price on the LME. Assume a ninety-day forward price of £7,100. He hedges by selling such a contract. If, by the date of sale of his physical tin, the LME spot price has risen to £7,200, he will make a gain of £200 on the physical sale (selling for £7,200 the tin he purchased at £7,000); but he will lose £100 on closing out his LME contract (paying £7,200 to terminate the obligation he assumed for £7,100). On the other hand, if the spot price has fallen to £6,800, he will lose £200 on his sale of physical tin and gain £300 on his LME contract. No matter how much the spot price rises or falls, his net gain will be equal to the size of the contango on the day the hedge was made. If, however, a backwardation existed, the hedger would have a net loss equal to the size of the backwardation regardless of the change in the spot price—a loss that Keynes described as a sort of insurance premium paid by the risk-avoiding hedger under what he viewed as the usual condition of normal backwardation. Tin traders, it may be noted, regard a backwardation as an aberration rather than as the normal situation.[35]

The above example is greatly oversimplified. Most important, the length of smelting and shipping time assumed is approximately correct for a purchase on the Penang market for subsequent sale in Europe or the United States; but the example does not take into account the difference likely to exist between the Penang price and the LME spot price on a particular day. Shipping charges are such that these differences can become quite large without any correction by arbitrage. Nevertheless, the basic point remains that a contango is a more attractive condition for hedging by a dealer or producer holding title to physical tin than is a backwardation. The size of a contango is limited by the possibility of buying tin spot on the LME, bearing the storage and interest costs, and closing out a higher-priced futures contract by delivery of the tin. But there is no such limit on the possible extent of a backwardation.

A dealer who wished to hedge a short position in the physicals market—perhaps one who had contracted to buy a specific quantity of tin for later delivery at the then prevailing price—would be in a reverse position from the one just described in that if he hedged he would be assured of a gain equal to the backwardation or a loss equal to the contango existing on the day the contract was first made. However, the nature of the tin trade is such that most hedging is done by those holding inventories of physical metal for later resale. Therefore, backwardation reduces the value to the trade of hedging.[36]

Over the 1977–79 period, there was a distinct relationship between the volume of trading in tin on the LME and the extent of contango or backwardation. On the 180 days with a contango, average turnover was 1,498 metric tons. On the 285 days with a backwardation of £100 or more, average turnover was only 1,361 tons. A regression of turnover on the difference between the current and the future price (with backwardation shown as a negative figure and contango as a positive one) indicated a positive relationship that was significant at the 99 percent confidence level. It appears, as one would expect from the higher cost to the typical hedger, that there is a reduction in the volume of hedging when a substantial backwardation appears. While the uncertainties generated by the changing fortunes of various stockpile disposal bills before the United States' Congress may have dampened speculation and thereby affected the volume of trade, these uncertainties characterized the entire 1977–79 period up until

December 1979, and do not appear to have varied systematically with levels of contango or backwardation.

Unlike the situation in hedging, there does not seem to be any reason why pure speculation should become more or less attractive when a backwardation develops. Therefore, the reduction in turnover associated with an increase in the size of a backwardation is probably also associated with a higher proportion of purely speculative trading. If this assumption is correct, there is evidence indicating that speculation in tin may have played a stabilizing role on the LME from 1977 through 1979. The daily price was regressed on time for one period of sustained contango (January 3, 1977 to July 28, 1977) and for two periods of sustained backwardation (August 2, 1977 to March 15, 1978 and April 15, 1978 to July 20, 1979). Again, the correlation coefficient was treated as an index of stability. For the one period of contango, $r^2 = 0.011$. For the two periods of backwardation, $r^2 = 0.163$ and 0.378. The difference between the r^2 for the contango and that for the first period of backwardation is significant at the 90 percent confidence level; and the difference between the correlation coefficients for the contango and the second backwardation is significant at the 99 percent level.

Over the entire three-year period, the coefficient of variation of daily prices for the days of contango was 0.061, whereas for the days of backwardation of £100 or more it was 0.048.

When the year 1980 is added to the period analyzed, the hypothesized relationships continue to be observable, but in a much weaker form. A regression of turnover on the extent of backwardation or contango still yields a positive coefficient, but a statistically insignificant value for r^2. There is virtually no difference in stability as measured by r^2 between periods of backwardation and contango in 1980; but during that year there were no prolonged periods in which to test the hypothesis, such as those in the 1977–79 period. The basic contention that speculation is stabilizing is given some added support, though, because for the entire four-year period the coefficient of variation for days of contango is 0.113, while for days of backwardation of £100 or over, it is only 0.048. While the ITC's buffer stock remained inactive, other events make it doubtful that it is appropriate to include 1980. GSA began sales from the United States' strategic stockpile around the middle of the year. Even though the dollar amount of these sales was minute, GSA certainly represents a major governmental force, potential as well as actual, impinging on the private market. Further, the price of tin dropped sharply in the last half of 1980 as recession appeared in the industrialized consuming countries. There was, thus, a sharp turning point in the series that is substantial enough to reject any analysis of the variations around an underlying trend that is assumed to be constant.

The above must not be taken as a definitive demonstration that speculation in tin on the LME was a stabilizing force during 1977–79, much less that such speculation is in general stabilizing. Other important forces operating on price, such as changes in uncertainty regarding disposals from the United States' strategic stockpile, have not been taken into account. The assumption that speculation becomes a larger proportion of turnover in periods of backwardation is based entirely on a priori reasoning; and even if the assumption is correct the calculations do not include any quantitative estimate of the size of this change in proportion. Although writers such as Rowe and C. P. Brown[37] have emphasized the effects of speculation on day-to-day prices, it is arguable that fluctuations in daily prices do not represent the kind of instability of most concern to producers and consumers. But at the very least, the 1977–79 experience gives no support whatsoever to the contention that speculation has destabilized tin prices in recent years.[38] To the contrary, the evidence suggests strongly that criticisms of speculation on that score should be viewed as unfounded.

7. Industry Structure: Patterns of Concentration and Ownership

Smelting

Tin smelters, acquiring tin concentrates from miners in the various local producing areas and subsequently selling tin metal on the closely integrated world market, are in a key position in the vertical chain of production and distribution of the metal. As noted in the simple supply and demand model in chapter 4, the degree of monopolistic or monopsonistic power at the smelting level, and the extent of vertical integration between smelters and miners, are crucial elements in the analysis of discretionary pricing ability in the industry.

Over the past half century, fundamental changes in ownership and location have drastically altered the structural characteristics of the industry at the smelting level. Essentially, predominance has passed from European to local interests, the degree of public ownership or control has grown, and the level of concentration has fallen.

The era of European predominance: the consolidation of 1929 through World War II. In 1929, three large British smelters and Eastern Tin Smelting Company, one of the two large Malayan smelters (later renamed Datuk Keramat), were all absorbed into a single organization, Consolidated Tin Smelters. The majority stock interest in Consolidated Tin Smelters was indirectly held by Patiño Mines and Enterprises, while Anglo-Oriental Mining Corporation held a substantial minority interest. Figures on annual output of tin metal by country are available from several sources, but there are no such series for the production of individual smelters. Further, published data on ownership of smelters are scattered, particularly for the earlier years of this century; and the accuracy of any depiction of the industry's interlockings or interest groupings is, at best, approximate. However, a reasonable reconstruction of the degree of concentration in smelting for the year 1929 does seem feasible, in large part thanks to estimates made by E. B. Scott for that year's edition of *Mineral Industry*. In his discussion of the establishment of Consolidated Tin Smelters and its 1929 acquisitions of the Cornish Tin Smelting Company, the Eastern Tin Smelting Company, Penpoll Tin Smelting Company, and Williams Harvey & Company, Scott noted:

> It will thus be seen that in the course of last year quite a new grouping was established in the tin smelting industry. First of all there was the Consolidated Tin Smelters, with an output of something like 77,000 to 78,000 tons of tin consisting primarily of the big English smelters, but with an important foot in the Eastern field through the Eastern Smelting Co., with an output for the

> year of about 29,500 tons of metal. Secondly there was the Straits Trading Co., with an output last year of about 63,000 tons of metal and cooperating to an important extent with the Dutch Government interests, Banka and Billiton, and smelting about a third of the production of the first-named island and the whole of the second. Thirdly there was the smaller and more or less unconnected independents such as Thames Metal Co. and Capper Pass in Great Britain, the new Arnhem Smelter in Holland, the Oolen Smelter in Belgium, the Berzelius Works at Duisberg, and the undertaking of the Société Electro-Metallurgique at Annency in Savoy.[1]

Scott closed his article with a warning to the reader that "the figures are of course approximate, but in the case of the larger producers represent official intimations received by the writer, and if the smaller totals are estimates, possible differences are not large enough to affect the general result."[2]

Table 7.1, showing smelting interests in 1929, is derived from Scott's study with additional information obtained from other issues of *Mineral Industry* and early issues of *Minerals Yearbook*. The table and calculations are based on Scott's estimates and some crude assumptions of my own. Not the least of the difficulties involved stems from the fact that Consolidated Tin Smelters did not make output figures public.

To the extent that the estimates underlying table 7.1 are accurate, Consolidated Tin Smelters accounted for about 43 percent of the total world output of tin metal in the first year of its formation. Only slightly behind came the remaining independent Malayan firm, Straits Trading Company, which owned one smelter at Butterworth in Malaya and a second, then the world's largest, at Pulau Brani in Singapore.[3] The two smelters of the Straits Trading Company accounted for another 33 percent of world output. While the largest stockholders in the Straits Trading Company were Chinese-Malay plantation owners and the Overseas Chinese Banking Corporation, the firm was founded by two Europeans, one of whom was associated with one of the British agency companies that managed a number of diverse Malayan enterprises; it was regarded as European-controlled as late as 1960.[4] Government-owned smelters in the Netherlands Indies ranked third, accounting for 8 percent. Thus, the recognized "big three" smelting interests handled almost 85 percent of the world's tin output.

In 1930, the Thames Metal Company ceased operation and the Mount Bischoff Mining Company closed its smelter. Over the following few years two continental smelters—Billiton's smelter at Arnhem and the Belgian smelter at Hoboken—expanded their outputs substantially. In 1932 Billiton enlarged its Arnhem smelter, which had been constructed only a few years earlier and which had been receiving virtually all of its concentrates from Bolivia; and in the following year the company began shipping ores to Arnhem from its Indonesian mines. (Billiton held a three-eighths share in these mines and acted as manager. The Netherlands Indies government owned the remaining five-eighths.) The Hoboken smelter, which relied on the mines of the Belgian Congo for its tin

Table 7.1. World output of tin metal, by smelting interest groups, 1929

Smelter or smelting group		Output (metric tons)
Consolidated Tin Smelters		
United Kingdom[a]	47,500	
Eastern Smelting Co. (Malaya)	29,500	
Zinnwerke Wilhelmsburg (Germany)[b]	3,500	80,500
Straits Trading Co. (Malaya)		63,000
Banka (Netherlands Indies)[c]		15,000
Kochiu Guild (China)[d]		6,400
Capper Pass & Son (U.K.)[a]		5,636
Thames Metal Co. (U.K.)[a]		5,635
Berzelius Metallhütten (Germany)[b]		3,500
Soc. Gén. Metallurgique de Hoboken (Belgium)[e]		1,000
Billiton - N. V. Hollandsche Metallurgische (Netherlands)[c]		1,000
Mt. Bischoff Mining Co. (Australia)[f]		834
O. T. Lempriere & Co. (Australia)[f]		833
Woolwich Tin Smelting Co. (Australia)[f]		833
Société Electro-Metallurgique (France)		700
F.M.S. Chinese[g]		3,400
Total		188,271

a. Scott's lower estimate of 77,000 tons produced by Consolidated Tin Smelters is used here, with Consolidated's British production shown at that figure less the 29,500 tons attributed to Eastern Smelting Co. British tonnage not attributed to Consolidated Tin Smelters is arbitrarily divided equally between the two British independent smelters, Capper Pass and Thames Metal. In general, equal division is the most conservative assumption to make in this sort of analysis of industry structure in the sense that it minimizes the indices of concentration. In 1930, Thames Metal closed its smelter, leaving Capper Pass as the sole independent British smelter.

b. Williams Harvey & Co., the largest British smelter and a component of Consolidated Tin Smelters, had acquired an interest of undetermined size in the Wilhelmsburg smelter shortly after the end of the First World War. German output of 7,000 tons is arbitrarily divided evenly between this smelter and the Berzelius works.

c. Although Billiton Mining Company owned mines in the Netherlands Indies as well as the smelter in Arnhem, Holland, in 1929 all Indonesian ores were either smelted locally at the government-owned smelter at Banka or shipped to the Straits Settlements for smelting. Therefore, Banka's smelting operation is shown as independent of Billiton's smelting in the Netherlands. Billiton began shipping its Indonesian ore to Arnhem in 1933.

d. Little was known of tin production in China other than that virtually all ore came from the Kochiu mines in Yunnan and was shipped to Hong Kong for smelting. In its 1918 edition, *Mineral Industry* described the "Kokiu Guild" as an entity or group, and noted that "the Chinese Guild or Corporation . . . is said to be closely related to the refineries in Hong Kong" (p. 701). This and other references to Kochiu Smelters seem to warrant treating all Chinese output as that of a single interest group.

e. Scott appears to be in error in attributing 1,000 tons of tin metal output to the Congo, as there was no recorded smelter in the colony at the time. In the 1930 edition of *Mineral Industry*, Scott attributes this same 1,000 tons to Belgium for the year 1929. The Hoboken smelter had been constructed to treat Congolese ore.

f. In the absence of information on actual output of the individual smelters, Australia's output is arbitrarily allocated equally among the three smelters operating in the country in 1929. In 1930, Mount Bischoff Mining Company closed its smelter.

g. The only information on Chinese smelting in the Federated Malay States at the time is that it was carried out by a number of small-scale operators utilizing simple and rather primitive equipment, but competing effectively with the large European-engineered smelters of Eastern Smelting Company and the Straits Trading Company in both ability to attract custom and in quality of tin metal produced.

Sources: G. A. Roush, ed., *Mineral Industry* (New York: McGraw-Hill), various issues. U.S. Bureau of Mines, *Minerals Yearbook* (Washington, D.C.: Government Printing Office), various issues.

concentrates, increased production rapidly in the early 1930s as the Belgian government refused to join the International Tin Committee or to agree to any restrictions on Congolese mining output. As a result, by 1935 Billiton produced 15,600 long tons of tin metal at Arnhem, and the Hoboken output had risen to 4,000 long tons.

The only other significant change in structure before the Second World War occurred in 1934 when Simon Patiño, the largest Bolivian tin mine owner, obtained control of Consolidated Tin Smelters through acquisition of a controlling interest in the British Tin Investment Company, which *Minerals Yearbook* described as having "extensive holdings in Malayan tin mining companies and in Consolidated Tin Smelters, Ltd."[5] Patiño already held a substantial minority interest in Consolidated Tin Smelters, as Patiño Mines and Enterprises Consolidated had acquired Williams Harvey shortly before the latter's 1929 merger into Consolidated.[6]

The largest entrant to the tin industry during the Second World War was the Longhorn smelter built at Texas City, near Galveston, Texas. This smelter, constructed by a Billiton subsidiary, was established by the United States' government's Reconstruction Finance Corporation, to process low-grade Bolivian ores. After operating under a peacetime subsidy, granted on grounds of national security requirements, the Longhorn smelter was sold to a private firm in December 1956. During the 1970s, under ownership of Gulf Chemical and Metallurgical Corporation, a subsidiary of Associated Metals and Minerals Corporation (ASOMA), the Texas City smelter processed from 4,000 to 7,000 tons of Bolivian tin-in-concentrates per year under a toll arrangement with COMIBOL—down from a peak rate of 43,000 long tons in 1946. As Bolivia's domestic smelting capacity increased, imports of tin-in-concentrates for the Texas City smelter fell, to 840 metric tons in 1980 as compared to 4,528 metric tons in 1979. Since November 1980 there have been no imports of tin concentrates to the United States. The Texas City smelter is now processing its own slag, and Gulf expects this operation to continue for several years. The company has also installed equipment for recovery of secondary tin.

The only other tin smelters constructed during the Second World War were those of the Belgian firm of Geomines in the Congo (now Zaire), built to process tin concentrates that had formerly been shipped to the smelter in Hoboken, Belgium.

The era of national independence: restructuring after World War II. The major structural changes in tin smelting in the postwar period reflected far reaching political changes in the producing nations, notably the Bolivian revolution of 1952, and the independence of all of the colonial producers—Indonesia in 1949, Malaya in 1957, and both Zaire and Nigeria in 1960.

The greatest obstacles to establishment of a domestic smelter were faced, and in large part overcome, by Bolivia. Prior to the First World War, two unsuccessful attempts had been made to establish smelters in Bolivia, including one by

Patiño.[7] There were three formidable and interrelated problems: technical difficulties of smelting tin at high altitudes; the low grade, or tin content, of Bolivian concentrates processed at the mine site; and the complex character of Bolivian ore.

Until the 1960s, experts in tin smelting were of the opinion that tin could not be smelted with any reasonable degree of economic efficiency at altitudes of over 12,000 feet, although two small smelters were in operation in Bolivia during the 1950s. Smelting at or near its mine sites (some of the most important of which are located on sites 12,000 to 16,000 feet above sea level) is of particular importance to the Bolivian tin industry, since the grade of concentrates produced is very low, ranging roughly from 20 to 60 percent tin content, as contrasted with 70 percent and over for the alluvial ores of Southeast Asia and Africa. Approximately one-fourth of Bolivian ore contained only 20 percent tin in the late 1950s,[8] and the average grade of ore mined since has been declining. While the differences between the costs of shipping Asian concentrates containing 70 to 76 percent tin to European smelters and the costs of shipping tin metal smelted in Penang or Banka to Europe or America are small and yield no significant advantage of location to smelters, it was estimated that if all Bolivian ores could be treated at the mine site to concentration levels of 60 percent tin metal the realization charges (mainly transportation and warehousing) could be reduced from 40 percent of gross value to 15 percent.[9]

Bolivia's smelting problems have been compounded by the complexity of its ores, or the extent to which the tin is combined with other elements and impurities in compounds that are difficult to separate. Smelters, even at sea level, must be specially designed to smelt and refine Bolivian concentrates.

In 1952, at the time of nationalization of the Patiño, Aramayo, and Hochschild mines, the only smelters in the world capable of treating Bolivian ores were Williams Harvey and Capper Pass in England, Berzelius Metallhütten in West Germany, and the RFC's Texas City smelter. COMIBOL had no choice but to continue to ship a substantial fraction of its concentrates to the British smelters, given Berzelius's limited capacity, unless it wished to become entirely dependent on a United States government-owned, subsidized smelter whose continued existence was uncertain. Williams Harvey was clearly under the ownership and control of Patiño; and Capper Pass was indirectly linked to Patiño through the Rio Tinto Zinc Corporation (see table 7.2). As noted above, COMIBOL experienced serious difficulties in running the nationalized mines; and Bolivians have contended that these difficulties were compounded by excessive and unfair charges levied by these smelters, including an obligatory 10 percent set-aside of the gross value of Bolivian concentrates shipped to Williams Harvey that was collected as part of the arrangements made for compensation to the tin barons for nationalization of their mines.[10]

Under these circumstances, it was inevitable that Bolivia would make great efforts to develop a domestic smelting capacity capable of handling its entire output of tin-in-concentrates. In 1967, Bolivia entered into a contract with the

West German firm of Klöckner Industrie for the design and construction of a smelter to begin production in 1970 with an initial capacity of 7,500 metric tons per annum, to be increased over time to 20,000 metric tons per annum. The smelter was built at Vinto, near Oruro, in the heart of the mountainous mining district. By 1975, despite severe and time-consuming technical problems, this smelter had grown to a capacity of 11,000 metric tons per annum, with a 20,000 metric ton capacity still projected; and a second contract was awarded to Klöckner for construction of another smelter of 12,500 metric tons per annum capacity.

In 1961–62, two smelters were established in Nigeria. Before independence, all Nigerian ore had been shipped to Williams Harvey; and the Ministry of Mines and Power had been urging that firm to establish a smelter in the newly independent nation. Williams Harvey was reluctant to do so, and was goaded into forming the Makeri Smelting Company only by the 1961 entry of a Portuguese firm, the Nigerian Embel Tin Smelting Company. In 1962, Makeri smelted 7,300 metric tons of tin metal compared to Embel's 300 metric tons; and in 1963 Embel declared bankruptcy.[11] Makeri has the capacity to handle 18,000 metric tons of tin-in-concentrates per annum, far above Nigeria's production. Nigerian exports of tin concentrates fell from 6,643 long tons in 1961 to 238 long tons in 1962, and Nigeria has not exported significant quantities of concentrates since.

In 1965, a large smelter, the Thailand Smelting and Refining Company (Thaisarco), opened on the island of Phuket in Southern Thailand. The firm was a joint venture, with 70 percent ownership by Union Carbide and 30 percent ownership by a Thai company controlled by the Minister of the Interior, General Praphas Charusathiara. Thaisarco, with an initial capacity output of 25,000 metric tons of tin metal per annum, was given temporary protection against competition ("monopoly privileges" in the Thai industrial promotion program), and all Thai miners were required to deliver their ore to Thaisarco for smelting. Since 1965, no tin concentrates have been exported from Thailand except for those smuggled out of the country. General Praphas and his confreres subsequently sold their 30 percent share in Thaisarco to Billiton. At the same time, Billiton purchased a 20 percent interest from Union Carbide, so that Billiton and Union Carbide each came to own 50 percent of Thaisarco.

In Indonesia, only one of the three prewar government-owned smelters was rehabilitated after the Second World War. In 1967 a new smelter, built by Klöckner for the Indonesian government-owned tin enterprise P. N. Timah, began operations. By 1971 this facility, known as the Peltim smelter, was operating at its capacity rate of 13,000 metric tons of metal per annum, with eventual expansion planned to 27,000 metric tons per annum. In 1975 P. N. Timah added three reverberatory furnaces to the Peltim smelter, raising its capacity to 33,000 metric tons per annum.

In what may well turn out to have been the last consolidation of European ownership in tin smelting, two Australian smelting firms and the Patiño-owned firm of Consolidated Tin Smelters entered into a joint venture to form Associated Tin Smelters of Australia in 1967, with each of the three parent companies holding a one-third interest. Until 1980, Associated Tin Smelters operated the only tin smelter in Australia. But this slight increase in concentration and in the European position was more than offset by the closing of Billiton's Arnhem smelter in 1971 and a 1973 decision by Consolidated Tin Smelters to put Williams Harvey into liquidation. With the establishment of Thaisarco and Indonesia's Peltim smelters, Arnhem had lost its principal sources of ore supply. Williams Harvey had lost all of its Nigerian concentrates in 1962, and had been operating at a loss in smelting and refining low-grade Bolivian ore. Capper Pass, which specializes in treating low-grade and complex ores, was left as the single British smelter handling Bolivian imports.

By 1975, the distribution of world tin smelting by interest groups was as shown in table 7.2.

Comparative scanning of tables 7.1 and 7.2 should suffice to indicate the decline in world concentration of tin smelting between 1929 and 1975. Commonly used statistical measures of concentration, shown in table 7.3, have been calculated from these tables.

Scott's comment that Straits Trading Company was "cooperating to an important extent" with Banka, and his inclusion of both in one grouping, suggest that it might be informative to calculate concentration indices with Straits Trading Company and Banka treated as one interest group in the 1929 indices, although there was no formal linkage between the two. In table 7.3 these figures are shown in parentheses. For purposes of calculating the 1929 Gini and Herfindahl indices, the output of the small Malayan smelters is arbitrarily assumed to have been produced by ten smelters of equal output. In light of the indirect ownership link between Patiño N.V. and Capper Pass, the various 1975 indices of concentration are calculated both with Capper Pass treated as an independent smelter and, in parentheses, with the output of Capper Pass included with that of the Patiño group.

The measures CR1, CR4, and CR8 in table 7.3 are simply the percentages of total output accounted for by the largest interest group, the largest four, and the largest eight, respectively. The most dramatic change is the decline in the share of the largest concern, Consolidated Tin Smelters (whose properties had passed to Patiño N. V. by 1975), to about one-half of its 1929 share by 1975.

The Gini Coefficient and the Herfindahl Index are measures used frequently in analyses of economic concentration. The Gini Coefficient is a measure of the degree of inequality within a given size-distribution, with 100 percent representing absolute inequality and zero representing perfect equality. The Gini Coefficient does not take into account the number of firms in the group under study. The Herfindahl Index is designed to measure concentration in terms of both number of firms and differences in their sizes, with the larger values representing greater concentration. Interpretation of the Herfindahl Index, or evaluation of the significance of the differences between two Herfindahl indices, is sometimes ambiguous since the index incorporates two aspects of the same size-distribution in one number. In table 7.3, the Herfindahl indices unquestionably indicate a decline in concentration of tin smelting.

Tables 7.1 and 7.2 also reflect the growth in importance of government-owned smelting facilities. In 1929, the only such smelter was that of the Netherlands Indies government on Banka Island, accounting for 8 percent of that year's output. In 1975, the Soviet, Chinese, Indonesian, Bolivian, East German, Zairoise, Vietnamese, and Czechoslovakian smelters, accounting for 35 percent of total world output, were public enterprises.

Not only was ownership of smelting capacity diffused over this period, but the physical location of smelters shifted, away from Europe and toward the tin mining nations. The extent of this shift is indicated by table 7.4 indicating that the fraction of tin-in-concentrates smelted in country of origin, excluding the socialist centrally planned economies, rose from 45 percent in 1960 to 70 percent in 1970, primarily reflecting the establishment of the Nigerian and Thai smelters, and to 84 percent in 1980, as Indonesia achieved smelting capacity matching its mine output and the Bolivian smelters' output continued to increase.

Since 1975, the trend toward relocation of smelting capacity in the producing nations has continued. The Bolivian government's smelting organization, Empresa Nacional de Fundiciones (ENAF), had a large volatilization plant, designed to handle highly complex ores, under construction by early 1982. The Soviet Union is providing technical and financial support for the project. When completed, this smelter will allow Bolivia to smelt large quantities of mine tailings and other low-grade ores, ultimately adding as much as 12,500 metric tons per annum capacity to that of the existing smelters. The new facility has been plagued with problems, including landslide damage, a breakdown of its electric furnace, and gas emissions that have been "killing animals and plants and causing sickness among local inhabitants."[12] It is five years behind an originally scheduled opening in 1979.

Two smelters, under single ownership by a local entrepreneur, were built in Singapore in 1978, evidently in response to the large volume of smuggled concentrates flowing to that island nation. By 1980, these smelters were producing at an estimated rate of 4,000 metric tons metal per annum, but with the capacity to double that output. Thailand has withdrawn the temporary monopoly privileges originally given to Thaisarco and has granted permits for construction of two new local smelters with a combined projected capacity of over 12,000 metric tons per annum. These Thai smelters are owned by two independent companies, Thai Pioneer and Thai Present, but a subsidiary of Metallgesellschaft constructed the smelter for Thai Pioneer, and Hoboken-Overpelt is serving as a technical adviser to Thai Present.[13] Thai Pioneer began operations late in 1981, with an initial capacity of 3,600 metric tons of metal per annum; Thai Present expects to open in 1983.

In late 1979, South Africa's largest mining company, Rooiberg Tin, opened a smelter that, along with the older Zaaiplaats smelter, gives South Africa the

Table 7.2. World output of tin metal, by smelting interest groups, 1975

Smelter or smelting group		Output (metric tons)
Patiño N. V.[a]		
Datuk Keramat (Malaysia)[b]	38,810	
Associated Tin Smelters (Australia)	5,254	
Makere (Nigeria)	4,677	
Estanifera do Brasil[c]	2,267	51,008
Straits Trading Company (Malaysia)[b]		44,260
Soviet Union		30,000
Peoples' Republic of China		22,000
P. N. Timah (Indonesian State Tin Enterprise)		17,825
Royal Dutch Shell (Billiton)[d]		
Thailand Smelting and Refining	16,630	
Kamativi Smelting and Refining (Rhodesia)	600	17,230
Capper Pass & Son (United Kingdom)[e]		11,520
Empresa Nacional de Fundiciones (Bolivia)		7,133
Gulf Chemical and Metallurgical (U.S.A.)[f]		6,500
Metallurgie Hoboken-Overpelt (Belgium)[g]		4,562
Metalurgia del Noroeste (Spain)[h]		1,993
Minero Metalúrgica del Estaño (Spain)[h]		1,827
Industrial Fluminense		
Industrial Amazonense (Brazil)[c]	1,600	
Industrial Fluminense (Brazil)	200	1,800
Berzelius Metallhütten (West Germany)[i]		1,306
Zinnhütte Freiberg (East Germany)[i]		1,120
Ferroaleaciones Españolas (Spain)[h]		930
Mitsubishi Metal Mining (Japan)[j]		866
Mamore-Mineracão (Brazil)[c]		800
Zaaiplaats Tin Mining Company (South Africa)		780
Zairetain (Zaire)[k]		575
Estaño Electro (Mexico)[l]		545
Electrometalurgia del Agueda (Spain)[h]		499
Neostana (Portugal)		409
Best Metais e Soldas (Brazil)[c]		400
Rasa Kogyo (Japan)[j]		346
Fundidore de Estaño (Mexico)[l]		228
Metales Potosi (Mexico)[l]		227
Vietnam		200
Bera do Brasil[c]		133
Soc. Min. Pirquitas, Picchetti y Cia. (Argentina)		120
Czechoslovakia		108
Morocco		12
Total		227,262

a. In 1974 Patiño N. V. held a 76.4 percent interest in Consolidated Tin Smelters Inc., which in turn had interests of 79 percent of Makere, 50.5 percent of Datuk Keramat, and 33.3 percent of Associated Tin Smelters. In July 1975, Amalgamated Metal Corporation became the surviving member of a merger between it and Consolidated Tin Smelters. Patiño N. V., following the merger, had a 51.4 percent ownership interest in Amalgamated Metal. Cia Estanifera do Brasil is 65 percent owned by Patiño N. V. In addition to the one-third ownership by Patiño, Associated Tin Smelters was owned one-third by Australian Iron and Steel and one-third by O. T. Lempriere & Company.

b. The figure for the output of Datuk Keramat is taken from the *1976 International Directory*. Although the reported output is not dated, it is presumably for 1975 since the directory's publication date is 1976. Output for the Straits Trading Company is the 1975 output for Malaysia as reported by *Minerals Yearbook* less the figure shown for Datuk Keramat.

capacity to smelt its entire output of tin concentrates. An Australian mine, Greenbushes Tin, began operating an electric arc furnace smelter in September 1980, with a capacity of 1,000 metric tons per annum reached in 1981.

In one important respect, the situation resulting from the energetic and uncoordinated efforts of tin producers to obtain their own smelting capabilities is an unstable one for the industry as a whole. The world's total tin smelting capacity in 1980 was estimated at approximately 400,000 metric tons of metal per annum, roughly twice as high as that year's world mine output of tin-in-concentrates.[14]

There have been three major changes in ownership of tin smelters since 1975. Symbolic, in terms of the passing of the old order, was the 1978 transfer of Patiño N. V.'s interest in Amalgamated Metal Corporation to a West German metals and engineering firm, Preussag A. G. In return, Preussag gave up a 29 percent interest in Patiño N. V. This transaction put Datuk Keramat in Malaysia, Makere in Nigeria, and Associated Tin Smelters of Australia into the Preussag group, leaving Patiño N. V. with a controlling interest in only one smelter, Estanifera do Brasil, and with an indirect link to Capper Pass through Rio Tinto Zinc.

Equally symbolic of the rise of the new national interests, the Malaysia Mining Corporation (MMC), 71.35 percent owned by the government corporation Pernas, agreed in March 1981 to acquire a large minority interest in

---

c. Brazilian output is distributed among smelters in proportion to smelter capacity at the end of 1974.

d. Billiton International Metals B. V. is a wholly owned subsidiary of Royal Dutch Shell. Thailand Smelting and Refining Company was 50 percent owned by Billiton and 50 percent owned by Union Carbide Corporation. Billiton also owns the Kamativi Smelting and Refining Company through the Industrial Development Corporation of Rhodesia. Billiton is described in the *International Directory* as consisting of "more than 50 operating companies conducted by four holding companies."

e. Capper Pass & Son is a wholly owned subsidiary of Rio Tinto Zinc Corporation. Rio Tinto Patiño holds a 5 percent ownership interest in Rio Tinto Zinc. Patiño N. V. owns 40 percent of Rio Tinto Patiño. It is not obvious whether such a link justifies inclusion of Capper Pass in the Patiño N. V. group.

f. Wholly owned subsidiary of Associated Metals and Mining Corporation.

g. 44.8 percent owned by Union Minière.

h. Spanish output is allocated among smelters in proportion to smelter capacity in 1977, as supplied in a letter from the U.S. Bureau of Mines.

i. Wholly owned subsidiary of Metallgesellschaft A. G.

j. Japanese output is distributed between smelters in proportion to smelter capacity at the end of 1974.

k. 50 percent owned by the government of Zaire, and 50 percent owned by Geomine Brussels.

l. Mexican output is distributed among smelters in proportion to smelter capacity at the end of 1974.

Sources: Output figures from: U. S. Bureau of Mines, *Minerals Yearbook, 1975* (Washington: Government Printing Office, 1977). Identification of companies and interlockings from: American Metal Market, *Metal Statistics, 1976* (New York: 1976); Engineering and Mining Journal, *E/MJ International Directory of Mining and Mineral Processing Operation, 1976* (New York: McGraw-Hill, 1976); International Tin Council, *Trade in Tin 1960-1974* (London: no date); and Metal Bulletin Books, *Non-Ferrous Metal Works of the World*, 2d ed. (1974). Minor discrepancies among these sources have had to be reconciled, sometimes arbitrarily.

Table 7.3. Statistical measures of concentration of world tin smelters, 1929 and 1975

Measure	1929	1975
CR1	0.43 (0.43)	0.22 (0.28)
CR4	0.88 (0.91)	0.65 (0.70)
CR8	0.96 (0.96)	0.81 (0.91)
Gini	0.78 (0.79)	0.73 (0.75)
Herfindahl	0.30 (0.36)	0.13 (0.15)

the Straits Trading Company's smelter at Butterworth. A new firm, Malaysia Smelting Corporation, is to be established, 42 percent of it owned by MMC and 58 percent by Straits Trading Company. The smelter will be transferred to the new corporation. The parties agreed that within three years Malaysia Smelting Company's shares should be distributed to the Malaysian public. At the end of 1981, MMC had entered into negotiations to acquire an interest in Malaysia's other smelter, Datuk Keramat.

Third, in a 1976 reshuffling of investments by international corporations Union Carbide sold its interest in Thaisarco to Billiton. Billiton, in turn, had in 1970 become a wholly owned subsidiary of Royal Dutch Shell.

Principal features of the current structure. In light of the developments reviewed above, the simple model of the smelter as a monopsonistic purchaser of tin concentrates and a competitive seller of tin metal must be modified, although the implications of the model do appear to yield insights into the structural evolution of the industry. The extent of vertical integration between mines and smelters will be reviewed in the next section of this chapter, in a discussion of the structure of the tin mining industries of the several producing countries; but recognition of the crucial strategic position of the smelter in the vertical chain of production, distribution, and pricing has clearly been the principal common element in the policies for development of the industry adopted by the independent producing nations.

Yet however essential control of smelting may be to an overall position of strength in the industry, it is also evident that the independent, nonintegrated smelter, no matter how large, is in a highly vulnerable situation. On the one hand, entry into smelting is for the most part easy from the technical and financial points of view, while on the other, access to supplies of tin concentrates is vital to both successful entry and subsequent survival.

Barriers to entry that are important in many other industries are of only negligible to moderate significance in tin smelting. As noted in the first chapter of this study, and confirmed by the entry patterns and size-distribution of existing smelters, there are no important economies of scale in tin smelting beyond the capacity of a single reverberatory furnace.

The basic technology of tin smelting is widely known, and entrants handling high-grade alluvial ores have not been disadvantaged by lack of access to the

Table 7.4. World production and export of tin-in-concentrates, 1960, 1970, and 1980 (metric tons)

Year	Country	Production	Exports	Exports as a percentage of production
1960	Zaire	9,202	5,885	64
	Rwanda	1,277	1,277	100
	Nigeria	7,675	7,860	102
	South Africa	1,284	796	62
	Bolivia	20,219	18,338	91
	Indonesia	22,596	18,184	80
	Malaya and Singapore	51,979	—	—
	Thailand	12,081	12,618	104
	United Kingdom	1,199	—	—
	Australia	2,202	—	—
	Brazil	1,556	—	—
	World total[a]	136,500	75,118	55
1970	Nigeria	7,959	—	—
	Zaire	6,458	4,333	67
	Bolivia	30,100	29,379	98
	South Africa	1,986	880	44
	South West Africa	1,044	—	—
	Rwanda	1,320	1,582	120
	Argentina	1,172	866	74
	Indonesia	19,092	13,517	71
	Malaysia	73,794	—	—
	Thailand	21,779	—	—
	United Kingdom	1,722	—	—
	Australia	8,828	3,657	41
	Brazil	3,610	—	—
	World total[a]	185,600	55,900	30
1980	Nigeria	2,527	—	—
	Rwanda	1,600	1,270	79
	South Africa	2,434	810	33
	Namibia	1,000	—	—
	Zaire	3,159	2,005	63
	Bolivia	27,271	7,085	26
	Brazil	6,930	—	—
	Peru	1,202	703	58
	Burma	1,100	—	—
	Indonesia	32,527	3,400	10
	Malaysia	61,404	—	—
	Thailand	33,685	—	—
	United Kingdom	3,028	1,375	45
	Australia	10,391	7,425	71
	Unspecified origin[b]	7,000	7,000	100
	World total[a]	199,500	31,711	16

a. World totals include unlisted minor producers.

b. Based on reports by importers and presumably smuggled.

Sources: International Tin Council, *Statistical Yearbook*, 1968; International Tin Council, *Tin Statistics, 1966–76* and *1970–80*.

best known techniques. Treatment of complex lode ores, however, does demand technical know-how and skills that are limited to a small number of smelters and that are in some cases regarded as valuable company secrets. The two Malaysian smelters, for example, handle large quantities of complex Australian ore that cannot be treated by Associated Tin Smelters—a situation that stimulated the construction of Greenbushes' smelter. Officials at both Straits Trading Company and Datuk Keramat view their superior technical production abilities as a major asset—and one won only through years of experience. The problems that Bolivia's COMIBOL faced as a result of its dependence on a small number of such smelters, and the successful efforts made to break that dependence, further illustrate both the problem and the opportunities for overcoming this barrier with the aid of technically sophisticated independent engineering firms such as Klöckner—or of governments such as that of the Soviet Union.

Capital barriers, either absolute or relative to cash flow, are also low. To compare two extremes, a 1969 report by the United Nations Industrial Development Organization (UNIDO) stated that Indonesia's Peltim smelter was being built at a cost of US$3.5 million for a capacity of 25,000 metric tons per annum, while the Bolivian smelter at Vinto would cost US$9 million to attain a capacity output of 20,000 metric tons per annum.[15] These capital costs are modest, given the value of the product. The Peltim smelter's cost, as cited by UNIDO, would be $140 per metric ton of tin produced per year, or $7 per metric ton of output if amortized over a twenty-year period. The figures for the Bolivian smelter work out to be $450 per metric ton of annual capacity or $22.50 per metric ton amortized. In 1970, the price of tin averaged £1,529 per metric ton on the LME, or US$3,363 at the exchange rates then in force. In other words, amortized capital costs would have amounted to two-tenths of 1 percent of value of the product at Peltim and six-tenths of 1 percent at Vinto, if calculated on the basis of then current tin prices. Given the purity and high grade of the "clean" Banka concentrates treated at Peltim and the contrasting complexity and low grade of the "dirty" Bolivian ores, the per ton capital costs of most smelters would presumably lie within the limits estimated for these two.

Since the basic tin smelting operation, as distinguished from the refining of complex ores, is a simple one, the operating costs of processing alluvial tin are low. While cost data from individual smelters are unavailable, the magnitude of these costs is suggested by the charge for smelting levied in March 1979 by Datuk Keramat. The total charge that month for smelting a picul of concentrates of 75 percent tin content consisted of a monetary charge of M$12.10, plus a so-called unitage deduction of 1 percent of the assayed tin content. This unitage deduction is taken by the smelter to account for loss of tin in the furnace. The monetary charge for smelting 1.33 piculs of such concentrates (needed to yield one picul of tin metal) was M$16.13, or roughly 0.83 percent of the value of the metal, calculated on the basis of the average Penang market price for tin metal for March 1979 of M$1,951 per picul; and the total real cost to the miner, including the unitage deduction, was therefore approximately 1.83

percent of the final market value of the metal he delivered in his concentrates. This percentage, it must be remembered, includes earnings for the smelters as well as the actual costs of the operation.

In all, there are no serious impediments to the entry of new tin smelters that are in a position to receive supplies of tin concentrates; and therefore in the long run the existing smelters are not in a strong monopsonistic position, even if they are sole purchasers, unless miners can in some way be prevented from offering their concentrates to others.

In contrast, those owning or controlling sources of tin concentrates may be in a position to block the entry of a new smelter or eliminate an existing one. As a consequence, the nonintegrated smelters outside the producing countries are becoming anachronisms. Metallurgie Hoboken-Overpelt announced the closing of its smelter in December 1980, "following the steady decline of Belgium's imports of tin-in-concentrates, principally from Zaire and Rwanda, for refining at Hoboken."[16] Williams Harvey, in liquidation since 1973, but thereafter producing at low levels using its own slag as well as concentrates purchased in spot lots and presumably consisting mostly of smuggled ores, closed in June 1981 and began selling off and scrapping its plant and equipment.[17] The future of Capper Pass is at best problematic as Bolivia develops a domestic smelting capacity to absorb that nation's entire mining output. Capper Pass may continue to survive, at a greatly reduced scale of operation, relying principally on the 2,000 to 3,000 tons of tin-in-concentrates produced by the British mines. Gulf Chemical, no longer receiving any Bolivian ore, is now processing slag and has recently installed equipment designed to reclaim secondary tin from recycled domestic metals.

These import-dependent smelters have been caught in the double bind of having no way of limiting entry of competitors, while at the same time having to rely for their raw materials on those who have the greatest interest in promoting such entry. While in a superficial sense the major source of discretionary economic power on the supply side of the tin market today still lies in control of smelting capacity, such control is in its turn dependent on control of supplies of tin ore; and this more fundamental form of market power, as will be discussed below, can be obtained only by governments and not by miners.

Mining

In mid-1981 there were nearly fourteen hundred tin mines operating outside of the socialist centrally planned economies, including dredging, gravel-pump, open-cast, and underground mines, but excluding such very small-scale operations as dulang washing, suction boats, and individual or family tributers. Working tin mines ranged in size from such giants as Renison Ltd. of Australia with capital assets of over A$50,000,000 and Billiton's Indonesian offshore dredge constructed at a cost of nearly US$35,000,000, down through hundreds of efficient-sized gravel-pump and open-cast mines with 1980 capital costs of a few hundred thousand U.S. dollars, to individual dulang washers panning the streams of Malaysia and Thailand and tributers combing the slag heaps of Bolivia.

The structure of tin mining reflects the diversity of sizes and types of operations in the industry as well as the policies of the governments of the various producing countries. Restrictions on the export of tin concentrates are pervasive, and thus there is economic logic as well as convenience in reviewing the mining sector country-by-country.

Malaysia. For the third quarter of 1981, Malaysia reported 730 tin mines in operation, consisting of 617 gravel-pumps, 33 open-cast, 57 dredges, 20 underground, and 3 other.[18] Dredges accounted for 29.6 percent of mine output, gravel pumps for 55.2 percent, and all other mines for 11.1 percent. Dulang washers were responsible for the remaining 4.1 percent.

Gravel-pump mining in Malaysia, since the beginning of the modern industry in the mid-nineteenth century, has been carried out almost exclusively by small-scale Chinese-Malayan entrepreneurs.[19] Writing in 1960, Puthucheary noted that very little information was available about ownership patterns in the Chinese-Malayan mining sector. Most tin mines were owned by individuals or families, or were organized as "kongsis" or partnerships. "At present," he concluded, "we are faced with about 600 mines which seem to be unconnected with each other." The problems of identifying any relationships among these mines were insuperable. Mines owned by the same or related kongsis in different parts of the country had different names. Individual and family mine owners operated under both Chinese and Malay family and given names. There were, Puthucheary observed, several of the wealthiest Chinese in Malaysia who were known to have interests in a number of tin mines.[20]

As far as I have been able to ascertain, no more is known about ownership patterns in gravel-pump mining now than in 1960. There are certainly individuals and families with interests in several mines, but there is no reason to assume that the gravel-pump sector is anything but highly competitive and diffuse.

Dredging, which requires a substantial initial investment in long-lived equipment, has commonly been carried out by firms organized along European lines as limited liability joint stock companies. From the first introduction of dredge mining in Malaya in 1912 until quite recently, this sector of the industry has been predominantly under the ownership and control of European investors. In the early years of dredge mining, a large number of unrelated concerns entered the industry, but in the late 1920s a consolidation movement began.[21] The single most important factor in increasing concentration, as described in chapter 3, was the acquisition drive of the Anglo-Oriental Mining Corporation.

D. R. Williamson noted that in 1974 there were thirty-five tin mining companies listed on the Kuala Lumpur Stock Exchange, twenty-two of which were managed by one of three agency houses. Anglo-Oriental, by far the largest of these, managed twelve concerns that had produced a total of 20,828 metric tons of concentrates in the most recent twelve-month periods covered by their annual reports. Associated Mines managed four companies producing 8,534 metric tons of concentrates; and Osborne and Chappel managed six firms which mined 5,004 metric tons per annum.[22]

The agency houses, headquartered in London, provided central accounting, sales, purchasing, shipping, statistical, and tax services, as well as on-site management and periodic visits by experts. The extent to which these agencies coordinated the operations of the separate firms they managed is not certain, but Stahl concluded that "while individual directorates are watchful of their independence, yet they almost invariably include a director of the agency house, and it would be exceptional for them to branch out in a new line of policy which diverged to any extent from that pursued by the group."[23] Rowe, noting the "common policies and combination" promoted within the Malayan tin-dredging sector by the managing agencies, described them as a form of "horizontal combination."[24]

In 1977, pursuant to the Malaysian government's New Economic Policy (NEP), there was a sudden and drastic realignment of that country's tin industry, with the government replacing European investors as the dominant group in the dredging sector. New Tradewinds, a subsidiary of the Malaysian public corporation Pernas,[25] acquired sole ownership of Anglo-Oriental's parent firm, London Tin Corporation. At the same time, Pernas exchanged the shares of New Tradewinds for the mining interests held in Malaysia by Charter Consolidated Ltd., resulting in the absorption of three additional mines into the New Tradewinds group, and in Charter obtaining a 28.65 percent interest in New Tradewinds. As part of the reorganization scheme, Pernas and Charter Consolidated formed a jointly owned subsidiary, Pernas Charter Management, to serve as a general management agency for all of the group's Malaysian mines. In 1978, New Tradewinds changed its name to Malaysia Mining Corporation (MMC).

The mines and managing agencies acquired by MMC, the countries of operation, and the percentages of ownership held in each, are listed in table 7.5.[26] MMC has since disposed of its non-Malaysian holdings, leaving the corporation with interests in mining properties that have been producing a substantial majority of the tin-in-concentrates accounted for by Malaysian dredges and approximately 22 percent of the country's entire output.

In addition to acquiring a 42 percent interest in Straits Trading Company's smelter in 1981, MMC underwent a further consolidation and reorganization later in the same year. In 1980, Malayan Tin Dredging had acquired full equity ownership of four of the companies managed by MMC—Southern Malayan Tin Dredging, Southern Kinta Consolidated, Lower Perak Tin Dredging, and Bidor Malaya Tin—as well as a majority interest in Kramat Tin Dredging. In October 1981, MMC merged with Malayan Tin Dredging, creating a new firm named Malaysia Mining Corporation Bhd. Pernas holds 56.6 percent of the

Table 7.5. Mines and mining agencies owned by MMC, countries of operation, and percentage of ownership

Tin mining		
Amalgamated Tin Mines of Nigeria (Holdings) Ltd.	Nigeria	31.5
Aokam Tin Bhd.	Thailand	4.6
Austral Amalgamated Tin Bhd.	Malaysia	37.4
Ayer Hitam Tin Dredging Malaysia Bhd.	Malaysia	13.0
Berjuntai Tin Dredging Bhd.	Malaysia	37.4
Bidor Malaya Tin Sdn. Bhd.	Malaysia	62.3
Kampong Lanjut Tin Dredging Bhd.	Malaysia	34.3
Kamunting Tin Dredging Malaysia Bhd.	Malaysia	23.8
Kramat Tin Dredging Bhd.	Malaysia	30.9
Kuala Kampar Tin Fields Bhd.	Malaysia	29.3
Lower Perak Tin Dredging Bhd.	Malaysia	43.2
Malayan Tin Dredging Malaysia Bhd.	Malaysia	39.2
Southern Kinta Consolidated Malaysia Bhd.	Malaysia	31.8
Southern Malayan Tin Dredging Malaysia Bhd.	Malaysia	10.1
The Sungei Besi Mines Malaysia Bhd.	Malaysia	4.0
Tongkah Harbour Tin Dredging Bhd.	Thailand	37.2
Tronoh Mines Malaysia Bhd.	Malaysia	26.0
Mining management and services		
A. O. (Australia) Pty. Ltd.	Australia	100.0
A. O. Nigeria Ltd.	Nigeria	100.0
Anglo-Oriental (Malaya) Sdn. Bhd.	Malaysia	100.0
Associated Mines (Malaya) Sdn. Bhd.	Malaysia	62.3

share capital of the new firm, down from 71.35 percent of the old MMC; Charter Consolidated holds 14.5 percent, Datuk Keramat holds 3.8 percent, and the remaining 25.1 percent is held by other private investors.

Government participation in mining has also increased as a result of the establishment of public corporations owned by Malaysian State governments, the most prominent of which has been Kumpulan Perangsang Selangor (KPS). Under the NEP, a limit of 30 percent foreign ownership has been set for new mining enterprises; and the State of Selangor announced in 1978 that, in accordance with its interpretation of that policy, KPS and Malaysian interests must henceforth hold a total share of 70 percent in any new mining venture within the state, including renewal of leases held by existing firms. The vast Kuala Langat deposits are located in Selangor, giving particular significance for the future to the actions of that state. Following the joint enterprise arrangements made by KPS with MMC, Berjuntai, and Pacific Tin, a State Development Corporation (SDC) was formed in Perak. SDC has entered into a joint enterprise with MMC and Tronoh Mines for the mining of a new tract, with MMC having a 40 percent interest, and SDC and Tronoh each having 30 percent.

The overall effect of these recent developments and the trend that they indicate are clear: the governments of Malaysia, both federal and state, now compose by far the largest tin mining interest in the country, and their role will grow over time. This projection must be qualified by noting that public corporations, such as MMC and its parent Pernas, are designed as temporary organizations, holding shares in trust for eventual sale to individual *bumiputra*

Table 7.6. Frequency of shared directors, MMC, Osborne and Chappel, and unaffiliated companies

Unaffiliated firms tied to other unaffiliated firms	12
Unaffiliated firms tied to firms in the MMC group	6
Unaffiliated firms tied to firms in the Osborne and Chappel group	5
Unaffiliated firms with no ties	5
MMC firms tied to unaffiliated firms	4
MMC firms tied to firms in the Osborne and Chappel group	14
MMC firms with no external ties	0
Osborne and Chappel firms tied to unaffiliated firms	4
Osborne and Chappel firms tied to MMC firms	3
Osborne and Chappel firms with no external ties	2

investors. But this ultimate objective for restructuring of the Malaysian economy is unlikely to be obtained in the near future.

In one respect—the pervasiveness of interlocking directorates—the substitution of MMC for Anglo-Oriental changed little, if anything. An examination of the statements of the thirty-eight tin mining companies listed on the Kuala Lumpur Stock Exchange by July 1978 indicates the extent of the interlocks shortly after the formation of MMC.[27] These thirty-eight companies were divided into three groups: the fifteen listed MMC companies (excluding non-listed Amalgamated Tin Mines and Bidor), the six classified by Williamson in 1974 as managed by Osborne and Chappel, and seventeen unaffiliated companies. All of the MMC companies were linked to each other by common directors, as were the six Osborne and Chappel firms. The numbers of instances of one or more common directors within the unaffiliated companies, between the two groups, and between one of the groups and independent firms, are shown in table 7.6.

Datuk Keramat is now linked to MMC, since it held 10 percent of the outstanding shares of Southern Kinta, Malayan Tin Dredging, and Southern Malayan Tin Dredging at the time of MMC's full takeover of these three members of the Malayan Tin Dredging group.

A further interlocking results from Straits Trading Company's holding of a minority interest in all six of the members of the Osborne and Chappel group. Straits Trading also holds minority interests in three firms listed on the Kuala Lumpur Stock Exchange that are among those included in the unaffiliated group in the above analysis of interlocking directorates.[28]

Thailand. In the first quarter of 1981, Thailand reported 39 inland and offshore dredges in operation, 365 gravel-pump and hydraulicking mines, 2,416 suction boats, and 309 other, including approximately 60 ground sluicing mines and a remainder comprising mostly small tin-tungsten mines. Dredging accounted for an estimated 13.6 percent of production, gravel pumping and hydraulicking for 21.9 percent, suction boats for 53.3 percent, dulang washing for 1.8 percent, and the category of other mines for 9.4 percent.[29]

As in Malaysia, tin mining other than dredging in Thailand has been and remains the preserve of small-scale local entrepreneurs, predominantly ethnic Chinese. There is no evidence of interest groups with any economic power among these miners, who can safely be characterized as individualistic and competitive. Dredging in Thailand was developed by British and Australian firms, the largest four of which, Tongkah Harbour, Aokam Tin, Southern Kinta, and Kamunting (the so-called TASK group), came under partial ownership and control by the Anglo-Oriental group. In 1939, thirty foreign firms, including Anglo-Oriental, operated forty of the forty-five dredges working in Thailand;[30] but in the postwar decades the independent inland dredges have come almost entirely under local ownership and management.

The primary impetus to recent structural changes in Thai tin mining (in addition to the impact of MMC's takeover and subsequent disposal of the Anglo-Oriental firms) has come from efforts to assign rights to dredging of rich undersea deposits off the shores of Phang-nga and Phuket in the Andaman Sea. Offshore mining has been carried out in Thailand for at least a century. Indeed, the first dredge ever used for tin mining was adapted to that purpose in 1906 by an Australian, E. T. Miles, and was put into operation in Phuket harbor the following year.

In 1968, Tongkah Harbour and Aokam Tin were working sea dredges off Phuket, and Southern Kinta had a near-shore shallows dredge in operation.[31] Sources of later controversy emerged during that year, when the Thailand Exploration and Mining Company (TEMCO) was established by the same consortium that had formed the Thaisarco smelting company three years earlier. TEMCO was granted a number of leases for exploration and mining of tracts off the shores of Phuket and Phang-nga. The details of the ensuing events are still unclear.

In outline, a second firm, EMCO, was formed and also granted leases to a number of offshore tracts. In 1970, Eastern Mining Company, a Thai firm controlled by General Praphas that was then holding a 30 percent interest in Thaisarco, sold all of its shares in Thaisarco, TEMCO, and EMCO to Billiton; and the following year Billiton purchased enough of Union Carbide's holdings in these three firms to obtain a 50 percent interest in each. EMCO was then absorbed into TEMCO. Following the share transfers, TEMCO held mining rights to a substantially larger area than an independent company would have been entitled to receive through making direct applications to the Thai government. The terms under which TEMCO operated were extremely generous, involving a twenty-year lease on a concession of nineteen thousand acres and a 24 percent royalty fee.

In 1974, following the overthrow of the Thanom-Praphas regime, the new government announced an investigation of the circumstances surrounding the awards of the TEMCO and EMCO leases; and in March 1975 TEMCO's concession was revoked, with an announcement from the Ministry of Industry that the leases had been obtained "by means of corruption and illegal ways."[32]

The matter was never adjudicated, and there is still disagreement in Thailand whether Union Carbide and Billiton knowingly engaged in corrupt practices justifying the action of the government or whether they were innocent victims of the 1973 revolution that drove out the government with which they had worked.

The TEMCO tracts, particularly off Phang-nga, proved to be extraordinarily rich. As the controversy over the mining rights grew, an increasing number of small boats began encroaching on the area. These boats operate by sending a diver equipped with a suction hose into as much as fifty feet of water. The tin-bearing silt is typically under about two feet of sand which the diver has to push aside to work the hose over the pay dirt. On board the boat there is a small and simple *palong* that retains some of the heavier tin concentrates before dumping the remaining silt back into the water. This method of mining is an exceptionally inefficient one in terms of the amount of tin spilled back, even though some of the boats are fitted out with quite sophisticated and expensive equipment; and the dredging companies have claimed that it reduces the amount of tin that can ever be recovered, since it disturbs the ocean floor and makes later dredge mining of the area far more difficult and costly.[33] Further, the unlicensed small boats became the first link in a smuggling chain costing the Thai government approximately US$5,000,000 a year in lost royalties and taxes.

A Bangkok business magazine reported that an estimated 3,000 small boats were operating offshore from Phang-nga in 1975, selling tin-in-concentrates on the beach at approximately one-third of the regular domestic price of tin. "From that selling point onward," the article continued, "an elusive network of middlemen and shippers, allegedly with the connivance of high-ranking officials and financial backing from Bangkok sources, conveyed most of the ore to smelters in Penang, usually with forged certificates of origin from Burma." Based on the discrepancy between Malaysian government figures for tin-in-concentrates received from Burma by the Penang smelters and official Burmese figures for tin exports to Malaysia, the tin smuggled out of Phang-nga amounted to about 3,600 tons in 1975.[34]

While the small boats were described in the Thai and foreign press as illegal and were obviously poaching on the TEMCO tracts, no serious efforts were made to curb their operations. Entirely apart from the alleged matter of involvement of highly placed government officials and Thai businessmen, the small boat operators had the open sympathy and support of local officials, most notably the Mayor of Phang-nga, who for some time before cancellation of the leases in 1975 had been a harsh and vocal critic of TEMCO. The central government recognized that the suction boats could not be eliminated without an armed struggle that would have been highly distasteful and politically unpopular, pitting the government against thousands of its own citizens on behalf of multinational business.

Under these circumstances, the government-owned Offshore Mining Organization (OMO) was created in July 1975. According to contemporary reports, OMO was formed in an emergency atmosphere involving the heated political aspects of the TEMCO affair, growing concern over the inroads of the suction boats, and the desire to have the matter resolved and the offshore tracts back in legal production before the opening of discussions on the Fifth International Tin Agreement. The last was deemed important since the Thai government hoped thereby to obtain a higher quota and greater voting strength in the ITC.[35] It can safely be said that the establishment of OMO was not the result of reasoned debate as to the appropriate long-run role of the government in the nation's tin industry. OMO took title to all existing and future offshore tracts and promptly began negotiations with Billiton for lease of the former TEMCO tracts off Phang-nga.

In early 1976 OMO lent official sanction to the small boats, which nevertheless continued to be referred to as illegal, by leasing to provincial administrations three offshore tracts in areas that had formerly been a part of the TEMCO concession, with the expectation that the suction boats would operate therein. In early 1977, following another change of government, OMO and Billiton agreed on a five-year lease for approximately one-third of the former TEMCO tracts. At the same time contracts were signed with the two other foreign offshore dredging firms. In response to criticisms leveled during the negotiations, the Ministry of Industry described the proposed contracts as "a temporary employment of the said three companies by OMO for ore production. The companies are merely employees of OMO which supervises and controls their operations. This temporary employment will provide OMO with data and facts in assisting OMO in its decision whether it should invest in mining operations by itself after the expiries of the hiring contracts."[36]

The terms of the Billiton agreement were far less favorable than those that had been given to TEMCO. While OMO agreed to pay a 24 percent royalty to the Thai government, Billiton's share of after-tax revenue was calculated on a sliding scale, ranging from 68.5 percent for output up to 1,300 metric tons of tin-in-concentrates per annum down to 40 percent for any output in excess of 1,700 tons per annum. Despite a continued failure of OMO and the Thai government to keep the small boats from poaching on Billiton's preserve, agreement has been reached to renew the lease on terms even more favorable to OMO; and OMO and Billiton are considering further offshore dredging on a joint venture basis.[37]

OMO has since awarded other offshore leases to a number of firms. In addition to Billiton, Southern Kinta, Tongkah Harbour, and Aokam, concessions have been granted to eight Thai-owned firms.[38]

In 1978 a mineral development policy was announced, requiring all existing foreign offshore mining companies to reorganize as firms incorporated in Thailand, with no more than 40 percent of the issued shares held by non-Thai investors, to be eligible for renewal of their leases. New offshore mining firms will be required to meet this standard within five years of the beginning of operations. An exception, currently applicable only to Billiton, is made for firms mining in water over 200 feet deep. Such a firm is given a ten-year grace period.[39] In 1980, the required Thai participation was raised to 70 percent.

Billiton is now the only large tin dredging firm in Thailand under foreign control. One of the four Anglo-Oriental firms, Kamunting, sold its dredges and went out of business in Thailand after exhausting the ores on the land it had worked in that nation. Southern Kinta reduced the scale of its operations outside Malaysia to the mining of a small area in Thailand on a sublease from a Thai company. Both Aokam and Tongkah Harbour have restructured themselves as Thai companies, with MMC's holdings in the new Thai firms being reduced to 30 percent. Pernas Charter Management, however, continues to act as general manager for both Aokam and Tongkah Harbour.

While the impact of the TEMCO affair and the formation of OMO has unquestionably been a drastic one, replacing foreign investors with a government organization, its effects when considered in the perspective of the entire Thai tin mining industry have not been overwhelming. Since the small boat operators are not in any practical sense under the control of OMO, it seems more realistic to view that organization as having effective regulatory power only over offshore dredging, which accounts for less than 15 percent of the Thai industry. The inland mines have not come under any new forms of regulation or control. In the opinion of several knowledgeable mining engineers and others with an interest in the Thai tin industry, offshore mining will prove to be a temporary bonanza—a rich strike that will be in large part depleted within a decade or fifteen years. Finally, the owner of a seagoing dredge is not nearly so vulnerable in negotiations for renewal of a lease as is the owner of an inland dredge. Once a dredge has been constructed in its place in a paddock, the cost of moving it to a new site is high and in some cases prohibitive, while a seagoing dredge is a highly mobile piece of equipment.

Nigeria. It is appropriate to survey the structure of the Nigerian tin mining industry immediately after having discussed those of Malaysia and Thailand since Nigeria, like the other two, has recently undertaken a major reorganization of the industry, resulting in the replacement of foreign mining interests by a government body.

Nonindigenous commercial tin mining began in Nigeria in 1906, and expanded rapidly under boom conditions. By 1928 there were eighty-three registered tin companies in the colony, forty-eight of which were small operations capitalized at less than £100,000. As in Malaysia, a period of consolidation ensued, and by 1938 the number of tin mining firms had fallen to thirty-one, of which eleven had stated capital in excess of £100,000. Yet over this decade, during which the number of firms decreased by more than a half, the estimated aggregate capital investment in Nigerian tin mining rose from £9,881,000 to £41,856,000.[40] By 1968, eleven firms, all incorporated in the United Kingdom, accounted for 66 percent of Nigerian production. By 1977, when Nigerian production had fallen sharply (from 9,881 metric tons in 1968 to 3,267 metric tons in 1977), the Anglo-Oriental affiliated firm of Amalgamated Tin Mines of Nigeria accounted for 45 percent of Nigerian tin output, and the five next largest firms accounted for another 30 percent.[41]

The difficulties of tin mining on the isolated Jos Plateau of Nigeria are reflected in recent annual reports of the tin mining companies. The 1977 report of Gold Base and Metal Mines, for example, noted, "It is becoming increasingly difficult to obtain essential spares and stores from authorised agents or traditional suppliers. However, our requirements can usually be obtained from other sources but at an inflated price, thus adversely affecting costs." The report went on to discuss increases in the cost of electric power and the need to shut down production from time to time because of the unavailability of diesel fuel. In the same year, Ex-Lands Nigeria reported to its stockholders, "Production was adversely affected by difficulties in obtaining adequate supplies of fuel oil, explosives, and vital spare parts for plant and machinery. In addition, water, essential for our mining operations, was in short supply as a result of the lack of rainfall which was the lowest ever recorded."[41]

In 1972 the Nigerian government established the Nigerian Mining Corporation (NMC), and in 1974 NMC announced its intention "to acquire reasonable equity participation in existing well-organised sizeable and profitable mining and smelting enterprises which would thereby become the Corporation's associated companies."[42] Two years later, a Nigerian Enterprises Promotion Decree was promulgated, requiring Nigerian participation by 1978 of at least 60 percent of the equity of mining and quarrying firms and 40 percent of smelters. In the instance of Nigeria's tin mines, it was decided that NMC should acquire a 54 percent interest, and an additional 6 percent would be set aside for distribution to the mines' Nigerian employees. In 1977, NMC acquired a 40 percent interest in the Makeri smelter and began negotiations for transfer of ownership with Amalgamated Tin Mines and four other British-owned mines (the sixth large mine had been transferred to majority Nigerian ownership in 1970). In 1978, as decreed, these transfers were completed.[43] Thus, the government-owned enterprise has come to own a substantial majority of Nigerian tin mining capacity and has totally displaced foreign control, as well as holding a 40 percent stake in the nation's single smelter.

Indonesia. The tradition of public ownership of tin mining and smelting in Indonesia goes back to at least 1816, at which time the Netherlands government obtained mining rights.[44] At the time of independence, the colonial government of the Netherlands Indies owned the nation's smelter, one of its mining concerns, and a five-eighths interest in the only other mining firm, the Billiton Joint Mining Company. The government of the new Republic of Indonesia assumed title to the mining and smelting properties of the Netherlands Indies government on coming to power; and in 1958 the Billiton firm was dissolved and ownership was transferred to the State Mining Enterprise, P. N. Timah. With this action the entire Indonesian tin industry, consisting of twenty-five dredges and sixty-three gravel-pump and hydraulicking mines as well as the smelter, came under the sole ownership of a government agency.

Following the overthrow of Sukarno, and in the face of a decline in tin output from 33,367 metric tons of tin-in-concentrates in 1955 to 12,526 metric tons in 1966, the Indonesian government began to consider opening its mining industries to foreign private enterprise. In 1969, foreign private investment in mining was permitted, on a joint venture basis with the appropriate government organization, and the Indonesian government entered into negotiations with a number of foreign firms for offshore exploration concessions. In 1972 a set of regulations for foreign mining enterprises was issued. However, over the following four years few new contracts were entered into between P. N. Timah and private tin mining concerns. In 1976, the mining regulations were liberalized, allowing duty-free import of materials and equipment as well as various tax breaks for foreign investors. In the same year, P. N. Timah, which had been a government department, was reorganized as a government-owned corporation and renamed P. T. Timah, in order to allow greater managerial and financial flexibility.

Three private mining firms have entered into joint ventures with P. T. Timah. Koba Tin, a joint venture with 25 percent participation by P. T. Timah and 75 percent by Kajuara Mining Company of Australia, has been operating both gravel-pump mines and two small dredges since 1974. Koba Tin's output has grown rapidly in recent years, from 1,610 tons of tin in 1977, when it operated five gravel-pump mines, to 5,263 tons in 1980, when it had fourteen gravel-pump mines working. In 1980, Koba's output exceeded that of Australia's Renison, Ltd., until then the largest privately owned tin mining company in the world. Another Australian firm, Broken Hill Pty., entered into a venture in 1971 to rehabilitate an underground mine that had been abandoned since the outbreak of the Second World War, but despite the expenditure of over US$14 million, the mine had not been brought into production by the end of 1981, and the continuation of the project is in doubt. The most ambitious of the three joint ventures, originally known as BEMI, is owned 65 percent by Billiton and 35 percent by P. T. Timah. BEMI, recently renamed P. T. Riau Mining, was granted a lease for an offshore area, and Billiton agreed to construct the largest and most expensive seagoing dredge yet designed to work the tract. The dredge was completed in 1979, at a cost of over US$50 million including supporting facilities; but at the end of 1981 it was not in regular operation, as problems of excessive vibration had not yet been corrected.

P. T. Timah does not manage and operate all of its other properties. About half of the output of its gravel-pump sector, exclusive of joint ventures with foreign firms, comes from mines operated by private contractors, drawn largely from Indonesia's ethnic Chinese minority. The output of these contractors is presumed to be the primary source of tin concentrates smuggled from Indonesia to Singapore.[45] This smuggling has been estimated to run as high as 3,000 to 5,000 tons per annum.[46]

In 1980, Indonesia closed its tin mining industry to further foreign private participation.

Bolivia. In 1952, when the revolutionary government of Bolivia expropriated the Patiño, Aramayo, and Hochschild tin mines, the total output of these "big three" represented approximately 75 percent of Bolivia's tin production. The remaining output was produced by around thirty so-called medium miners (a category that has been defined by Bolivian law since 1934 in terms of capitalization, degree of mechanization, and technical and engineering capability), and as many as 3,000 small mines, about 500 of which were full-time operations. The remainder comprised individuals, often Indian farmers' wives and children, and families who worked mine tailings or abandoned shafts on a part-time basis. A recent study described these small mines as follows.

> Small mines and cooperatives are in the main, characterized by primitive technological methods (both in exploitation and in concentration), low productivity and generally abysmal health and safety standards. Commonly, small mines and cooperatives engage in reworking the tailings from the large concentrators, particularly those of COMIBOL. Others direct their efforts to such hazardous operations as extraction of safety pillars and minor ore blocks left by previous operators when the mines were closed down. Finally, other small mines and cooperatives operate extremely narrow and short veins which cannot be mined economically at medium or large scales of operation. These undertakings generally are economically worthwhile only if costs are kept low through the use of inferior safety, health or working conditions for staff and workers.[47]

By 1965, COMIBOL's share of Bolivian output had fallen to roughly two-thirds of the country's total. In recent years the share of the medium mines has risen, as shown in table 7.7. Several of Bolivia's so-called medium mines are quite large by world standards. As Fox noted, the average output of the nineteen medium mines operating in 1966 was approximately five times as large as that of the average Malaysian gravel-pump mine, and four of these medium mines were producing at the level of the average Malaysian dredging company.[48] Gillis et al. described these mines as being in general more efficient than the COMIBOL mines, with more modern equipment, better ventilation, and higher safety standards. Several of the medium mines are foreign owned or controlled, although the extent of such control was reduced substantially when a Bolivian firm, International Mining Corporation, purchased control of five medium mines from the United States firm of W. R. Grace and Company.[49]

The output of the small miners must, by law, be sold to the Bolivian government's Banco Minero (BAMIN), although COMIBOL acquires tin ore directly

Table 7.7. Shares in Bolivian tin output, 1966-80 (percentage of total Bolivian output as reported to the ITC)

	1966	1980
COMIBOL	66.9	68.3
Medium mines	17.6	21.6
Small mines	15.6	10.1

from its tributers. About 30 percent of the output of the medium mines is also sold to BAMIN. When the Vinto smelter came into production the government decreed that all ore not sold to BAMIN must be offered first to ENAF, and also required permits for any export of tin concentrates.

In late 1975 the Bolivian delegate to the ITC made an announcement that "Corporacion Minera de Bolivia (Comibol) is willing to form mixed companies with private capital in dredge operations and in areas where large exploration investment is needed. The State Enterprise would keep majority equity participation but is prepared to give management to the private sector. Mining industry's private sector is avidly seeking to increase its participation either through new ventures or by increasing the capacity of existing operations."[50] Only one such venture was formed, ironically not with a private firm but with Malaysia's MMC, for a dredge mining operation to involve one large or two small dredges. No commitment to construction of such dredges has yet been made.[51]

Other producers. In the second quarter of 1981 Australia had seven tin-mining establishments, comprising five underground mines and two dredges. Less than 5 percent of Australia's output came from the dredges. Australian mining is highly concentrated in structure. Renison Ltd., which operates the largest single tin-mine site in the world, accounted for approximately 49 percent of Australia's output in 1980. The second-ranking Australian firm, Aberfoyle Ltd., added another 26 percent of the nation's 1980 output. Aberfoyle is the product of a 1978 merger of three tin-mining firms, including those that had ranked second and third in previous production.

Australia possesses extensive deposits of tin. In 1971, noting active exploration in that country, *Minerals Yearbook* commented, "It is believed that Australia may possibly become a tin producer as large as Malaysia because of the vast potential of deposits in the Pilbara region of Western Australia."[52] These deposits, however, are of low grade and in remote areas that are waterless and among the hottest regions on earth. There is general agreement that at current tin prices and under current production techniques, involving large quantities of water for on-site concentration, these deposits cannot be mined economically.

In the early 1970s there was the prospect that Brazil as well as Australia might develop into a major tin producer, perhaps rivaling Malaysia. In the same issue in which it had reported on the Pilbara deposits, *Minerals Yearbook* observed,

"Alluvial tin deposits, richer in tin content than those of Malaysia but more inaccessible than any other in the world, have been discovered in the Pôrto Velho area of Rondônia by the Brazilian subsidiary of Billiton, Companhia de Mineração Ferro-Union."[53] *Minerals Yearbook* noted that, in addition to Billiton, three other companies were active in mining tin in Brazil, including W. R. Grace and a Canadian consortium. In 1972, *Minerals Yearbook* reported that twelve groups, including ninety-six companies, had applied for permits to prospect and mine in Rondônia.[54] The previous year the Brazilian government had banned small-scale nonmechanized tin mining and had announced its intention of issuing mining permits only to firms capable of undertaking mechanized operations.

In early 1975 the Brazilian Ministry of Mines and Energy issued the first government report on the situation in Rondônia, which it described as a "frank disclaimer of earlier wild estimates." The nation's tin reserves were estimated at approximately 600,000 metric tons, down from earlier estimates as high as 5,000,000 tons. The report noted that recovery of the tin would be difficult and protracted, observing that "as we can see, the deposits are characterized by small ore beds spread over several areas of the immense region along the little rivers and waterways." The report stated that there were eleven major tin mines in operation throughout Brazil at the time.[55]

By 1978 the small miners were still active in Rondônia, and the Brazilian government announced that it intended to crack down on these illegal operations and at the same time coordinate the exploitation of Brazil's tin deposits through a government-owned mining corporation, Companhia do Vale do Rio Doce. In 1980 there were nine companies reported to be mining in Rondônia, working nineteen mine sites.[56]

The largest enterprise, the Empresas Brumadinho group, operates mines at five sites, producing approximately 25 percent of Brazil's tin. In 1979, it established its own smelting and marketing subsidiaries. In 1981, the British Petroleum Company acquired a 50 percent share of the second largest Brazilian mining company, Brascan Recursos Naturais. The year before, Brascan had purchased Brazil's largest smelter, Estanifero do Brasil, from Patiño N. V.

In Zaire, since a 1976 merger of four companies, there have been only two active tin mining companies. Sominki, the successor firm emerging from the 1976 merger, mines approximately 80 percent of Zaire's ore. The remaining 20 percent is mined by Zairetain, which also owns Zaire's one smelter. The government of Zaire holds a 28 percent ownership interest in Sominki and a 50 percent share of Zairetain. The private partner in Zairetain, the Belgian firm of Geomines, is also the owner of the largest mining concern in Rwanda.

There are three underground mining companies in South Africa. Two of these, Rooiberg Minerals and Union Tin Mines, are subsidiaries of Consolidated Gold Fields Ltd. The third, Zaaiplaats Tin Mining, is an independent company. Zaaiplaats has owned and operated a smelter for many years, and Rooiberg opened a smelter of its own in 1979.

Table 7.8. Exports of tin metal, Peoples' Republic of China, 1966–80 (metric tons)

Year	Total	To U.S.A.
1966	5,205	—
1967	3,046	—
1968	4,786	—
1969	4,307	—
1970	5,423	—
1971	7,684	—
1972	8,425	163
1973	10,036	1,755
1974	10,176	3,336
1975	13,018	6,378
1976	7,094	1,727
1977	3,396	381
1978	5,415	1,571
1979	3,196	185
1980	2,574	858

Sources: International Tin Council, *Tin Statistics, 1966–76*; International Tin Council, *Monthly Statistical Bulletin* (June 1979 and March 1982).

In 1981, there were four underground tin mines owned by three firms operating in Great Britain, all in Cornwall. Geevor Tin Mines, 14.63 percent owned by the Union Corporation Group which holds extensive gold mining facilities in South Africa, but with no ties to any other interests in the tin industry, operated one of these near Penzance. Saint Piran Ltd., which also owned or controlled three dredge mining companies in Malaysia and held a 43 percent share in a Thai dredging company, owned a mine at South Crofty. Two other mines, the Mount Wellington and Wheal Jane mines, had been closed in 1978. Both were subsequently acquired by Rio Tinto Zinc (RTZ)—Mount Wellington being sold by Billiton and Wheal Jane by Consolidated Gold Fields. By the end of 1980, RTZ had driven new shafts connecting the two mines and had put the combined property back into operation at an annual rate of output of 1,500 tons per annum of tin-in-concentrates. A number of other properties in Cornwall and Devon are being explored and tested, including offshore sands that might be dredged, but none was in commercial operation by the end of 1981.

The Peoples' Republic of China, one of the world's major tin producing nations, experienced a rapid increase and an even more sudden decrease in its exports of tin metal over the 1970s, as shown in table 7.8. The increase, as can also be seen in table 7.4, was stimulated by exports to the United States, which began importing tin from the Peoples' Republic in substantial volume following the thaw in relations between the two countries in the early 1970s.

In 1977, the Malayan Tin Bureau reported that a long-term contract between the USSR and China, canceled seven years before, had been renewed, under which China would export 6,000 metric tons of tin per annum to the Soviet

Union. According to the Tin Bureau, execution of the terms of this contract would preclude any future Chinese exports of tin to the noncommunist world.[57]

A 1978 report by the United States Central Intelligence Agency (CIA) estimated a sharp drop in Chinese production (which is not reported to the ITC), by approximately 40 percent from 1975 to 1976. The CIA attributed this decline to low levels of investment in the industry, high development costs, inadequate transportation facilities, and aging equipment, most of which had been supplied by the Soviet Union twenty to thirty years previously.[58] No mention was made by the CIA of exports to the Soviet Union.

Any such reduction in investment in tin mining must reflect a deliberate decision by Chinese planners to husband the resource and channel investment funds to meet more pressing needs of the nation. According to a recent study by K. P. Wang, new tin ore was being discovered as rapidly as old ore bodies were being depleted during recent years when the rate of production was about 20,000 tons of tin-in-concentrates per annum; and this ore is of high grade. Wang notes that in 1976 there was an earthquake in Yunnan, where the tin mines are located, although he does not speculate on what, if any, damage might have been done to the mines or infrastructure; and he points out that Chinese domestic consumption of tin may have risen to 8,000 or 9,000 tons of tin per annum.[59]

A Malaysian delegation visiting Yunnan in 1980, led by Minister of Primary Industries Datuk Paul Leong Khee Seong, concluded that there is no prospect of Malaysian smelters obtaining Chinese tin concentrates or of joint production ventures, since "tin mining was accorded a low priority compared with gold, copper, zinc, and lead" in China's plans for development of its mineral resources.[60] Indeed, according to Datuk Leong, Chinese spokesmen told their visitors that their country might well become a net importer of tin.[61]

International holdings. Discussion of ownership patterns in the world tin mining industry must take account of international holdings. Prior to the independence of the producing nations, such holdings were a predominant element in the structure of the industry. Puey, for instance, identified three major groups, the London Tin Corporation (LTC) or Anglo-Oriental, the Patiño, and the Dutch, the last comprising both the Netherlands Indies government and Billiton, as controlling among them 40 percent of the world's mine output in the years just before the outbreak of the Second World War. Anglo-Oriental owned interests in mines in Malaya, Nigeria, Thailand, and Great Britain. Patiño's mining ventures embraced Bolivia and Malaya. Further, LTC and the Patiño holdings were linked through the British Tin Investment Company, a firm in which both LTC and Patiño held shares.[62]

Private international combinations still exist in the tin mining industry. Several such holdings have already been identified. Saint Piran Ltd. has interests in mines in Cornwall, Thailand, and Malaysia. Geomines Brussels is involved in tin mining in both Zaire and Rwanda. The largest of these combinations, until its 1977 takeover by MMC, was the Anglo-Oriental group with its holdings in Malaysia, Thailand, and Nigeria.

The most pervasive group now in existence is that of the Billiton companies. One subsidiary, Billiton (Thailand) Ltd., is engaged in offshore mining in that country while another subsidiary, Thaisarco, remains the largest smelter in Thailand even after losing its monopoly privileges. In Zimbabwe, still another Billiton subsidiary, Kamitivi Tin Mines Ltd. owns mines as well as the country's one smelter. Billiton companies are active in offshore dredging in Indonesia, pioneered in tin mining in Brazil's Rondônia, and sought briefly to restore the Mount Wellington mine in Cornwall after the failure of an American firm to revive it. Billiton owns the smelter at Arnhem and operates a research and development facility there. Billiton markets tin worldwide through Billiton Handelgesellschaft in Lucerne, Switzerland, and Billiton Trading Corporation in New York, having sold approximately 47,000 tons, or 25 percent of the world's total in 1980, according to a company spokesman.[63] Through Billiton Enthoven Metals it holds a seat on the LME. The Billiton structure consists of forty-six operating companies around the world in addition to exploration, metallurgical, and service units, most of which are concerned with nonferrous metals and minerals other than tin.

Consolidated Gold Fields Ltd. is the ultimate parent of the giant Renison mine in Australia and of both the Rooiberg and Union Tin mines in South Africa; it owned Wheal Jane in Great Britain until 1978. Consolidated Gold Fields is active in tin trading through another subsidiary, Tennant Trading Ltd., which was a founding member of the LME.

In addition now to owning the Wheal Jane and Mount Wellington mines, RTZ is the majority owner of one dredging company in Malaysia and holds a 30 percent interest in another, in partnership with Selangor's KPS. RTZ has subsidiaries engaged in exploration for tin in both Australia and Malaysia. RTZ is the sole owner of the Capper Pass smelter. Through Rio Tinto Patiño, RTZ has indirect links to the modest tin holdings left in the hands of the Patiño group after the sale of Amalgamated Metal Corporation to Preussag and Estanifera do Brasil to Brascan.

By the most generous estimates of their outputs, drawn from reported ore production and capacity figures given in financial statements, the combined output produced or controlled by all five of the extant private international mining companies mentioned above (Billiton, Consolidated Gold Fields, RTZ, Saint Piran, and Geomines), has been less than 9 percent of total world tin-in-concentrates production (excluding the centrally planned socialist economies) in recent years. This figure seriously overstates the market position of these firms because Billiton is operating on contract to government-owned organizations in both Thailand and Indonesia; all of the Malaysian outputs of Saint Piran and RTZ must be sold to the Malaysian smelters in open bidding on the Penang market; and Geomines shares control of a substantial minority of its production with the government of Zaire. These international mining firms, like the European-owned smelters, have become a relatively minor factor in the world tin industry, both singly and collectively.

Far overshadowing the large private firms are the state mining enterprises. Assuming that MMC and the state government mining companies account for 25 percent of Malaysian production, and that comparable figures are 95 percent for P. T. Timah in Indonesia, 70 percent for COMIBOL in Bolivia, 10 percent for OMO in Thailand, 70 percent for NMC in Nigeria, and 30 percent for Zairetain in Zaire, these six government-owned mining organizations now account for approximately 36 percent of the world's mine output of tin, again excluding the socialist centrally planned economies.

The total output of the public enterprises is almost matched by that of the least organized and most highly competitive operators. The aggregate output of the world's small mines—the hundreds of operators of gravel-pump, hydraulicking, and open-cast mines, the small-boat suction dredges of Thailand, dulang washers, and the small mining sector of Bolivia—is approximately 33 percent of world production. Roughly 23 to 24 percent is produced by what may be called the medium-scale sector, including the privately owned inland dredge mining companies of Thailand and Malaysia, the Australian mining companies with the exception of Renison, the medium mines of Bolivia, the mechanized mines operating in Rondônia, and the output of scattered mines in the minor producing nations. These totals add to more than 100 percent because the estimates are approximate and, more important, because of double counting—Billiton's offshore dredging output in Indonesia and Thailand, and Zairetain's production being included in the percentages of both the international corporation and state enterprise groups.

In all, there is no identifiable firm, government enterprise, or existing coalition that exerts economic power in the tin market in the sense of being able to affect the world price of tin metal in a significant and sustained way by curtailing its output of tin concentrates. On the other hand, the structure is not so diffuse that effective coalitions cannot be imagined (MMC, COMIBOL, and P. T. Timah, for example).

There are other organizations, such as the Malaysian smelters and the trading subsidiaries of Billiton and Consolidated Gold Fields, with the power to influence the day-to-day price fluctuations of tin metal on the organized markets and with conceivable interests in doing so growing out of their vertical linkages with tin mining concerns.

Until recently, such problems appeared to be potential rather than actual. But the mystery buyer episode of 1981–82, although a failure if its objective was either to maintain a higher price level or to earn short-run profits for the financial backer, may be a harbinger of future efforts by or on behalf of producing governments. There is a clear trend toward vertical integration with the decline of the European and American smelters, the rapid growth of Bolivia's government-owned smelting capacity, acquisition or construction of smelters by private mining concerns in Australia and Brazil, and, most recently and significantly, MMC's direct involvement in smelting. P. T. Timah and COMIBOL have had their own marketing organizations for some time: in 1980 MMC decided not to renew a marketing agreement with Anglo-Chemical and Ore, but to set up its own London-based subsidiary, MMC Services Ltd.

Nevertheless, in the absence of overt collusion or governmental intervention in the production and trading activities of private companies, the current structure of the tin industry seems compatible with efficient pricing and distribution on a world market. The structure, it must be added, has changed in fundamental ways since 1929, and is currently undergoing rapid transformation.

8. Mining Costs and Market Pricing

The Cost and Technical Structure of Tin Mining

Overall cost levels by type of mine and by country. As types of tin mining vary widely, so do costs of investment and operation. Converting limited information on representative or average mines in a number of countries to a composite index, if fully allocated costs per ton of tin-in-concentrates produced are set at 100 for dredging, such costs would be around 130 for gravel-pump and opencast mining, and 145 for underground mining. These differences in the costs of various types of mining are reflected in differences in the average costs of the producing countries. Table 8.1 indicates these differences, as reported to the ITC for the first half of 1980.

A survey of principal types of mining. Approximately 40 percent of total world mine output of tin, from other than the centrally planned socialist countries, is accounted for by gravel-pump, hydraulicking, and open-cast mining, which may be regarded as one general type of alluvial and eluvial tin mining involving excavation of ore-bearing dirt from above ground or an open pit and transportation of the paydirt to on-site concentrating and dressing facilities. This type of mining is carried on around the world. Some of these mines are highly mechanized, with investments in large and costly machinery for digging and earth moving, while others rely more heavily on hand labor. Sizes vary, depending on the nature and extent of the ore to be mined as well as the degree of mechanization.

Another 25 percent, roughly, of tin output is produced by dredges. Where terrain permits, dredging is almost always the most efficient method of tin mining, but it requires a large flat tract of alluvial tin-bearing ground, ample water in which to float the dredge, and an absence of underground impediments such as rock pinnacles, boulders, or large stumps. Only in Southeast Asia are these conditions common. Dredges, which are basically floating platforms containing a chain-bucket excavating mechanism and an onboard concentration facility, range in size and expense from large stabilized seagoing vessels to small installations dredging mine tailings in artificial ponds. Dredging and gravel-pump mining are commonly regarded as technically complementary in that a suitable area can be dredged first and later mined again by gravel-pump techniques, since the gravel-pump mines can excavate paydirt in and around rock formations and in odd corners that are inaccessible to the dredge.

John Thoburn has questioned this conventional view of complementarity. Noting that the rate of return on investment criterion favors capital-intensive forms of production at relatively low prices but capital-saving forms at higher prices, he has shown in a 1978 article that under cost and price conditions pre-

Table 8.1. Weighted average unit costs of tin production, January–June 1980, by producing country

	Excluding royalties, export duties, and tributes		Including royalties, export duties, and tributes	
Country	M$/picul	As a percentage of average price	M$/picul	As a percentage of average price
Australia	1,318	60.0	1,366	61.1
Bolivia	1,700	76.1	2,215	99.1
Indonesia	1,279	57.3	1,492	66.8
Malaysia	1,120	50.1	1,865	83.5
Nigeria	897	40.2	1,487	66.6
Thailand	1,086	48.6	1,755	78.6
United Kingdom	1,907	85.4	1,978	88.5
Zaire	1,649	73.8	1,649	73.8

Source: *Metal Bulletin* (May 12, 1981). Reprinted in *Notes on Tin* (June 1981).

vailing in the previous decade dredging was generally the preferable investment alternative in Malaysia, but that at higher price levels there was a greater average return on investment in gravel-pump mining.[1] In his 1981 book, Thoburn observes that there are substantial economies of scale in gravel-pump mining, and that four or five efficient-sized gravel pumps might work a new tract at lower unit costs of extraction than a dredge. He does warn, however, that "further work, which would best be done by specialist mining consultants, would be necessary to establish what would be the costs of a deep multiple gravel pump operation shifting the same yardage as a dredge."[2] An experienced tin mining engineer with whom I discussed this idea was skeptical, pointing out that, among other things, the disposal of tailings from a number of gravel pumps would pose a much greater problem on most sites than would be the case with a dredge that simply fills in the back of its paddock as it works new ground.

About 25 percent of tin output comes from underground mines, primarily those of Bolivia, Australia, South Africa, and Great Britain. The remaining 10 percent or so is accounted for by such small-scale and individual operations as dulang washing in Southeast Asia, tributing in Bolivia and Nigeria, and suction-boat mining off of the coast of Thailand.

Cost structures and mining technologies. As a result of this technological diversity, comparisons of cost structures are difficult to make, and may be misleading if generalizations about broad types of mining are applied to specific countries.

In 1977, the ITC requested member producing nations to provide its newly established Economic and Price Review Panel with cost information based on analysis of model investments or costs for hypothetical new mines under specific assumptions as to size, location, and grade of ground at the mine site as well as choice of the most efficient mining technique. The model investment exercises were subsequently distributed in an ITC internal working paper, "Model Studies of Investment Required for New Mining Ventures," issued January 27, 1978, but with circulation restricted to the ITC and delegations of member nations. Some results of the model investment study have since been released, however, and published both by the ITC[3] and in Thoburn's 1981 book. There is virtually no other published information on the cost structure of tin mining. I have, however, been able to obtain a scattering of data from other sources—from actual accounting records, cost sheets of project evaluation studies, and rough estimates made in the course of interviews—virtually all provided on a confidential basis. Thus, only rough and broad comparisons can be made.

Yet it is important to recognize technical and cost characteristics that may be significant determinants of the industry's conduct and performance. Yip, for example, observed, "Since gravel-pump mines require less expenditure on capital equipment than dredges, they can be more quickly rehabilitated and brought into operation. On the other hand, since they employ a comparatively larger number of persons, the problem of unemployment becomes more serious during export control."[4]

To cite another example, Puey noted what he referred to as the "non-reversibility" of the long-run supply curve for tin. In response to a rise in price, new tracts would be opened up, new infrastructure such as roads and bridges would be installed, and schedules for planned rates of depletion of existing mines would be changed. If the price later fell to the earlier level, production would not revert to its previous rate nor would proportionate disinvestment occur, since the actions that had been taken in the period of higher prices would have altered the conditions determining costs and optimal output. One manifestation of this problem, Puey noted, was that capacity created in a boom period might continue to produce in a following slump, increasing the severity of the drop in price.[5]

In certain respects dredging is far more capital intensive than gravel-pump or open-cast mining. In terms of prices prevailing in the late 1970s, the initial expenditure on capital and equipment for a large modern offshore dredge capable of mining six to seven million-cubic meters of ore per annum, including supporting dredge-to-shore boats and shore facilities but excluding initial costs of survey and exploration, would have ranged from US$20 to US$30 million. A large inland dredge of similar capacity would have cost from $7 to $10 million to build and install, without taking account of land acquisition and development costs. In contrast, the capital cost of a large and well-equipped gravel-pump mine handling in the neighborhood of 500,000 to 600,000 cubic meters of ore per annum would have been in the range of $300,000 to $400,000, again excluding site costs.

Very crudely, initial capital investment in plant and equipment for a tin dredge is twenty to one hundred times as great as for a gravel-pump mine, assuming firms that are large enough to enjoy economies of scale in both instances, although the amount of ground treated by such a dredge is likely to be only ten times as high. Gravel-pump and open-cast mines, being of smaller size and more flexible in their ability to choose the ground worked than are dredges, tend on the average to mine higher grades of ore than do inland dredges. For this reason, comparisons of capital costs per ton of ore handled may slightly overstate the relative capital intensity of inland dredging in terms of capital per unit output of tin-in-concentrates. Offshore dredges, on the other hand, have typically mined higher grade ores than either type of onshore mine.

An inland dredge requires about ten to twenty times as much land for a working tract as does a gravel-pump mine, so the relative costs of land acquisition would be that much greater for a dredge. Costs of providing access to the site, however, need not vary with size of the tract alone, nor need prospecting costs.

In Malaysia, reported employment in 1975 and 1976 averaged roughly 130 persons per inland dredge, and approximately 55 per gravel-pump mine. In Thailand, there were estimated to be approximately 175 employees per dredge and 35 per gravel-pump mine. Despite the diversity of these figures, it seems clear that in terms of the capital/labor ratio, as well as in absolute size of initial outlay, dredging is unquestionably the more capital intensive method of operation.

There is, however, one difference in mining technologies that is crucial to recognize in comparing the capital intensities of dredging and gravel-pump mining. The capital and equipment used in gravel-pump and open-cast mining —such items as pumps, tubing, wooden *palongs*, high-pressure hoses, trucks, bulldozers, and lifts—have much shorter useful lives than do dredges. It is customary in Southeast Asia to assume an average five-year useful life for real capital invested in a gravel-pump mine, and a fifteen-year useful life for an inland dredge.

There are two concepts of capital intensity that must be distinguished in making these cost comparisons. One is the size of the capital stock employed at any one time. Capital intensity in this first sense can be measured either in absolute amount, or in relative terms, such as capital invested per unit of output or per employee. From this balance sheet or static point of view, dredging is a much more capital intensive form of tin mining than gravel-pump or open-cast mining. Absolute size of the capital stock needed is one important factor in determining the number of firms that a given market can support, and may comprise a barrier to entry in instances where not all potential entrants to the industry have equal access to sources of financial capital. One common explanation for the predominance of European capital in the development of tin dredging in Malaysia and Thailand, after the failure of Europeans to establish themselves in the gravel-pump sector, is the superior opportunities that European investors had to incorporate in their home countries as limited liability joint stock companies and to make use of the organized capital markets of Europe in financing their companies. In general, capital stock requirements play an important role in long-run economic analysis that addresses itself to such matters as numbers of firms and conditions of entry to and exit from an industry.

But if the analysis takes the existing number and sizes of firms in an industry as given, and considers how these firms will adjust their output to price changes, a second concept of capital intensity, emphasizing flows rather than stocks, is needed. It becomes necessary to distinguish variable costs such as those for labor and raw materials, that are functions of output, from the fixed costs that are independent of current output, such as long-term interest costs and that part of depreciation which depends on time rather than use. These fixed costs are irrelevant to the short-run production decision in that if a firm can earn enough revenue to cover its variable costs and make some contribution to fixed costs, no matter how small, it should do so rather than shut down while continuing to incur the inescapable fixed costs.

The firm with high fixed relative to variable costs is limited in its short-run production options, continuing for some time to produce at prices well below its full costs. Capital costs—or the flow of costs over time necessary to maintain a capital stock—comprise a large element of fixed costs. In the comparison of dredge and gravel-pump mines, if the capital used in the dredging sector is approximately three times as long-lasting as that in the gravel-pump and open-cast sector, then over time the gravel-pump and open-cast mine owners will have to make capital expenditures equally as large as those of the dredge owners in order to have a stock of capital of one-third as great a value at any one time.

As can be seen in table 8.2, as a result of the differences in the relative sizes of firms and the useful lives of plant and equipment, dredging appears to be somewhat more capital intensive than gravel-pump or open-cast mining in terms of cost flows, or in a rough comparison of fixed and variable costs, but not overwhelmingly so. The data used and assumptions made to obtain the figures in table 8.2 are, however, extremely crude.

The relative costs shown in table 8.2 are reasonably consistent with a calculation reported by Engel and Allen that the capital cost per ton of tin produced over the mine's lifetime is US$1,111 for the hypothetical Malaysian gravel-pump mine in the model investment study, and $1,939 for the study's Indonesian offshore dredge.[6] Table 8.2 is also given support by a paper, presented at the 1981 World Conference on Tin, noting that the energy cost of producing a ton of tin-in-concentrates by gravel-pump mining was twice that of producing a ton by dredging.[7]

The largest underground mine, the Renison mine in Australia, was capitalized at A$50,903,000 (approximately US$58,238,000) in 1978. Pahang Consolidated, Malaysia's only large underground mine, listed total assets exclusive of quoted investments and trade credit at M$13,364,000 (approximately US$6,900,000) in its 1977 statement. In both cases, the figures understate the actual costs of establishing similar mines. Geevor was capitalized at £1,870,994 (US$3,836,000) in 1977. Costs of opening a new underground lode mine of average size in Bolivia have been estimated at around US$12,000,000.

Length of useful life of capital invested in underground mining varies greatly. A newly discovered vein unrelated to an existing mine is commonly considered

Table 8.2. Estimated percentages of total costs, exclusive of royalties and taxes, by types of tin mining[a]

Type of Cost	Gravel-pump and open-cast	Dredge	Underground
Salaries and labor	30%	24%	35%
Power	28	15	5
Materials	18	20	19
Realization costs[b]	5	9	19
Total variable costs	81	68	78
Depreciation	7	9	8
Amortization of exploration and development costs	3	1	5
Other charges	9	22	9
Total fixed costs	19	32	22

a. These percentages are subjectively weighted averages of both reported and estimated costs in a number of countries rather than actual cost structures of specific mines. The categories are too broad to be entirely suitable to the purposes of this table. Materials, for example, may include purchases for maintenance of plant and equipment that would have to be made regardless of whether the mine was in production. Depreciation must be viewed as an accountant's schedule for amortization of the original cost of owned assets rather than the actual loss of value over time. Other charges are not broken down in the original material. The bulk of these charges appear to be long-term carrying charges on plant and equipment financed by borrowing, as in the original data they vary inversely with reported depreciation. In all, the categorization of variable and fixed costs is only approximate. The figures can be taken, however, as indicative of the actual cost structures of the three types of mining.

b. Transportation, concentration, and storage of ores.

worth opening only if it is believed that it can be worked for three or four years. While a mine shaft is certainly a long-term investment, such shafts will be extended in the day-to-day operations of extracting the ore. Write-off rates for underground mining tend to center around ten years. Output of tin-in-concentrates per employee varies widely, from one to seven tons per annum in estimates I have obtained.

Supply responsiveness. The discussion of supply elasticities in chapter 4, while stressing the difficulties of making accurate estimates and of translating observed short-term and long-term responses into the economic concepts of short-run and long-run supply functions, nevertheless led to a tentative conclusion that these elasticities are higher for gravel-pump and open-cast mines than for dredging. If this is actually the case, the short-term phenomenon cannot be attributed solely to differences in capital intensity or to the variable/fixed cost ratio. But there is a marked difference in how rapidly the different types of tin mines can make long-run adjustments in the sense of putting new capital in place. Once the site has been selected, a gravel-pump mine can be constructed and the overburden cleared in less than a year. Construction of a dredge takes three to four years from placing an order to on-site installation. Output, therefore, can respond more rapidly to favorable price or cost stimuli in the gravel-pump sector. It is also more difficult for dredges to exit from the industry, given the longer economic life of the equipment and the physical immobility of inland dredges. Occasionally, dredges are disassembled and put back together again on a new tract, but this is a costly procedure and most dredges spend their entire working lives at one site.

There are heavy costs incurred in closing down all types of tin mines for a short period of time. In the case of a gravel-pump mine, the *palong* begins to rot as it gets dry. Further, many Chinese miners continue to give financial support to their regular work force in periods of idleness. In the case of dredging, the paddock will silt up when the dredge is not in operation, and it may take months just to restore the site to its original condition when work starts up again. Underground mines are subject to flooding if unmaintained. The strong and militant miners' union in Bolivia makes it difficult for COMIBOL to lay off its workers.

P. P. Courtenay made a study of the rates at which mines in Malaysia closed down during the time of the ITC's most severe export restrictions, from December 15, 1957 to September 30, 1960. By February 1958 the number of gravel-pump mines in operation had declined by 23 percent below the December 1957 figure, while the number of operating dredges had fallen by only 7 percent. By June the percentages were roughly equal and remained so until near the end of the year.[8]

Sources of change in cost structure: mechanization, grade of ground, and power. A 1965 undergraduate honors thesis written at the University of Malaya gave the initial capital requirements for a standard-size gravel-pump mine as being on the order of US$50,000 to $75,000, in contrast to the $300,000 to $400,000 cited above as an estimate for the late 1970s.[9] One evident reason for the discrepancy is the change in the general price level. Another is that the 1965 estimate is for a smaller and more labor intensive mine, with no provision for vehicles or mechanized earth-handling equipment. Use of such capital items has become more common in Malaysia in recent years, not only adding to capital intensity but also increasing the optimal scale of operation. But there is a third reason. Because of differences in assumptions as to the grade of ground to be mined—0.18 kilograms of metal per cubic meter of paydirt in the earlier study and only 0.12 in the later one—the mine of 1965 had to handle much less earth to produce a quantity of tin-in-concentrates similar to that of the 1977–78 mine. These sample figures are closely in line with estimates of actual declines in the average grade of ground worked by gravel-pump mines in Malaysia.

The decline in grade of ground is a worldwide phenomenon, according to data collected by the ITC. In a 1977 estimate, Thailand reported that the average grade of ground dredged on inland sites fell from 0.26 kilograms per cubic meter in 1962 to 0.13 in 1977. Comparable figures for gravel-pump mining in that country were from 0.32 to 0.19, and for offshore dredging from 0.52 to 0.32. The weighted average grade of Bolivian ore fell from 1.5 percent tin content in 1951 to 0.72 percent in 1977. The average grade of ground dredged in Malaysia fell from 0.18 kilograms per cubic meter in 1966 to 0.12 in 1975. Estimates of average ore grade in Malaysian gravel pumping before 1972 were not considered reliable by the mining authorities of that country, but in 1977 it was 0.13 kilograms per cubic meter.[10]

Table 8.2 indicates that power is a large component of expense in both gravel-pump and dredge mining. Tin mining costs are therefore strongly affected by changes in the prices of petroleum products and electricity. The latter has been a source of great concern to Malaysian miners, particularly gravel-pump mine operators, in recent years. The Malaysian government's National Electricity Board raised rates for nondomestic consumption of electricity by 60 percent in 1975, by another 20 percent in 1978, and again by 30 percent in 1979. Then, in 1980, the rate schedule was revised so that the unit charge now rises rather than falls with usage. This change in rate structure raised the cost of electric power to Malaysian tin miners by approximately 50 percent.[11]

Summary: cost characteristics and performance. To summarize, fixed and short-run shutdown costs are high enough in all major types of tin mining so that output may remain quite steady in the face of temporary price declines causing severe losses to most miners. Despite differences in capital/labor and capital/output ratios, there do not appear to be marked differences among types of mines in this respect. If short-run price declines or export controls are severe enough to lead to temporary shutdowns, start-up costs will be high, particularly for dredge and underground mines.

If price increases or decreases are sustained for long periods of time, so that miners perceive a change in the long-run average or trend price great enough to either attract entrants and new investment by existing firms or to cause some firms to leave the industry and others to withdraw part of their capital, the gravel-pump and open-cast sector will be able to expand or contract output more promptly than the dredging sector. Since dredge operators are likely to have to bear losses for a longer period of time before being able to disinvest and to suffer greater losses from temporary shutdowns, they should benefit more from the ITC's defense of floor prices than gravel-pump miners, even though the magnitudes of the losses relative to sales revenue would be about the same for operating mines of both types.

Tin mining costs are rising not only because of worldwide inflation, but also because of the declining grade of ground in all of the major producing countries and rising costs of fuel and electricity.

Tin Markets and Pricing

Several of the most basic structural characteristics of the tin industry—a large number of independent mines no one of which could hope to manipulate the price of tin by varying its output, low transportation costs for concentrates as they leave the mine relative to their value, ease of entry and an absence of major economies of scale in smelting, and an essentially homogeneous product sold on a worldwide market with active trading on an organized commodity exchange—are in combination conducive to more highly competitive marketing and pricing than would be possible in any other major industrial metal. Until recently, another aspect of structure—the absence of vertical integration between mines and smelters—would have been added to the list of features favorable to competitive market behavior. But as we have seen, even as tin smelting became less concentrated worldwide over the past half century, an ever greater percentage of smelting capacity came to be located in the producing nations; and within each of these countries the formal or informal ties between the domestic mines and smelters grew closer. Today, only a small fraction of the tin concentrates produced moves in international trade.

The crucial underlying structural element behind this development is the small number of countries producing the overwhelming majority of the world's tin ore, coupled with the political fact of the national independence of all of them in recent decades. Not only have the tin producing nations achieved predominance in smelting, but also they have bolstered the domestic positions of their smelters by limiting or banning the export of tin ore. In addition, the governments of these countries have become important factors in tin mining, either supplanting or regulating private enterprise. These features of the industry, which must be recognized as limiting the scope of competitive economic behavior no matter what their other positive aspects may have been within the various countries, constitute the principal qualification to the assertion that the fundamental structure of the tin industry is competitive.

Current markets and world pricing. Current marketing channels and practices have evolved to adapt to this structural framework. There are three recognized markets for tin, two of which, Penang and the LME, can be described as organized exchange markets allowing for interactions among large numbers of buyers and sellers. The third, New York, is an informal center for individual transactions in tin metal.

The Penang and LME markets, particularly the former, exert pervasive influences on tin trading and distribution throughout the world, including transactions in the large amounts of concentrates and tin metal that bypass open markets altogether. While Thai law forbids the export of tin concentrates, for example, the miners of that country have been protected from potential exploitation by the monopsonistic smelter through a requirement that Thaisarco pay the going Penang price for its tin-in-concentrates. Associated Tin Smelters in Australia quotes prices based on those in Penang less an adjustment for transportation costs. The price received by Bolivian private miners from ENAF is geared to the Penang price. The Indonesian State Tin Enterprise, vertically integrated from mining through smelting to a worldwide sales agency, sells approximately 70 percent of its tin metal on negotiated long-term contracts with settlement prices based on either Penang or LME prices, usually depending on the location of the buyer.[12] Since 1972, the ITC's floor and ceiling prices have been expressed in terms of the Penang price, and the standard before that was the LME price.

The single most important market in establishing freely determined competitive prices for both tin-in-concentrates and physical tin metal is the Penang market. On the face of the matter, it is remarkable that a truly competitive market could ever have developed under the structural circumstances existing in Malaysia, with a ban on the export of ore and only two smelters to which the nation's miners could bring their concentrates.

The nature of trading on the Penang market. The Penang market is a one-way auction market, with the smelters acting in concert as a clearing agent. Miners throughout the country are required early each morning to inform the smelters or their various collecting branches and agencies how much ore they wish to deliver that day. At the time the miners so advise the smelters, they do not know what price they will receive. By 10:00 A.M. the quantities being offered around the country are transmitted to the head offices of the smelters and totaled. Buyers from around the world and Penang tin dealers are required to submit bids by letter, telex, or in other written form by 10:30, specifying quantities of tin bid for and a specific maximum offer price. Alternatively, the offer price may be "at market," meaning that the bidder will pay whatever the price turns out to be on that day. Buyers are allowed to bid only in units of five metric tons, with a maximum bid of twenty-five metric tons, although a single buyer can submit a number of bids at either identical or different prices.

After receiving the day's bids, the smelters communicate by telephone, reporting to each other both the quantities of tin-in-concentrates being offered and the bids received. All of the bids are ranked, with the "at market" bids listed first and the others following in descending order of offer price. Bids are filled in order of this ranking until all of the tin offered that day has been taken up. All successful bidders pay the same price, that quoted by the lowest bidder whose offer clears the market. The sales contracts issued to the buyers call for delivery of the tin metal within sixty days of the sale, but require the buyer to pay in full on the day he is notified that his tin is deliverable. The smelted metal is normally delivered within two to four weeks of the date of purchase.[13]

Under this system, total bids and offers are matched, but they need not match for each of the two smelters. Unless bidders state a preference for one smelter, bids may be transferred from the smelter with an excess of bids over offers to the other. The smelters are willing to cooperate in this manner because under the Penang market system their revenue is earned from the charges they levy for smelting the tin delivered to them, rather than from sales of tin metal. Miners are paid the day's price for tin, less the smelting charge, based on an assay of the tin content and analysis of impurities in the ore delivered.

The process, as described, would appear to be automatic, with the role of the smelters being merely the clerical function of counting bids and offers and ascertaining the market clearing price. In fact, there are elements of discretion. There may, for example, be twenty-five or thirty metric tons bid for at an identical price, whereas only seven or eight metric tons bid at that price would clear the market. In such an instance, there may be rationing, or an allocation partially to fill the requests of the low bidders. On occasion, "at market" bids exceed the quantity offered. In such cases the senior smelter official may decide to set the price at the highest specific bid and ration the tin offered among that bidder and the "at market" bidders, or he may set a price slightly higher than the previous market price and supply only the "at market" bidders. Occasionally, not enough tin is bid for to absorb the day's offerings at any price, or perhaps only at a price that the smelter officials regard as excessively low. In such an event, according to an advisory statement issued by the Straits Trading Company: "The Smelters have discretion to withhold or to limit sales if a realistic market cannot be made. In practice exercise of this discretion has seldom been necessary."[14] This is referred to as placing limits on miners.

The operation of such a system, in which a price with worldwide ramifications is set each day in private by the pronouncement of a single smelting official (the current practice being for the senior of the two representatives conferring by telephone to make the final determination), and in which the details of the prices offered and the buyers involved are held confidential by the smelters, has inevitably given rise to questions as to the integrity of the process. The Penang market works mainly because the smelters, who run the system, have to a large extent removed themselves from an interest in the level of the price of tin. Rather, their revenue is maximized by generating the greatest possible volume of smelting business that can be obtained at a given smelting charge; and this, in turn, requires an open market clearing system. The smelters realize that the survival of the Penang market depends on the confidence of the participants that it is being run fairly and impartially.

To this end, not only is the daily price-setting done with scrupulous care, but the smelters also seek to limit manipulation of the market by traders. One manipulative practice a dealer might employ would be to put in a reasonable bid for his actual requirements of tin and a second nominal bid at a very low price, hoping that the volume of bids on a particular day would be so small that his low bid would set the price. To counter this tactic, the smelters reserve the right to reject unreasonably low bids.[15] Smelter officials are also concerned that a trader might accumulate a large stock of tin, say two to three hundred tons, over time and then put in a very high bid for twenty-five tons on a particular day, hoping thereby to raise the Penang price at a time when he could dispose of his large stock to customers at a price he had helped inflate, or in the expectation that the LME price would rise in response to the Penang price and he could sell his accumulated tin at a profit on that market. The smelters, therefore, also retain the right to reject bids at prices that they consider to be excessively high.

The origin and sources of survival of the Penang market. The question most in need of explanation is not whether the Penang market is a fair and workable auction market: virtually every person connected with the tin trade with whom I talked agreed that it is managed honestly and efficiently by the smelters and that manipulation, while it exists, is only a minor problem. A more fundamental question is why such a system ever arose in the first place, or why the Malaysian smelters have chosen not to exercise the market power inherent in their duopolistic position vis-à-vis the competitively structured mining sector. The supply and demand analysis in chapter 4 led to the conclusion that profit-maximizing smelters in such an asymmetric market should either buy tin at monopsonistic prices and sell it on their own account or integrate backwards into mining.

The answer does not lie in mutually harmful competition between the smelters. Stahl noted that at one time "keen competition" did exist but that it was ended in 1940 "by a mutual agreement to adjust smelting charges."[16] Since that time the two firms have quoted identical schedules of charges for smelting, and whatever competition exists between them takes the form of competition for concentrates through provision of credit to supplying miners, ease of access to local buying agencies, and service to small ore dealers who collect ore from dulang washers and very small mines.[17]

There are a number of plausible explanations for the smelters' voluntary establishment of and desire to continue the Penang market. First, and perhaps the most convincing, is that the system shifts nearly all of the risks stemming from price fluctuations back to the miners and forward to the dealers. Not only are both miners and buyers uncertain as to the price that will prevail when they submit their offers and bids, but also the miners are obligated by contract to deliver the ore pledged regardless of the price later announced, while the buyers are committed to payment of the previously agreed-upon price at the time the tin is ready for delivery no matter how much the market price of tin may have risen or fallen in the interim. The smelters hold large stocks of tin concentrates as work in process, carried at far more than the value of plant and equipment on the recent balance sheets of both smelters; but they are completely insulated against any inventory losses or gains resulting from price changes. Sir Ewen Fergusson, former chairman and managing director of the Straits Trading Company, commented on the price risk imposed on the miners, and dismissed the problem by explaining rather airily that they "are encouraged to sell regularly with the object of achieving the average tin price over a period." Such behavior, he continued, benefited all concerned because "as a result of this orderliness an evenness is achieved which avoids large daily fluctuations in the quantity to be sold and, consequently, helps to avoid large daily fluctuations in price."[18]

It should be noted in passing that the smelters do enjoy a gain from a rise in the price of tin and suffer a loss from a decline, since a substantial portion of the overall smelting charge is the unitage deduction set at roughly 1 percent of the current price of tin for high-grade concentrates, and nominally levied to compensate the smelter for metal lost in the furnace in the process of smelting. In actuality, according to a number of my informants, the loss in furnace is far lower than this figure, perhaps around 0.4 percent. But this is the one, and minor, qualification that must be made to the assertion that through establishment of the Penang market the smelters have removed themselves from exposure to the price risks they would have to assume if they bought and sold tin for their own accounts. It has long been recognized that risk-averse businessmen are behaving rationally in forgoing some obtainable profits in order to increase the degree of certainty as to future costs and revenues.

A second reason for the smelters' continued support of the Penang market system is that they are financially involved with the miners as their principal suppliers of credit. Both smelters will pay a miner up to 80 or 90 percent of the current Penang price less smelting charges on delivery of tin concentrates to the collecting agency, with the remainder being paid after the ore is assayed and the price on the day of sale is known. The smelter, in turn, is not paid until the buyer is notified that the metal is deliverable. "Whether the miners realize it or not," one smelter official pointed out, "the cost of these cash advances is built into the smelting charge." The smelters are also the major source of credit to established miners of good reputation for loans to open new mines and acquire new equipment, with the mine frequently being tied to the smelter by an agreement that the loan will be repaid out of the proceeds from sales of tin-in-concentrates subsequently brought to that smelter. Although these loans comprise one of the major competitive devices used by the smelters to attract and hold suppliers of ore, interest is charged and represents a substantial portion of the smelters' revenue. The cash advances and outright loans, as well as the heavy capital investment tied up in work in process, involve the Malaysian smelters as deeply in the business of financing the tin trade of that country as in smelting its ores. These financial ties both make it possible for the smelters to obtain some of the benefits of vertical integration under the present system and lessen the net advantages to them of monopsonistic restrictions on the level of tin mining activity.

Third, it may be unrealistic to assume that the two Malaysian smelters have a high degree of monopsonistic power in the long run even though they do not compete in setting smelting charges. The source of such power is the ban on the export of tin concentrates that assures that all tin ore mined in the country will be delivered to one of the two Malaysian smelters—with the quite serious exception of smuggled concentrates. Yet this situation may not give the two smelters the power to raise the long-run level of smelting charges to monopolistic levels, much less depress the price received by miners through collusive buying power, both because the Malaysian government could easily counter such practices by eliminating or reducing the control on export of concentrates and because there is no protection against entry of other smelters if the two existing ones exploit their potential short-run position. Whatever independent monopsonistic power may have existed in the past has been further diluted by the establishment of MMC and its partial integration with Straits Trading Company.

With only a modicum of cynicism, it must be pointed out that consumers—in this instance the industrial users of tin in the developed countries—benefit at least as much from competitive pricing and levels of output as do producers from the absence of monopsony. It is, therefore, pertinent to note that the tinplate industry in Great Britain was the world's largest prior to the First World War and second only to that of the United States until surpassed by Japan in the early 1970s. Because of the existence of this industry in the home country, the Penang market and the associated restraint on exercise of the smelters' market power were clearly in the interests of the empire as a whole; and the smelters' adoption of this system must have exerted a positive influence on the colonial administrators and the legislature of the Federated Malay States in their willingness to protect the interests of the local smelters by curbing the export of tin ore. Whether the result was intended or not, and I have no evidence whatsoever that it was, the Penang market was close to an ideal device for exploitation of a colonial resource to the benefit of the mother country, especially during the period when the majority of tin mining in Malaya was carried out by dredges owned by British investors.

Finally, the Penang market grew out of the particular institutional history of the Malaysian tin smelting industry. I asked several people, with both current and past associations with one or the other of the smelters, why the Penang system had been established and why the smelters did not instead simply buy tin ore and sell tin metal. The most frequent answer was that the risk of buying and reselling would be great, and since high and quite steady profits can be earned under the Penang market this risk was not worth taking. Another related answer was that the smelters do not have the marketing skills and sophistication necessary to engage in commodity trading, nor any interest in acquiring such a capability. These responses are consistent with ownership by risk-averse investors, but also with a degree of "managerialism," the ability of the managers to earn a high enough rate of return to satisfy the owners while at the same time preserving for themselves such benefits as a less stressful set of duties than would be required to run the company entirely in the stockholders' best financial interests, or the discretion to pursue objectives other than profit maximization. Another common answer, however, was that the Penang market was set up in response to an episode in the early 1920s that nearly bankrupted the Straits Trading Company.

As recounted by Tregonning, before the appointment of W. F. Nutt as managing director in 1918, Straits Trading Company had made daily offers to purchase tin ore based on that day's selling price of the metal and had tried each day or as soon as possible thereafter to sell an amount equal to that purchased, thus minimizing its price risk. In 1919, Nutt held metal off the world market in anticipation of a price rise and succeeded through this maneuver in making a large profit for the company. In so doing, however, he incurred the displeasure of at least two directors who warned him never to repeat such speculation. Nevertheless, in the spring and summer of 1921, with tin prices falling, Nutt withheld so large a volume of tin that it was necessary to pass over a semiannual dividend. This was the first time in the history of the Straits Trading Company that a dividend had been missed. When his action became known to the directors, Nutt was asked to resign. He appealed to the general meeting of the company in January 1922, but failed to obtain support for his position from the shareholders, who thereby indicated their concurrence with the risk-averse attitudes of the directors.

In the meantime, large stocks of tin had also been accumulated by the governments of the Federated Malay States and the Netherlands Indies, and the depressing effect on price of the overhang of all of these stocks had been an important contributing factor in the creation of the Bandoeng Pool. The shrinkage in the cash flow resulting from Nutt's accumulation of tin, heightened by the subsequent inability to dispose of this stock promptly at a profit in the face of the other stocks, placed the Straits Trading Company in a critical financial situation that was eased only gradually by the operation of the Bandoeng Pool. The tight cash situation was particularly painful to the shareholders as the Straits Trading Company had increased its capital by M$9,000,000 in December 1920, in order to finance a major postwar expansion of facilities. In 1923, Nutt's successor ended the company's practice of buying tin ore throughout the day at a posted price and inaugurated, in Singapore, the system still in use and now known as the Penang market.[19]

In the opinion of most of those with whom I discussed the matter, this near-disaster is so seared into the institutional memory of the Straits Trading Company that it will never voluntarily abandon the Penang market system in favor of holding tin ore and metal at its own risk.

It is doubtful that any commercial practice such as the Penang market can survive for long unless it proves to be consistent with the actual economic interests of the firms establishing it, as these interests are conditioned by the objective structural conditions of the industry; and therefore the incident just recounted cannot be regarded as a sufficient explanation for the continued existence of the Penang market. On the other hand, an examination of basic structural factors that neglects the particular historical circumstances under which a given pricing mechanism was created is incomplete and in many cases, such as this, is incapable of explaining the form taken by the system.

The future of the Penang market and the Kuala Lumpur Commodity Exchange. While there are a number of plausible and convincing reasons why it has been in the best interests of the Malaysian smelters to conduct the Penang market as they have done, and while this market did serve broad governmental purposes as long as the Federated Malay States were viewed as part of a colonial system, it does not follow that perpetuation of the Penang market is in the best interests of the independent nation of Malaysia. Certainly,

Malaysia has no interest in providing its large industrial customers from the developed countries with a market that gives them competitive production levels and prices for the tin they buy, other than the fear of losing these customers to other suppliers. The producers' surplus that the Penang market preserves for the predominantly Chinese-Malaysian miners may not be of great significance to the current government in light of the restructuring of the economy and the priority being given to enlarging economic opportunities for the *bumiputras* under the NEP. Indeed, this surplus may not be large or visible enough for the miners themselves to desire the continuation of the Penang market. The risk imposed on miners through the requirement that they commit themselves to delivery of a definite quantity of tin-in-concentrates, when they do not know the price they will receive for it, is a source of widespread discontent, according to several miners with whom I talked about the matter. (The requirement is somewhat less onerous for the larger miners, who are permitted to deliver ore to a smelter or collecting agency but designate in advance any day up to four weeks after delivery as the day of sale.) In any event, the continued survival of the Penang market is in doubt.

The August 1, 1978 decision of the Malaysia Mining Corporation (MMC) to withdraw from the Penang market was one blow. KPS (the Selangor state tin organization) has since made a similar decision to have its ore smelted on toll and to market the metal for its own account through an agent. Since the withdrawal of MMC, concern has been expressed as to the impact of the decline in the volume of tin traded on the day-to-day stability of the Penang market price. A thin market and the volatility of price that could result might both make the Penang market much less useful to the miners and dealers who use it and might undermine the validity of the Penang price as a benchmark for other transactions.

The performance of prices on both the Penang market and the LME in recent years provides no support whatsoever for this concern. To the contrary, day-to-day price stability on the Penang market increased after MMC's withdrawal, both absolutely and relative to the stability of the LME price. Over the nineteen months from January 1, 1977 through July 31, 1978, while the mines in the MMC group continued to use the Penang market, the monthly coefficient of day-to-day variation of the Penang price averaged 0.0222. The figure for the LME over the same period, 0.0210, was slightly lower but quite close. From August 1, 1978 through December 31, 1979, the average monthly coefficient of variation of prices on the Penang market was down substantially from the previous period, to 0.0148. Over this later seventeen-month period forces other than the absence of MMC's concentrates and their buyers acted as major influences on the stability of tin prices, as evidenced by a like drop in the coefficient of variation for the LME, which declined to 0.0158. Even were it possible to isolate the effects of MMC's action from other market influences on the variability of tin prices, which is most doubtful, it is not necessary to do so here, since the day-to-day Penang price shifted from being slightly less stable than that on the LME to being somewhat more so. Unless there were additional but unidentified phenomena that had a significant impact on one market but not the other, the conclusion is inescapable that the withdrawal of MMC's business did not destabilize the Penang market.

Far greater strain will be put on the viability of the Penang market if and when trading begins in tin on the Kuala Lumpur Commodity Exchange (KLCE). The Exchange opened on October 23, 1980, trading only in palm oil futures. As described in the bill presented to the Malaysian Parliament that spring, the KLCE will add rubber, tin, pepper, cocoa, and timber as experience is gained and the Exchange grows in volume of business and reputation. Each commodity added will require a separate act of Parliament. It was not specified when trade in tin might begin.

The proposal to establish a Kuala Lumpur commodities exchange did not arise out of any expressed dissatisfaction with the Penang market. The chief impetus for the idea came from what one Malaysian government official described as traumatic experiences with fluctuations in rubber prices coupled with the belief that organized exchange trading in rubber in Kuala Lumpur would stabilize that price. Further, the exchange was viewed as an instrument of economic development in its own right, contributing to the growth of Kuala Lumpur as an international financial center. Trading in tin and other commodities, excepting rubber, was to be introduced not so much in the interests of those commodities as to promote the visibility and use of the exchange itself and to obtain economies of scale in supporting facilities such as international communications and financial services.

A few days before the opening of the KLCE, States of Malaya Chamber of Mines President Rahim Aki, who is also the chief executive of MMC, stated that futures trading in tin on that exchange would mean the end of the Penang market. This statement was in contradiction to an earlier announcement by Minister of Primary Industries Leong that his Ministry had made a study of the feasibility of trading tin futures on the KLCE and had concluded that if such trading were established the Penang market would remain in operation.[20]

Rahim's view was in accord with the findings of a report submitted in 1977 by Commodities Research Unit, Ltd. (CRU), which had evaluated the proposed exchange as a consultant to the Ministry of Primary Industries. CRU, although noting that the Penang market "performs a particular function outstandingly well, and fits into the special conditions of the Malaysian tin industry in a way which would be difficult to improve,"[21] had concluded that "it is difficult to see how the existing daily market can continue to exist in its present form."[22] CRU noted that the smelters could continue to advance cash to miners, but identified two major problems. First, with the Exchange in operation, some miners would prefer to keep title to their concentrates and smelt on toll charges, limiting their risk by hedging in tin futures. Second, current prices would be influenced by near futures prices, and it would therefore be far more difficult to establish a current price by matching daily bids and offers for physical tin-in-concentrates and metal. CRU concluded that the smelters would not want to operate such a "half-sized" market.[23]

By July 1981, Rahim had changed his mind, or at least was cited as saying that he "saw no reason why the current Penang physical market could not exist side by side with a futures contract, with futures providing hedging possibilities that currently do not exist locally."[24] A similar conclusion was reached in a paper by Ismail Ahmad and J. H. Wilson which was presented at the Fifth World Conference on Tin at Kuala Lumpur in October 1981.[25] Since Ahmad holds the post of Commodities Trading Commissioner, the conclusion suggests a significant change in official attitudes.

Most of those with whom I discussed this proposed commodities exchange are not so optimistic about the prospects for survival of the Penang market. Dealers and traders might well shift the majority of their business to the Exchange. While the competitive market clearing process now established at Penang is appreciated, traders have long noted certain shortcomings. Some would prefer to be able to buy metal at a known price at any time during the day. Under the Penang market system, a buyer either submits a bid "at market" and assumes the risk of paying an unknown price, or quotes a price and runs the risk of not being awarded any tin at that price. The one-way nature of the market, in which the smelters are the sole sellers of tin and other traders can only buy the metal, is regarded as a limitation. Dealers have indicated that their risk exposure in Penang, on a market in which they have contracted to pay a definite price for metal to be delivered at some uncertain future date, is often an unsatisfactory one to them and requires hedging. There is no forward trading or any other facility for hedging on the Penang market, and it is awkward to hedge a purchase on that market with an offsetting transaction on the LME because of both the time differential and the imperfect correlation between prices on the two markets.[26]

Trading in tin futures on the KLCE would meet these objections. The consensus of my informants, including smelter officials, is that as a result of these advantages to buyers as well as the desire of many miners for known prices at the time they offer their ore for sale, enough trade in physical tin would be diverted to a well-run and active commodities exchange in Kuala Lumpur to make the Penang market too small to function effectively or to sustain the smelters' interest in conducting it.

If the Penang market were displaced by tin trading on the KLCE, the Malaysian smelters would face two choices. They could buy tin concentrates and sell tin metal, or they could smelt on a toll basis. My discussions with officials of both of the existing smelters made it perfectly clear that their preference for risk avoidance is such that the latter option is the only one they would consider acceptable.

As toll smelters, Straits Trading Company and Datuk Keramat would not continue to bear the costs of carrying work in process. If only metal and not ore is to be sold on the Kuala Lumpur Commodity Exchange, the miners will have to assume this financial burden, as well as forgo the cash advances they now receive, unless some new institutional arrangement evolves. Further, it is doubtful that the smelters will continue to have as great an interest in supplying loans to miners. The large miners, both dredge and gravel-pump, do not appear to be unduly worried about this eventuality, but there is widespread agreement that many small mines do not have the capital resources to survive without the financing now being provided by the smelters. The problem in and of itself does not appear to be an overwhelming one in the long run. Miners should benefit from a fall in smelting charges as the smelters disengage from financing. New local dealers can be expected to emerge to profit from new demands, engaging in the purchase of ore from small miners who cannot or do not wish to wait for payment until after smelting. However, given the past reluctance of private banks in Malaysia to finance tin mining, some public intervention such as guarantees of loans or provision of credit through the national bank will probably also come about, increasing still further the role of the Malaysian government in the tin trade.

A more serious concern is with the ability of the Kuala Lumpur Commodity Exchange to provide a genuine competitive market for tin metal. The proponents of such an exchange have expressed the hope that it would attract tin that is now marketed privately from both Thailand and Indonesia. MMC, as a government body, would probably have to give up its present practice of marketing its own tin directly through a subsidiary in order to support the exchange. It is not clear how much speculative trading would be attracted, either new or from the LME.

An equally crucial matter is whether private traders on both sides of the market will have as much confidence in the neutrality and integrity of MMC as they have reposed in the past in the two private smelters. As one expressed it, "MMC will be able to play the K. L. tin market like a piano," either in its own financial interests or in the economic and social interests of the Malaysian government, if it so desires. There are ominous signs that just such interference with, and even manipulation of, the KLCE tin futures market may be contemplated. Rahim, in his initial statement that the Penang market could not survive, said that the method of setting the Penang price should be altered so as to take account of producers' interests. According to *Comtel Reuter*, "He mentioned the possibility of a pricing panel whose responsibility would be to strike a balanced price with physical availability satisfying all relevant bids."[27] *Business Times* of Malaysia, in a November 1980 article on the new exchange, commented that "the Malaysian government—mainly through the Primary Industries Ministry—would in fact be pioneering the drive by the developing countries generally in reorienting the traditional pattern of international trade in commodities in favour of the producers rather than the consumers as at present."[28] In their paper presented at the 1980 Kuala Lumpur Tin Conference, Ahmad and Wilson noted that "the system of the Penang tin market does not provide the sellers of the concentrates, who are the miners, with much scope for influencing the prices they are paid. . . . According to modern standards of marketing of the world's major commodities it is unusual that so small a role should be played by the producer."[29]

Whether or not either Malaysian government agencies or MMC was behind the mystery buyer's 1981–82 manipulation of tin trading on the LME, the widespread suspicion of such involvement can only heighten doubts among traders as to the manner in which the marketing of both physical tin and tin futures in Malaysia will be carried out if the Penang market is eliminated or changed in any fundamental way.

World market linkages. The Penang market, as also would be the case for any successor, is pervasively but imperfectly linked to the LME and to trading in New York. Most of the world's tin is acquired from producers and resold to industrial users through a network of intermediary dealers and brokers. Virtually all of these traders operate on both the Penang market and the LME, and can readily shift their activities from one to the other whenever it is in their own interest or that of their clients to do so.

Prices can and do differ on the tin markets, however, and on occasion move in opposite directions. Arbitrage is constrained not only by the one-way nature of the Penang market, which makes it impossible to bring down a high price there by the sale of lower-priced tin from elsewhere, but also by the time and costs of transferring metal from Penang to London or New York. Transportation charges and carrying costs such as insurance and forgone interest, that are incurred during the two months or so that it usually takes to ship tin from Penang to Europe, or the seventy-five days required on the average for shipments to North America, amounted to approximately £220 per metric ton in 1979. Divergences between Penang and London prices have been heightened in recent years by the growth of Japan as a major tin consumer, since locational factors dictate that purchases for shipment to Japan are made solely on the Penang market. A speculator who needs to buy tin within the next few days to cover an earlier forward sale on the LME cannot buy the needed tin on the Penang market merely because it is cheaper there. Under extreme conditions, however, such as those which developed on the LME in June 1979, when a severe shortage of metal deliverable under contracts coming due forced a backwardation of up to £1,000, tin metal can be and is shipped from Penang by air freight.[30]

Tin traders, acting both as dealers on their own account and as commission brokers, operate on a basis of confidentiality; and little information is available regarding the details of market structure and conduct at this level. In 1972, ITC Buffer Stock Manager R. T. Adnan expressed alarm over a decline in the number of dealers and brokers bidding for tin on the Penang market, and a concurrent increase in the purchasing power of recent entrants to the market. Older dealers, "serving the consumer and the producer alike in making the daily price," were giving way to "other more powerful and bigger concerns with more influential bids." In his reference to "influential bids," Adnan was presumably alluding to the power to manipulate the market.[31]

There is undoubtedly some ground for Adnan's concern. At the time he made his observation Thaisarco was marketing all of Thailand's output of tin metal through Philipp Brothers, a subsidiary of Engelhard Minerals and Chemicals Corporation. Engelhard, described by the *New York Times* as "a sizable but secretive company," ranked thirteenth in total revenues for all United States' nonfinancial corporations in 1979.[32] Several years later, when it decided to bypass the Penang market, MMC obtained the services of Philipp Brothers as its marketing agent. Thaisarco has since shifted from Philipp Brothers to its own affiliates, the Billiton trading companies, to market its metal; and MMC now has its own marketing subsidiary. All of Indonesia's tin is marketed through a state trading arm of P. T. Timah. COMIBOL has a marketing office in New York. Datuk Keramat holds a seat on the LME through its affiliation with Amalgamated Metal Corporation. In explaining his firm's acquisition of Amalgamated Metal, the chairman of Preussag stated that one of the principal motivations for the acquisition was Preussag's desire to obtain access to the LME through one of its newly acquired subsidiaries Amalgamated Metal Trading, which was a ring-dealing member of the LME. Data are not available to show the volume of trading or market shares of these and other leading tin trading organizations; but the general structure of the trade lends little support to Adnan's fear of undue manipulation of price by powerful buyers. Despite the existence of a number of large organizations, both private and public, with at least some ability to influence tin prices in the short run, available evidence strongly supports the contention that this sector of the world tin market is workably competitive.[33]

Approximately one hundred buyers are registered with each of the two Malaysian smelters, undoubtedly with a good deal of duplication. On a typical trading day, I was told by officials of both, there would be in the neighborhood of twenty bidders (although probably a larger number of bids since multiple bids are allowed). I was able to corroborate this estimate to some extent, as on one day during which I was permitted to watch the operation of the Penang Market from the office of one of the smelters, there were seventeen identifiably separate bidders.

Officials of Datuk Keramat and Straits Trading Company realize that there are individual attempts by traders, from time to time, to influence the day's price of tin on the Penang market, but they do not believe that there is any collusion among dealers for this purpose. A survey of recent buyers' directories and lists of brokers and agents in *Tin International*, American Metal Market's *Metal Statistics*, and *Engineering and Mining Journal*, after eliminating obvious duplications such as subsidiaries of Billiton and Amalgamated Metal, indicated fifty-five apparently independent firms trading in tin either in London or in New York, and often in both. No one of these firms has the power to dominate the trade. Herkstroeter noted that most tin traders deal in several metals, and are owned by or have ties to large international firms that not only provide financial support but also, often, supply their affiliates with one or more of the metals in which they trade.[34]

There is a powerful disincentive to the rise of collusion in the tin trade. The industrial buyers include a number of the world's largest corporations. United States Steel Corporation alone accounted for 18.5 percent of the world's 1980 output of tinplate, and for 42.2 percent of the tinplate production of the United States. In that same year, the world's four largest tinplate producers, two United States firms and one each from Great Britain and Japan, supplied 49.5 percent of the world's output.[35] The traders in tin metal, therefore, are frequently dealing with powerful and sophisticated customers, capable not only of switching their business but of integrating backwards and purchasing for themselves on the Penang market and the LME if not satisfied with the overall functioning of the existing dealer mechanism. Mitsubishi has already established a tin trading subsidiary with offices in Penang and New York that not only purchases the metal for its associated companies but also acts as a general broker-dealer.

To date, however, almost all of the tin-using firms have preferred the convenience of being able to buy the precise grade and quantity of tin desired, in their own currencies, to be delivered at the time needed, relying on the dealers to provide the specialized knowledge of the tin market necessary to buy skillfully on the Penang market and the LME and to hedge or to carry both price and exchange rate risks while managing an inventory of the metal from smelting to final sale. The dealers have also developed expertise in the details of booking shipping and arranging for warehousing of tin metal in the distributive pipeline. One explanation for this lack of vertical integration is that since tin is a commodity with such a high unit value and low use volume, most customers, even the large ones, buy only a few tons at a time, at infrequent intervals. Thus, there are economies of scale in having specialized dealers handling purchases for a large number of buyers, with these buyers paying brokerage fees only when they make purchases, rather than having each buyer maintain underused expert agents and office facilities of its own at continuing expense. Herkstroeter pointed out that "another consideration is that the margin on the tin trade is so small . . . that even these large consumers have found it more convenient and cheaper to buy from the trade rather than . . . having to take all the necessary steps themselves with regard to shipping, insurance, currencies, etc."[36]

The New York market is not an organized exchange facility, but takes its name from the fact that a large number of tin dealers maintain offices in that city. The so-called New York price quoted in the financial press is obtained by *American Metal Market*, whose tin editor telephones the major dealers each day following the second round of tin trading on the LME (4:15 P.M. in London, which is 11:15 A.M. in New York) and asks each the lowest price it is quoting for twenty-five tons of Straits tin forward and for fifteen tons spot. These quotations, not identified by source, are disseminated to all of the New York dealers. Any dealer can challenge a price by calling *American Metal Market* back within ten minutes and requesting the dealer quoting the lowest price to deliver a lot at that price. Following the challenge period the price is announced to the public. A similar procedure is followed in the late afternoon. *American Metal Market*, in describing this system, warned, "It is recognized that these prices represent offering levels only, and offering levels at which dealers are willing to sell to other dealers. They are not necessarily the same levels at which consumers would get metal, as dealers have different prices for consumers, often discounting METAL MARKET prices by ¼¢–½¢, and sometimes more."[37]

An overview of market performance. Viewed as a whole, the worldwide marketing of tin comprises an efficient and highly competitive pricing system and distributive network. An effective and smoothly functioning market may well have been a critical factor in the sustained support given to the postwar international tin agreements by producing and consuming nations alike. At least the performance of the world market has been such that it has never become a source of serious contention within the ITC; and consequently the council has not been a forum for demands for public action, either national or international, to eliminate or restructure elements of the trade that one country or another found objectionable. In contrast, commodity agreements in other metals, such as copper and aluminum, have been torn by controversy over such matters of market structure and conduct as the alleged monopolistic or monopsonistic power of multinational corporations, producer pricing systems, and the internal transfer pricing of integrated transnational producers. In general, issues of reform of market structures and practices or substitution of nonmarket channels of distribution pit the interests of primary producers against those of consumers and are thus highly divisive.

The world tin marketing system, as effective as it is, is vulnerable in that it is based almost entirely on that of the largest producer, Malaysia. It will continue to yield satisfactory results to both producers and consumers only so long as the price set on the Penang market, or on the KLCE, is regarded as acceptable not only within Malaysia but by those around the world who use it as a reference price for nonmarket transactions in both concentrates and metal. There appear to be two major threats to this system. First, competitive pricing with its attendant benefits to foreign customers may not be regarded as in the national interest by the Malaysian government which is in an increasingly strong position to modify domestic pricing as its direct involvement in the tin industry grows. Second, recent issues within the ITC, notably those connected with conflicts between the United States' and Bolivian delegations, have given new emphasis to the question of the extent to which the ITC is to become concerned with the basic levels of tin prices rather than exclusively with their fluctuations, with all that implies for reducing the perceived worldwide validity of prices set on a market in which the ITC operates directly.

9. Economic Performance and Public Policy

The economic performance of an industry is best understood as the result of the conduct of its buyers and sellers within a particular structural context. But before considering the implications for public policy of the structure and conduct of an industry, it is essential to identify the kind of economic performance desired. The model of pure and perfect competition provides one such norm or standard for evaluation of performance, in that at a theoretical competitive equilibrium, with price equal to full or social marginal cost in every market, resources are allocated in such a way that the sum of producers' and consumers' surplus is maximized. In the case of tin, as in many other industries, the competitive norm provides an initial—although abstract and incomplete—insight into criteria for optimal performance and for assessing the actual outcome of the industry's activities. The most crucial aspects of the tin industry that must be dealt with in formulating a workable and realistic performance norm are the nonrenewable nature of the product and the international separation of producers and consumers.

Global Aspects of the Theory of the Mine

In pure theory, the nonrenewable character of a commodity does not affect the optimality of competition in its market. It is well established in the literature dealing with the economics of the mine that a competitive miner, faced with constant costs of extraction, will mine all of his ore in the current period if he expects the future price to fall, stay constant, or rise by less than the discount rate reflecting his preference for present over future income—i.e., if $P_{t+1} < P_t(1 + r)$. If he expects the price to rise by even slightly more than the discount rate, r, and either his expectation is certain or he is not risk averse, he will defer mining until the next period, $t+1$, at which time he will make his production decision on the basis of the then expected price in period $t+2$. If all miners behave this way, total output of the mineral will fall when $P_{t+1} > P_t(1 + r)$ and, given the demand schedule, this decrease in supply will cause P_t to rise. Conversely, if $P_{t+1} < P_t(1 + r)$, total output will rise, causing P_t to fall. Thus, in equilibrium, the price will rise over time by the discount rate.

In calculating costs, the miner will take into account both his extraction costs and the reduction in the value of his mine caused by removal of some of its nonreproducible ore. The latter, referred to as "user cost," is the present discounted net value of the ore if sold later, where net value is measured by expected future price minus expected future marginal extraction cost. Total marginal cost in the current period, therefore, equals marginal extraction cost plus marginal user cost, with the latter assumed constant for the individual miner.

If, realistically, marginal extraction costs are assumed to be rising, in equilibrium each competitive miner produces an output where marginal extraction cost plus user cost equals price. Price minus marginal extraction cost will rise over time by the discount rate. It has been shown that if the discount rate used by miners is equal to the social discount rate (reflecting society's willingness to forgo present consumption to provide for the future consumption of either existing or future generations), and if both miners' and consumers' expectations of the future are correct, the competitive outcome is optimal, in the sense that resources are allocated both in every period and over time in such a way as to maximize the present discounted value of the sum of producers' and consumers' surpluses.[1]

In the competitive model's treatment of nonrenewable resources, the price of a mineral may ultimately rise to a level where the demand for it is zero. At and above such a price, consumers would resort to substitutes or simply forgo consumption of products containing the metal. Such an eventuality implies that the resource is not essential for all time. A mineral may be completely used up if the highest price at which it will be consumed is greater than the marginal cost of extracting the last unit. Otherwise, it will never be entirely exhausted. If there is no upper limit to the price, or demand remains positive at any price, reflecting the economy's inability to continue to function without the use of the resource, the ever-higher expected future prices would induce miners to limit the rate of extraction more and more severely, reserving dwindling supplies for the most pressing and costly uses. In theory, if there is no ceiling on the price, the resource will never be totally depleted.

The economic theory of the mine is complicated once the possibility of discovery of new deposits of the mineral is introduced, so that the resource, while still nonrenewable and ultimately available only in some finite amount, is not assumed to be fixed at some currently known quantity for all time. In a model assuming undiscovered deposits, miners invest in prospecting on the basis of estimated prospecting costs and probabilities of success, as well as expected prices and extraction costs. Each miner must take into account that the amount of successful prospecting by all miners will have a feedback effect on future prices. Models that incorporate uncertain opportunities for exploration, compound probability distributions for expected costs and prices, risk aversion on the part of miners, and risk aversion coupled with ability to stock the metal on the part of consumers, are complex; and the intertemporal welfare effects are uncertain. For example, an aversion to risk raises the discount rate applied by individual miners and thus encourages them to increase the rate at which they deplete existing mines; but risk aversion also discourages exploration and new investment and may therefore slow down the rate at which the total available amount of the resource is exhausted.[2]

Perhaps the two most dubious assumptions made in a formal demonstration of competitive optimality in mining are first, that the miners' rate of discount is equal to the social rate of discount, or even that the present generation can and will determine an appropriate discount rate to apply to the welfare of future generations, and second, that miners have perfect foresight with respect to future prices and costs. Geoffrey Heal has shown that where miners' forecasts are in error, there can be either exponential movements away from equilibrium, or oscillations over time around the equilibrium levels of price and output, with the possibility that these oscillations will be explosive. Heal argues, however, that the basic problem is that the absence of adequate opportunities for hedging on forward markets forces miners to forecast in situations where there can be no basis for making accurate estimates of future conditions and that organized commodity exchanges therefore have a very important role to play.[3]

Despite such difficulties in modeling the performance of an extractive industry over time, the views of most writers on the economics of the mine are summarized in Peterson and Fisher's conclusion that "extractive resources are probably allocated by competitive markets about as well as other resources, subject to the usual variety of imperfections."[4] The most serious and widely held objection to this viewpoint is that a monopolistic miner, who will set a higher price and sell less of his ore in the earlier years of mining, will in almost every conceivable case exhaust his mine at a later date than would be the case under competition. Conservationists have argued that the current private discount rate applied by miners is bound to be higher than the appropriate social discount rate both because private risk is higher than social risk and because individual members of the present generation do not put a high enough value on income streams accruing to future generations. According to this line of argument, the tendency of monopoly to prolong extraction may provide a welfare-augmenting offset to the opposing tendency of an unduly high private discount rate to accelerate depletion.[5]

Consumer Criteria for Performance of the World Tin Industry

Consumers benefit from the competitive structure of the tin industry at both mining and smelting levels. They should welcome the market-clearing price determined on the Penang market, or alternatively on an active and unmanipulated commodities market in Kuala Lumpur; and they should regard domestic taxes and the various constraints put on the free export of tin concentrates by the producing countries as detrimental to their interests.

In a high-risk industry such as tin, in which major determinants of performance are future price and ore discoveries, both of which are unknown and must be estimated, competition will be more effective if risk and uncertainty can be reduced, and if unavoidable risk can be transferred to those who have the knowledge and willingness to bear it. Thus consumers as well as producers enjoy the benefits of improved market efficiency from whatever success the ITC achieves in stabilizing tin prices and from the opportunities for hedging provided by the LME. This joint and unambiguous gain from reduction and transfer of risk is, therefore, a different phenomenon from the increase in consumers' surplus at the expense of producers under conditions of fluctuating demand and steady supply, or the reverse trade-off under stable demand and variable supply.

Representation on the ITC is undoubtedly of value to the consuming countries. But such representation may not be used to promote the best interests of final consuming households, as prescribed for domestic welfare by the competitive norm. As discussed in chapter 4, the economic interest of final consumers in the price of tin is small, and their awareness of that price may be minimal or nonexistent. As a result, those with more immediate and substantial interests, such as dealers and manufacturers, may be the domestic groups whose views are best known to and represented on a consuming nation's delegation to the ITC. This phenomenon, as Rowe and others have pointed out, is likely to give a greater weight to price stabilization and a lower weight to the level of price than would optimize the nation's economic welfare, especially if the final demands are inelastic, as is the case for many tin-bearing products, so that cost increases can be passed on by fabricators.[6]

More recently, Anthony Edwards observed that "many industrialists have suggested that more stable prices for commodities, even if by implication somewhat higher ones, would have helped their operations." Among the advantages to industrial consumers that could induce them to accept higher prices, Edwards cited the following: dampening of fluctuations in demand; reduction in average inventory holding requirements; amelioration of the marketing problems growing out of instability in end-product prices; lessening of the incentive for substitution of synthetic materials with more stable prices; and reduction of general uncertainty.[7]

Today, tin is essential in some uses, notably in high-quality solders and bearing metals. In the absence of technological advances in the development of substitutes, a protracted shortage or exhaustion of tin would be a more serious matter for the economies of the consuming nations than for those of the producers, with the exception of Bolivia. There do not seem to be any major problems, however, with the economic performance of the tin industry with respect to the rate of resource depletion.

Throughout the present century, the rate of growth of consumption of tin has not been nearly so rapid as consumption of the other nonferrous metals. As the price of tin rises over time, the world will have to resort to more costly sources of supply, including lode ores, the extensive low-grade deposits of Australia, deeper and less accessible alluvial ores such as the deposit at Kuala Langat, and offshore deposits with high costs of discovery and recovery. Mining engineers and geologists are confident that there is a great deal of undiscovered tin in the world. As one put it, the vast alluvial deposits of Southeast Asia came from some lode, somewhere.

The problem is not physical exhaustion of the resource before substitutes can be found, or even, as suggested in the competitive market model, a rapid and drastic rise in price to forestall this eventuality. It is, rather, that the optimum rate of depletion cannot be determined in the absence of knowledge of the extent of undiscovered ore bodies. But there is no reason to believe that under the existing structure and market conduct of the industry the rate of depletion is too high to be in the best interests of consuming nations.

Technical progress over time, either reducing costs or developing new and improved products, may be as important to an industry's performance or even more important than competitive levels of price and output at any one time. The basic modern technologies of tin mining were introduced when the first gravel-pump mine was established in Malaysia in 1906 and the first dredge in Thailand the following year. Technical progress in mining has since taken the form of improvements in these methods, such as better and more economical pumps and motors, greater use of mechanized dirt-moving equipment, on-site jigs for ore treatment, and larger and more efficient dredges, most notably off-shore dredges capable of mining ocean areas hitherto inaccessible because of depth of the ocean floor or roughness of surface waves. Underground tin mining has shared in advances in mechanization common to that mode of mining.

It is not surprising that tin mining companies have engaged in very little if any industrial research of their own. At least since the work of Joseph Schumpeter,[8] it has been recognized that relatively small firms in highly competitive industries are disadvantaged relative to larger monopolistic or oligopolistic firms in conducting research and development. On the other hand, firms in competitive markets must be adaptive to survive, and with the apparent exception of the government-owned mines of Bolivia, the record of tin mining in utilizing and diffusing new technology is excellent.

In response to a question regarding research work done by tin mining companies, posed from the floor during the discussion session of a conference on tin consumption, B. K. T. Barry of the Tin Research Institute pointed out that the individual mining concerns were involved only in producing tin concentrates for sale to others. Unlike the much larger and vertically integrated firms both mining and processing other metals, the tin miners could not be expected to engage in research on processing techniques, new alloys, or new uses for tin. He added that it was "precisely because the various governments of the tin producing countries had sensed that the tin industry was so different from other metal industries in this respect, that they had set up the Tin Research Institute." Further, Barry pointed out, the industry itself did lend "considerable support" to TRI.[9]

There may be areas of conflict between consumers' and producers' interests in new technology. Rowe, for example, was particularly concerned that technical progress posed a severe threat to commodity control schemes such as the international tin agreements. Major technical change, he noted, requires prompt adaptation within an industry, including the elimination of high-cost and obsolete operations. Control schemes, to the contrary, tend to protect existing capacity, even if inefficient. Further, a commodity agreement that succeeds in raising prices may thereby unintentionally stimulate the very research and technological change that puts such strains on the agreement. Finally, Rowe observed, new technologies may promote the entry of new firms and thus increase the output of those not participating in the agreement.[10]

Rowe feared that "the likelihood is either that technical progress will be deliberately stifled by the control, or that the stabilised price will not be reduced, an excess of capacity will develop, and semi-permanent restriction of output will result."[11]

There are two international tin research and development organizations, the Tin Research Institute (TRI) and the South East Asia Tin Research and Development Centre (SEATRAD). TRI, located just outside London, England, is not supported by nor a part of the ITC, but is funded entirely by the producing members of the council. But ties between the two organizations are close. TRI representatives serve as advisors to the ITC and as observers at ITC meetings. The executive chairman of the ITC, similarly, sits as an observer at TRI meetings. TRI engages in three main areas of work: applied research on new products that will increase tin consumption; developmental work such as publications, conferences, and exhibitions to promote use of its research discoveries; and provision of technical advisory services to tin users. Its avowed overall objective is to promote greater use of tin. TRI has made important advances in metallurgy and development of new tin alloys, and in organotin chemistry.[12]

SEATRAD is financially supported and directed by the major three Southeast Asian tin-producing nations, Indonesia, Malaysia, and Thailand. It is a new organization, discussed within the United Nations for many years but established only in 1978. SEATRAD's program is planned to complement that of TRI in concentrating on research in all aspects of tin production. It also plans to collect data on various aspects of tin production, conduct conferences, seminars, and training courses on tin mining technology, and serve as a clearing house to facilitate the exchange of technical information among its members.

Both Malaysia and Bolivia have national institutes engaged in tin-related research—the Mines Research Institute Malaysia in the former country and the Institute for Mineral Investigation in the latter. The Mines Institute Malaysia, which shares laboratories with SEATRAD, has divisions dealing with prospecting, mining technology, mineral processing (primarily ore treatment), and basic research.[13] Much of its time and budget—too great a portion in the opinions of several with whom I discussed research and development in tin—is devoted to current problems such as assaying ore from potential new mine sites and investigating mine accidents.

These international and national institutes are engaged in programs related to the interests of producers—new uses for tin and new and improved methods of mining. But their activities do not appear to have been constrained by fears that technical progress would destabilize the industry, as suggested by Rowe.

Consumers' interests, in opposition to those of producers, are furthered by research into substitutes for tin. The overall structural context is favorable to such research, since the extremely high cost of tin relative to other industrial metals provides a powerful incentive, and since the users of tinplate, tin alloys, and solders include many of the world's largest industrial firms with capacities to carry out large-scale and sophisticated research. The growth of substitutes as well as tin-saving processes discussed in chapter 2 attests to the continuing vigor and success of this line of research.

Smelters regard the techniques they use as confidential information, but it is evident that a great deal of progress has been made by some, notably Gulf Chemical & Metallurgical's Texas City smelter, Bolivia's Vinto smelter, and Capper Pass, in treatment of complex ores. Even more important in terms of creation and diffusion of new technologies is the contribution made by firms such as Klöckner, Metallgesellschaft, and Hoboken-Overpelt in design and construction of technically advanced smelters to be owned and operated by others.

In all, despite the lack of incentives and opportunities for the primary producers to engage in research and development, the rate of technical progress and its diffusion throughout the industry must be regarded as a positive rather than negative aspect of the tin industry's performance.

Producer Criteria for Performance of the World Tin Industry

The above discussion of economic performance from the point of view of consuming nations took the appropriate standard and policy objective for such countries to be maximization of the present value of consumers' surplus. An equally simple and widely accepted standard—i.e., maximization of the present value of producers' surplus—cannot be posited for the producers.

Individual private mines and smelters may be assumed to be maximizing a preference function involving a trade-off between profit and risk. The price risks involved in the tin trade look quite different to the producing nation as a whole, however, than they do to individual producers. There is no net national advantage or disadvantage to Malaysia, for example, in having risk shifted from the smelters to the miners but in no way reduced, unless for some political reason the government wishes to favor the individuals involved in one sector or judges the other sector to be more capable of bearing the risk. It might appear that the forward price risk shifted by the smelters to the buyers of tin should be welcomed by the Malaysian government, except in those instances in which the dealers holding title to the tin are Malaysians, since it is a shift from Malaysians to foreigners. However, the Penang market price of tin paid by these foreign buyers is quite likely to be reduced so as fully or in large part to compensate them for bearing the risk. In light of this unavoidable consequence of a competitive market structure, it would be in Malaysia's interest to eliminate or reduce the risk. But it does not follow that maximization of producers' surplus, the economy-wide analogue of profit maximization by the individual firm, is an appropriate performance norm for the government of a producing nation even if its concerns are strictly domestic and its exposure to perceived risk is small.

In a developing country, particularly one such as Bolivia that is dependent on imports for current consumption, gross foreign exchange earnings may be more important than the net of such earnings minus the domestic costs of production. The demand for foreign exchange may be heightened by pressures to fulfill targets in national plans for growth and development that are dependent on imported materials and machinery. Maximization of producers' surplus would occur at a rate of sales at which the marginal revenue to the nation from the sale of one more unit of the commodity equaled the full marginal social cost of production, while maximization of total gross revenue would occur at a lower price and higher level of sales, with output expanded to the point at which the nation's marginal revenue was zero. Export revenue would be maximized at an even higher rate of output and lower price than those of a competitive equilibrium if the market-clearing price lay in the elastic range of the nation's demand curve (since marginal revenue is positive at rates of output at which demand is elastic).

A producing nation might also be hesitant to reduce output in order to raise price and thereby producers' surplus if the commodity in question is one that employs large amounts of domestic labor and capital. In such a case, even if it could be shown that the economic gains obtainable from output restriction would yield net economic benefits to the specific industry and to the nation as a whole, the hardships imposed on those displaced from the industry might make the overall economic gains seem undesirable. The political power of the Bolivian tin miners, for example, has exerted a major influence in raising that country's desired level of sustained tin production.

There is no reason to believe that the discount rates of the producing and consuming countries will be equal. A priori, one would expect the developing producing nations, with more urgent needs for both current consumption and immediate growth, to have a substantially higher social discount rate than the wealthier industrialized consuming countries. An exception to this generalization would be the case of a nation, such as some of the OPEC members, which is incapable of making efficient use of all of its exports receipts for either current consumption or prompt investment in productive capital; but none of the tin producing nations has the problem of absorbing its foreign exchange earnings.

A concern for foreign exchange earnings, the desire to promote domestic employment, and a high premium placed on current income, all serve to reduce the optimum price and raise the optimum level of output relative to those that would maximize producers' surplus. Further, to the extent that the producing countries take such considerations into account but differ in the importance they attach to them, the likelihood of their cooperating on an effective sustained collective policy is reduced.

There is a more formidable and fundamental barrier to collective producer action designed to maximize joint producers' surplus. As noted in chapter 4, a problem with private cartels and collusive oligopolistic arrangements is that the costs of the participants may differ, both in general level and with respect to changes at various rates of output. Such differences in costs, to the extent that entire nations may be categorized as low- or high-cost producers, may lead to marked differences in the world market price and rate of output desired. During the 1920s and 1930s, Malaya was generally recognized as the world's lowest-cost producer of tin. Yip, after reviewing the agreements reached by the International Tin Committee during the prewar period, concluded, "But when demand had returned to normal, the continuation of restriction as a weapon to raise the price of tin to a level that would keep in operation even some of the highest-cost producers tended to work to the disadvantage, in the long run, of low-cost producers such as Malaya."[14]

Yip regarded the objections of Malayan miners to the First (prewar) International Tin Agreement as "quite justifiable." These miners contended that under free competition they could drive the high-cost Bolivian and Nigerian mines out of business. Malaya's participation in the agreement, therefore, amounted to imposing on Malayan miners the responsibility of "carrying the burden of these high-cost producers on their backs." In addition, the Malayan opponents of the agreement feared that success in holding tin prices at artificially high levels for any length of time would encourage the development and use of substitute products.[15]

Today, Thailand, with its newly discovered offshore deposits and with a higher average grade of ground for inland mining than Malaysia, is the lowest-cost major producer. Differences in the costs of mining the high-grade alluvial ores of Southeast Asia and those of mining the complex and low-grade lode ores of Bolivia remain the basic source of potential conflict among producers.

Producers share consumers' interests in market stabilization, particularly in cases of commodities that are important enough to the domestic economy to have perceptible effects on overall price and employment levels. Because of foreign exchange needs and concern for the welfare of nationals engaged in the domestic industry, the governments of producing nations are likely to be more interested in stabilization of revenue than of price; but reduction in the price risk imposed on domestic producers should also be a matter of concern to governments seeking to promote the domestic economic welfare, reduce costs of production, and encourage an optimal rate of resource depletion. In the case of tin, the evidence reviewed in chapter 5 indicates that the ITC's efforts to stabilize price probably have had somewhat of a stabilizing effect on revenue as well.

Producers' concern for stability may be excessive. Several writers who have studied both the earlier experiences of the developed nations and the current growth rates of the developing countries have pointed out that there is no evident correlation between stability of either price or export revenue and rates of economic growth.[16] From a long-run competitive perspective, stability may be no more in the interests of the more efficient and lower-cost producing nations than are prices high enough to accomodate the high-cost producers. In a recent study of a domestic oligopoly, R. G. M. Sultan noted, "There is evidence that cyclical oligopolies are more profitable, on the average, than are noncyclical oligopolies; in other words, the disruption which attends varying prices and output probably brings its own profit reward, in terms of causing a 'shake-out' of extra costs and of marginal competitors, both of which tend to drag down profitability."[17] Analogous reasoning could apply to the attitudes of the Southeast Asian tin producers toward the efforts of the ITC to mitigate fluctuations in tin prices. But no such views have surfaced yet within the ITC. To the contrary, the positive value that the tin producers put on price stability as well as on a floor price is indicated by their willingness to finance the ITC's buffer stocks with no assistance from the consuming members at the outset and very little since.

A little-noted but serious problem with tin mining is that there are starkly visible external diseconomies of production arising from the large quantities of dirt that must be excavated and sifted in order to obtain the cassiterite. The primary reason why tin is an expensive metal is that its natural concentration is low. The result is that in regions where there is alluvial mining, large areas of land are stripped of topsoil, dug and churned to depths of fifty feet and over, and left for decades as barren tracts useless for agriculture and devoid of vegetation.

As Malaysia has developed and grown in population this problem has become more serious and in recent years has been recognized as an important factor inhibiting the government from opening up new land to exploration and mining. Conflict-of-use problems have arisen as a result of the expansion of roads, power lines, and railways across tin-bearing lands. The Mines Research Institute Malaysia has identified research on how to reclaim mined land for agricultural use as one of the most important items on its agenda. Underground mines pose a similar problem of disposal of tailings. Prospecting for tin has been prohibited on moorlands in Cornwall, and issues of environmental deterioration are of growing importance in consideration of the feasibility of opening new British tin mines. Offshore dredging in Thailand has fouled rich fishing grounds.

Future efforts by producing countries to either mitigate the effects of these external diseconomies or to shift the costs back onto the tin miners—by such means as requiring miners to restore mined sites to a condition where the land can be used for other purposes before abandoning them, or simply through user fees and taxes—may raise prices paid by consumers; but to the extent that prices reflect full social costs of production including external diseconomies or economies, global resource allocation will be improved, even though consumers will not enjoy any of the benefits for which they are paying higher prices.

In Situ Value and Taxation of a Nonrenewable Resource

In addition to the question of the extent to which maximization of producers' surplus represents an appropriate or plausible norm by which a producing nation should evaluate the performance of its tin industry, or any other, the issue of distribution of whatever surplus is obtained must also be addressed. The matter of appropriate distribution involves considerations of ethics and politics as well as economic incentives, and hence models of competitive markets alone cannot generate any sort of distributive norm with which actual distribution can be compared to evaluate its propriety.

In the case of a natural nonrenewable resource, the problem is complicated by the existence of two sources of value. First, there is the effort put into discovering and reclaiming the resource, representing the allocation of labor and capital that could otherwise have been devoted to other productive uses; and second, there is the inherent value of the mineral in the ground, which is a function of its scarcity and useful properties in consumption entirely apart from any human endeavors. The proposition that this latter in situ value should accrue to the nation claiming sovereignty over the territory in which the mineral is found, rather than to any private individuals who discover ore bodies or own the lands on which they are found, has become widely accepted throughout the world. In a recent survey of mining legislation, the authors concluded, "Generally speaking privately owned mineral rights may now be regarded as an anachronism, and where they still exist it is reasonable to anticipate the Governments will seek in one way or another to extinguish them."[18] Once the principle is accepted, difficulties arise both in ascertaining this in situ value and in appropriating it through taxes or royalties.

A producing nation might well estimate the in situ value as equal to the maximum producers' surplus obtainable from the resource, or whatever portion of that surplus it was deemed in the nation's best interest to try to obtain. The rationale for such an evaluation would be that while neither local nor foreign-owned mining enterprises should earn more than a competitive rate of return for locating and extracting the nation's mineral wealth, whatever price over and above the miners' full costs foreigners are willing to pay for the quantity of the resource the producing nation chooses to offer for sale indicates its in situ value.

In instances where the producing country is a developing one, particularly with historical experience as a colony, application of the willingness-to-pay principle may be defended as ethical on grounds of recompense for past exploitation, a more equitable global distribution of income and wealth, or an offset to payments for the import of goods produced in the developed countries by firms with monopolistic power. It is beyond the scope of this study to discuss the merits of this line of argument: rather, it can merely be accepted that most nations, both in setting norms and in prescribing policy, are concerned with augmenting their own national interests, and that there is nothing untoward about a small developing nation behaving in this manner.

There are two basic ways in which a government may collect the in situ value of minerals from private miners. It may levy a royalty on minerals extracted or exported, or it may tax the miners' profits. The royalty may be a flat fee based simply on the quantity of the mineral removed or exported, or it may be ad valorem. The royalties on tin charged by Bolivia,[19] Indonesia, Nigeria, Malaysia, Thailand, and all but one of the states of Australia are based on current prices of tin, with those of Bolivia, Malaysia, Nigeria, and Thailand being progressive in the sense of taxing a higher percentage of the price as the price rises. Zaire is the only ITC member that does not charge a royalty on tin. All of the major producing countries except Bolivia charge business income taxes, but Malaysia is the only one to have imposed an additional profits tax on tin mining.

In a 1975 book, *Negotiating Third-World Mineral Agreements*, David N. Smith and Louis T. Wells, Jr. pointed out that while royalties have in the past been the primary source of government revenue from private mining ventures, they have been declining in importance since the early 1970s and play only a minor role in most mining concession agreements today.[20] But no such decline has occurred in the case of tin. Smith and Wells also noted that most royalties still being levied today are based on value of the product, since worldwide inflation has reduced the value of real revenues obtained from royalties based on physical units. One of the major disadvantages of ad valorem royalties, they continued, is the difficulty of ascertaining a value for the mineral when much of the product is shipped to affiliates of the mining ventures rather than being sold at an arm's-length price. Widespread and heavy reliance on royalties may have persisted in tin in part because such a problem does not exist in that industry, since there is no integration between miners or smelters in the producing countries and fabricators in the consuming nations. Indeed, to the contrary, the Penang market, or a Kuala Lumpur commodities exchange, may be useful not only to Malaysia but to other producing nations as well in arriving at a freely determined and generally accepted market price on which ad valorem royalties can be based. As an additional practical consideration, a royalty or export tax is administratively easier to collect and less vulnerable to evasion than is a business income tax.

Gillis, likewise, noted and discussed the shift away from royalties in the mid-1960s. "The general shift from heavy reliance on royalties to income taxes was based on sound reasons," he wrote. "The average revenue that could be obtained from an income tax was higher than that which could be expected from royalties. The shift of risk away from the entrepreneur, implicit in the change to income taxation, enabled him to accept a higher rate of taxation. . . . And the income tax gave the desired incentive for the rational exploitation of marginal ores."[21]

Abdul Rahim Aki has made a similar argument, contending that royalty charges levied on tin mined, without regard to costs of extraction, discourage both full recovery of metal from ore bodies worked and exploitation of risky sites.[22] A tax on either a constant or a progressively increasing percentage of profits, on the other hand, encourages efforts to recover all of the metal that can be taken from the ore at a cost of less than the minegate price, and imposes no additional penalty on losses associated with risk-taking. At the 1981 World Conference on Tin in Kuala Lumpur, a Billiton executive, F. J. Blok, stressed the waste that accompanies the lessened incentive to maximize recovery from certain ores under a royalty system.[23]

Thoburn emphasizes the effects of high royalty charges on exploration and development of new mines, noting that, in both Malaysia and Thailand, the problem lies in the structure of taxation rather than the level. Not only does heavy reliance on royalties promote inefficiency and wasteful mining techniques, he argues, but it is also inequitable in failing to allow for cost differences and is insufficiently progressive in its claims on profits in periods of high price.[24]

The policy implications of this line of argument must at best be drawn with caution. Royalties have no effect on the risk stemming from the possibility that a new mine will prove barren or yield concentrates that cannot be sold at a price covering variable operating costs. If no metal is extracted, no royalties are paid (all of the tin-producing countries levy their royalties or export duties on assayed amounts of tin-in-concentrates, and not on the amount of dirt dug up). High-cost, marginal miners, like their more efficient competitors, are removing irreplaceable wealth from the nation's soil.

The point that prospective miners will be willing to pay more taxes in total if they are relieved of the added risk associated with royalties follows from the idea that royalty payments add to variable costs of production and thus increase the possibility that a miner will decide not to work his site after having incurred exploratory and development costs, or will be unable to recover fixed as well as variable costs. The effect on existing miners of a new or higher royalty would be similar: it would reduce their aggregate output and eliminate some mining on lower grades of ground. But this reasoning fails to recognize that total variable cost—including user cost as well as operating costs—may not rise at all with the imposition of a royalty, or may increase by much less than the full amount of the royalty charge.

Since an income tax has no effect on marginal cost, it will lead miners to increase their production relative to the rate at which they would produce under a royalty system only to the extent that the unit royalty charge raises marginal cost. And it is incorrect, in general, to treat this as a "desired incentive" for "rational exploitation." If the in situ value of the metal to the nation has been calculated correctly and reflected accurately in the royalty, that value should be included in the full marginal social cost of mining. Unless the miners have already calculated a private user cost as high as the social in situ value, they will overproduce and exhaust the resource too rapidly in the absence of an appropriate royalty, even if the equivalent of the full in situ value is collected from the inframarginal mines in business income taxes.

Gillis, Rahim, Blok, and Thoburn are correct in pointing out that if miners have calculated a user cost equal to the in situ value, they will be producing at an optimal level in the absence of any royalty. But they fail to realize that miners' user costs will be reduced by the imposition of a royalty that is expected to be permanent, since the private value of any ore mined in the future is reduced by the royalty that will be collected on it. The net disincentive effect of the increase in marginal extraction cost resulting from a royalty payment and the offsetting decrease in marginal user cost will be nil if the miners and the government apply the same discount rate as well as arrive at the same valuation of metal in the ground. If the government sets a royalty rate in excess of miners' zero-royalty user cost, either because it is more interested in maximizing public revenue than in allocating resources efficiently over time or because it has applied a lower discount rate than that used by miners, the reduction in output identified by Gillis will result. But there is nothing inherently wrong with such a reduction. It will be suboptimal only if the government is in error or has deliberately set a royalty higher than the in situ value. Robert F. Conrad and R. Bryce Hool have shown the effects of a royalty, or "severance tax" in their words, more formally. Conrad and Hool begin by noting that the discounted profit for a mine in any period, assuming no taxation, is given by the expression

$$\pi_t = \frac{P_t \alpha_t X_t - C_t(X_t)}{(1 + r)^t}$$

where

π_t = discounted profit in period t,
P_t = price of output in period t,
X_t = quantity of ore extracted in period t,
$C_t(X_t)$ = total cost of ore extraction and processing in period t,
α_t = average grade of ore extracted in period t, and
r = discount rate.

If an ad valorem tax on total revenue (i.e., a royalty) is imposed, the after-tax profit becomes

$$\pi_t = \frac{(1 - \beta) P_t \alpha_t X_t - C_t(X_t)}{(1 + r)^t}$$

where β = the proportion of revenue collected as tax.

Conrad and Hool note two sources of "distortion" associated with such a tax. First, the higher the price the greater the tax per unit. (Note that this effect is heightened if the tax is progressive or is based only on the portion of prices above a certain level, as is the case with most tin royalties.) This would induce the firm to reduce output in periods of higher discounted expected prices and increase it in periods in which lower prices were expected. Second, the cutoff grade of ore extracted would be raised, leading to "tax-induced high grading." This second distortion can be shown as follows. In the absence of taxation, profit maximizing output is set where

$$P_t \alpha_t = MC_t,$$

where MC_t = the marginal cost of ore extraction and processing in period t.

The cutoff grade, α^*, is therefore

$$\alpha^* = MC_t / P_t.$$

With an ad valorem tax, the cutoff grade becomes

$$\alpha^* = MC_t / P_t(1 - \beta).$$

Since $0 < \beta < 1$, the cutoff grade is raised with imposition of the severance tax.

Conrad and Hool, however, err in the normative implications they attach to their analysis, if the criticisms I have made of other writers are correct. They assume that the outcome in the absence of any taxation is optimal, and that changes in either rate of exploitation or grade of ore mined induced by taxation therefore represent distortions. But they fail to include user cost in their marginal cost, and regard the miner as maximizing profit by equating the price of the processed ore at the mine gate, $P_t\alpha_t$, to the marginal cost of extraction and on-site processing. If they had included unit user cost in their expression for marginal cost, and had noted that this user cost falls as the royalty rate, β, increases, their policy conclusions would have been different and, it would appear, similar to mine.[25]

The only argument against the use of a royalty of appropriate size that appears to have substance and must be taken seriously is that the royalty encourages wasteful methods of mining and leaves too much tin that could have been recovered in the tailings. Yet even this argument is valid only if it assumes that the metal is lost forever or is rendered more expensive to recover later. In light of the widespread practice of reworking grounds mined earlier as prices rise and average grade of ground falls, the problem is evidently not a serious technical one. As Thoburn himself noted, even the operation of suction boats does not preclude later mining of the same area by dredges, with very little if any loss in the amount of tin metal ultimately recovered. From a social point of view, the problems noted by Gillis, Rahim, Blok, and Thoburn would appear to be severe only to the extent that the level of royalties exceeds in situ value.

The foregoing discussion leads to the conclusion that a theoretically optimal combination of producing-country taxes on tin would be a royalty set just equal to in situ value plus an income tax to reduce the remaining profits of inframarginal mines to whatever level the government considered proper. It also follows that the concern of the consuming nations that current tin prices have been forced too high by the taxes levied on tin mining to justify continued ITC support ought to be directed only to the difficult but fundamental question of how much higher, if at all, are royalties than reasonably estimated in situ values. Income taxes will not reduce output, provided they allow returns on capital invested and risks assumed that are high enough to attract private entrepreneurs—returns that are included in the concepts of full average and marginal costs. Income taxes are, therefore, instruments involving issues of domestic income distribution within the various producing nations and not a matter that should be of any concern to the consuming members of the ITC.

International Cooperation and Global Performance

Public policies toward the tin industry are formulated and implemented in a setting characterized by an underlying economic structure conducive to effective worldwide competition at both mining and smelting levels, but with the conduct of the industry shaped by powerful external factors as well as its own industrial structure. The most potent of these outside features are an international commodity agreement, a small number of producing nations dominating smelting (and thus the trade in tin metal through their control over shipments of concentrates), and a vast stockpile in the hands of the largest consuming nation.

Issues dividing consumers and producers: the 1980-81 debate on the sixth agreement. Inevitably, the ITC has become the primary forum for identifying and venting policy issues that affect the global trade. The issues dividing consumers and producers were articulated pointedly in 1980 and 1981 in the meetings of the conference to negotiate the Sixth International Tin Agreement.

On May 14, 1980, the conference on the sixth agreement adjourned sine die in a deadlock. "It has been a failure," stated ITC Executive Chairman Peter Lai, "and we must not disguise that."[26] The major issues dividing the delegates were the size of the buffer stock, the nature of members' contributions to the stock, the use of export controls, and the role of the ITC in disposals from the United States' strategic stockpile. The United States was calling for a buffer stock of 70,000 metric tons, relying heavily on the Smith and Schink study as a basis for the figure, and for elimination of export control from the agreement. The Malaysians and Bolivians, on the other hand, were pressing for a buffer stock of 30,000 metric tons, with mandatory equal contributions from consumers and producers. The Bolivians, in addition, urged that ITC approval be made a requirement for any GSA sales of tin from the United States' stockpile. Shortly after the adjournment, spokesmen for the producing nations made it clear that export control was not negotiable; and the United States made it equally clear that no binding control over stockpile disposals would be ceded to the ITC or any other international body.

The emphasis put on a large buffer stock by the United States, and the contrasting insistence by the producers on preservation of export control, are at the core of the differences dividing the delegates. A buffer stock is the instrument of price stabilization. The basic effect of export control is to support or raise the price level. The relative importance placed on the two is thus indicative of the primary purpose envisaged for the agreement and the council.

During the months between May 1980 and the final adoption of the text of the sixth agreement over the opposition of the United States and Bolivia in June 1981, the two sides came closer together, although obviously never close enough. The United States accepted export control in principle, but only providing that it was designated as an instrument of last resort and accompanied by a buffer stock large enough to ensure that it would be used rarely. Somewhat inconsistently, in the opinion of the producers and most independent commentators, the United States continued to oppose mandatory contributions to the buffer stock by consuming nation members. The EEC delegates proposed a 35,000 ton buffer stock to be financed by direct contributions from both consumers and producers, and authority for the buffer stock manager to borrow if necessary in order to acquire an additional 25,000 tons.

But at a December 1980 meeting, the gulf dividing the United States and Bolivia was illustrated by a proposal from the former that export control be considered by the ITC only when 90 percent of a 60,000 ton buffer stock had been accumulated in metal, and a counterproposal by the latter that export control should be imposed automatically (not merely brought to the council for vote) whenever the buffer stock manager had acquired more than 15,000 tons of tin. In March 1981, Executive Chairman Lai proposed that a buffer stock of 50,000 tons be established, with export control to be considered by a two-thirds distributed majority whenever the buffer stock held over 35,000 tons in metal, and by a simple majority when the stock of metal rose to over 40,000 tons. Both the United States and Bolivia opposed this compromise.

There was also controversy over the width of the price band and the financing of borrowings to supplement the buffer-stock fund. A majority of the delegations to the conference were prepared to accept a 30 percent price range, with the lower or "may buy," middle or "nonintervention," and upper or "may sell" bands each set at 10 percent of the floor price. The United States, however, argued for a 20 percent middle range. The United States also contended that loans made by the buffer stock manager should be guaranteed by the governments of the member states, although others argued that such guarantees were unnecessary and would not reduce interest costs given the acceptability to the London banks of tin warrants as collateral. The matter was of substance since it would be necessary to get these guarantees enacted by the legislatures or other appropriate authorities in all member countries.

In the agreement as finally drafted for signatures, Article 21 set the buffer stock at 30,000 metric tons, to be shared equally by producing and consuming members. In Article 24, the buffer stock manager was given authority to borrow funds to acquire an additional 20,000 tons, on the security of tin warrants and, "if necessary, government guarantees/government undertakings." In Article 27, the range between the floor and ceiling price was set at 30 percent of the floor price, and divided into three equal bands. Under Article 32, export control could be imposed by a two-thirds distributed majority whenever the buffer stock holds 35,000 tons of tin, and by a simple majority when holdings rise to 40,000 tons. There was, however, a provision for an automatic increase in permissible exports if, at any time after the first quarter of a control period, the fifteen-day moving average price is at or above the top of the lower sector of the price range for twelve consecutive market days.[27]

A Bolivian spokesman, in explaining his government's reservations about signing the sixth agreement, described the features found most objectionable. From Bolivia's point of view, the size of the buffer stock is excessive, and the provision that export controls cannot come into effect until the stock holds 35,000 tons of tin is unacceptable, particularly in comparison with the 5,000 ton figure for that contingency in the fifth agreement. He further cited the provision for borrowing by the buffer stock manager as a weakness, commenting that "if the price of tin on the market is down governments would find it hard to get bank credit for purchase of the additional amount."[28] With the decision of the United States not to join the sixth agreement, Bolivia's concerns about the distribution of voting power and the need for the ITC to exercise control over stockpile sales became moot.

The draft of the sixth agreement provided that if enough signatures were not obtained by June 1, 1982, the Secretary of the United Nations was to convene a conference of signatories to determine whether to put the provisional agreement into effect, or to modify it as they might determine. This conference met and on June 23 the sixth agreement was adopted by four producing nations (Australia, Indonesia, Malaysia, and Thailand) and fifteen consuming nations (Belgium/Luxembourg, Canada, Denmark, Finland, France, Federal Republic of Germany, Greece, India, Ireland, Italy, Japan, the Netherlands, Norway, Sweden, and the United Kingdom) that attended the conference. This agreement came into force on July 1, 1982. At the first meeting of the ITC under the new agreement, on July 1 and 2, the council adopted a recommendation from the conference that the buffer stock be reduced from 30,000 to 19,666 metric tons, in light of the reduced membership. Authorization for the buffer stock manager to borrow, if needed, to acquire an additional 20,000 tons was retained. No adjustments to Article 32, evidently made necessary by the reduction in the size of the buffer stock and the concomitant greater reliance on export controls, had been announced by the close of this meeting.

Evaluation of the ITC's contribution to the tin market. Discussion of the merits of the arguments made in the debate on the sixth agreement is beside the point, as is any effort to ascribe good or evil motives to the participants. Realistically, the ITC must be viewed as an organization all of whose members expect to obtain some net benefit from membership; and in their participation in the council's affairs the delegates must be expected to promote the interests of their own nations. The ITC, therefore, should be evaluated in terms of how well it serves the mutual interests of its members and whether it mitigates or exacerbates the conflict engendered by their opposing interests.

Both consumers and producers seem convinced that they benefit from price stabilization and prevention of the delayed investment booms and ensuing periods of excess capacity to which the tin industry is undoubtedly prone under modern mining techniques. Tin is particularly amenable to buffer-stock control, because its high value and resistance to physical deterioration make it relatively simple and cheap to store, because it is traded on a small number of organized markets on which the buffer stock manager can operate, and because it is a highly although imperfectly homogeneous commodity.[29] Since it is impossible to quantify the asserted net benefits of stabilization to either producers or consumers, it is conceivable that the ease and low cost of stabilization have prevented both tin-producing and tin-consuming nations from making a thorough assessment of these benefits. It is, though, more likely that the benefits reviewed above do substantially exceed the costs; but the success of the tin agreements may not be replicable for commodities with much lower values per unit of weight or volume, or for perishable commodities.[30]

ITC officials have repeatedly stressed that the ITC does not purport to protect miners and dealers from day-to-day fluctuations in tin prices. "It is commonly believed in trade circles the world over that the Tin Agreement aims merely at stabilizing the tin price," R. T. Adnan commented at the 1972 Conference on Tin Consumption. Rather, he continued, the stated objective of the tin agreement is not "stabilization," but the prevention of "excessive fluctuations in the price of tin." The phrasing is significant, Adnan continued, since the ITC does not attempt to maintain or restore any one ideal price for tin. Instead, the council seeks to keep fluctuations around the market-determined trend of prices within the sectors demarcated by its buy and sell ranges.[31]

In a later paper at the same conference, H. W. Allen reinforced Adnan's point when he stated, "The Council aims to prevent excessive fluctuations in the price of tin. In practice this means the prevention of any excessively steep falls in the lower sector of the price range or excessively steep rises in the upper sector, with free movement in between in accordance with day-to-day conditions of supply and demand."[32]

More recently, B. C. Engel has stressed the point that the tin agreements are designed "to work within the limitations of a free market world economy for tin, and are not intended to replace that system. What the agreements set out to do is to keep the tin price within a fairly broad range between a floor and ceiling price." Thus, the "excessive fluctuations" referred to in the texts of all of the postwar agreements should be interpreted as referring to price oscillations outside of whatever ceiling and floor prices the ITC has set at the time.[33]

A criticism often leveled at commodity stabilization schemes is that well-functioning commodity markets serve the purpose of protecting producers, dealers, and fabricators from price fluctuations by providing hedging facilities, and at the same time reduce the severity of such fluctuations through the activities of informed professional speculators. The effective operation of a buffer stock, this argument continues, may discourage speculation to the extent that it becomes impossible to maintain adequate hedging opportunities on the commodity market. Thus, a government or international stabilization mechanism may merely supplant a private one of perhaps equal or greater efficiency. In light of Adnan's, Allen's, and Engel's observations, it is clear that this criticism misses its mark if applied to the ITC. There is an evident division of function between the ITC and tin trading on the LME. The buffer stock manager, while he must operate on the basis of day-to-day prices and may react to excessively steep price movements toward the floor or ceiling, is primarily concerned with confining price movements within the range set by the council. Speculation on the LME, if effective, should dampen day-to-day variations in price. The vigor of trade in tin on the LME, coupled with the Malaysian government's interest in expanding such trade to a Kuala Lumpur commodities exchange, makes it evident that the ITC's buffer-stock operations have not unduly discouraged private speculation.

In addition to its implicit partnership with the LME, the ITC has found itself in unplanned collaboration with another organization, GSA. Virtually every writer who has commented on the world tin trade shares the conclusion reached in this study that the ITC has been successful in reducing long-term or annual price fluctuations. But a number have asserted that this success was made possible only because the ITC was able to defend its floor price by export control while GSA defended an undisclosed and perhaps undefined ceiling by sales from the United States' strategic stockpile. As long as GSA's ceiling was equal to or above the ITC's ceiling, and the latter was a price arrived at by consensus and acceptable to the producing nations, this arrangement could work. But stockpile sales cannot be expected to continue to play such a generally acceptable role now that a coalition of consuming members has exercised its combined voting power to hold the ITC's ceiling well below the market price for over two years, and the United States' delegation has made clear its belief that tin prices are not justified by costs but are inflated by excessive taxes imposed on tin miners in the producing nations. Thus, the vehemence of the current reaction to authorized stockpile sales of 10,000 tons a year over the next three years, in contrast to the relative equanimity with which earlier and much larger disposals were accepted, is an understandable response to the changed circumstances.

Another general criticism of buffer stocks is that they may merely replace private stocks and thus make no net contribution to price stability. Tin users, it is said, will buy for inventory when prices are low, forcing prices up, and will withhold their purchases and let inventories run down when prices are high, thereby reducing the price. If a buffer stock is established, private tin users may reduce their stocks accordingly and no longer exert any stabilizing influence of their own on price. The argument is, however, a two-sided one. To the extent that the buffer stock's holdings are offset by reduced private stocks, the global real net carrying costs resulting from creation of the buffer stock are lowered. Further, a buffer stock may be a more efficient stabilizer than the operations of private inventory managers. Adnan addressed himself to this issue by noting, "There is no private trader in the world to whom the profit motive comes second to his other objectives. To the Buffer Stock, profit-making is not a goal. . . .

This basic difference in philosophy between the trade and the Buffer Stock leads sometimes to market situations which make the whole concept of a Buffer Stock and its intervention undesirable and unacceptable to the trade."[34]

One major difference is that a private businessman in a competitive market will decide whether to increase or decrease his inventory on the basis of his price expectations without taking into account the effects of his own actions on the price level, while the buffer stock is operating at a magnitude at which it can affect the price, and indeed buys and sells solely for the purpose of doing so. Buffer stocks were used only sporadically and were clearly regarded as subordinate to export or production controls in the years before establishment of the ITC. Puey, in discussing the 1920s and 1930s, noted that consumers' holdings of tin stocks had a "tremendous influence" on price, but that this effect was hard to predict since the overall level of these holdings was unknown, presumably to individual businessmen as well as to the governments operating the Bandoeng Pool, the Tin Producers Association, and the International Tin Committee.[35] Further, to the extent that inventory decisions are made on the basis of day-to-day prices rather than on calculations of carrying costs at average prices over a longer period, or to the extent that tin users speculate on their inventories, the issue is irrelevant in light of the ITC's exclusive concern with dampening longer-run swings in price.

In sum, the basic stabilization objectives of the ITC are to promote more desirable levels of investment in tin mining and facilitate long-run adjustments in production, rather than to substitute for private institutions in dealing with problems of short-run fluctuations. Successful stabilization of this type is also beneficial to private industrial consumers of tin in planning the nature and level of their capital investments, particularly in instances where long-run considerations include the possible shift to substitutes for tin. Since capital intensity is high and shutdown costs are large in all major tin-mining methods, and since adjustment lags are long in dredging and underground mining, successful long-term stabilization improves the performance of the market and is thus in the interests of consumers as well as producers.

Whether a much larger buffer stock and the elimination of export control would achieve the ITC's stabilization objectives is uncertain. Some who advocate a large buffer stock and less reliance on export control have argued that the power to impose export control should not be eliminated until experience is gained with the larger stock. Rowe ended his 1965 book with an analogy that seems apt today: "Excessive price fluctuations are . . . like an illness of the human body which does not kill but greatly impairs its efficiency. Commodity control is like a drug which can largely neutralise such ills if administered correctly, but which can produce extremely dangerous results if the right dosage is exceeded, or if it is administered in the wrong ways. . . . [T]o give up experimenting with such a drug is a counsel of despair." The experiment will be costly, "but it should be well worth while at the end."[36]

Whatever may be said about price stabilization, consumers most certainly do not obtain any economic benefit from the current objective of the International Tin Agreement which reads:

> (c) To make arrangements which will help to increase the export earnings from tin, especially those of the developing producing countries, so as to provide such countries with resources for accelerated economic growth and social development, while at the same time taking into account the interests of consumers.[37]

No such objective had been specified in the first two agreements, yet it had been noted from the beginning that export control injected a major asymmetry into the ITC's effectiveness in defending its floor and ceiling prices. Yip noted, "It was clear that the mechanism of control as set out in the 1953 Agreement was designed to be more effective in keeping prices above the set floor in the event of a surplus than in preventing them from rising above the ceiling in the event of a shortage. . . . For all intents and purposes, the 1953 Agreement, although it had changed substantially in character and purpose since the pre-war days, was still primarily a producers' scheme."[38] Another Malaysian economist, Siew Nim Chee, writing in 1957, offered the following comfort to consumers. "While the Agreement seems to benefit the producers more than the consumers, the latter will now at least have some means of preventing undue exploitation by the former."[39]

The objective of increasing the earnings of tin producers could be introduced, for the first time, into the third agreement without undue consternation among the consuming members in part because by that time the ability to overlook inconsistencies or sources of conflict in the formal language of the agreement and the willingness to make decisions by consensus on a case-by-case basis had become ingrained in the ITC's procedures, and in part because the majority of the consuming members still had economic or political ties to the newly independent producing nations that made them willing to support producers' interests. But by the time the fifth agreement came into force, no such easy accomodation was possible. It was the militancy with which demands for such arrangements were pressed on the industrialized countries by the third world nations that led the United States to join the ITC, but the United States at the time of its entry indicated in the clearest possible way its opposition to commodity agreements designed to "raise prices above market trends."[40] These sentiments were shared by West Germany, which had first joined the ITC during the fourth agreement, and by Japan, which had been a member since the second agreement but had been steadily increasing its tin consumption and thus its voting power.

In assessing the ITC's effects on consumers and producers, a basic question must be whether it has succeeded in raising tin prices. A comparison of indices of annual average prices of tin and other nonferrous metals, shown in table 9.1 with 1956 = 100, suggests that it has been able to do so.[41]

From 1956, the initial year of operation of the First International Tin

Table 9.1. Indices of annual average prices of tin and other nonferrous metals, 1956–80 (1956=100)

	Tin	Copper	Lead	Zinc
1956	100	100	100	100
1972	162	130	103	154
1974	294	266	220	537
1978	450	216	295	314
1980	484	243	275	339

Agreement, until the eve of the commodity boom in 1972, tin rose more in price than any of the other three metals. The effect of the commodity boom is shown by comparisons of the price indices for 1972 and 1974. The prices of all four nonferrous metals fell during the 1975 recession. Only the price of tin, and to a much lesser extent that of lead, managed to recover the gains of the boom and continue to rise to new heights. The 1956–80 experience ought not, however, to be interpreted as an indication that tin is now overpriced relative to the other three metals as a result of the activities of the ITC. The copper, lead, and zinc industries are all more highly concentrated than tin and vertically integrated at an international level; and they may have been overpriced relative to tin in 1956.

Controversy within the ITC has centered on export control because that is the one device possessed by the ITC designed solely to raise the price of tin. But recently consuming nation delegates and some disinterested observers have been attributing a secondary, supportive role rather than a primary one to export control and to the ITC in general. The concern expressed is that the producing nations took advantage of the commodity boom to raise their royalties and export taxes steeply, confident that the ITC would validate the new levels by setting higher floor and ceiling prices and would defend the floor by export control. This pattern, the argument goes, was continued until the 1977 constraint on further increases in the floor and ceiling prices by the United States, Japan, West Germany, and the Soviet Union.

The result was a one-way upwards ratchet effect, in which royalties would be increased to absorb the bulk of any market-induced price increase, followed by representation to the ITC from the producing members that the new level of after-tax costs required a higher floor price. The schedules of royalties, export taxes, and tributes levied by Bolivia, Indonesia, Malaysia, and Thailand all roughly doubled between 1972 and 1974, and then doubled again by 1978, despite modest reductions in 1977 by Malaysia and in 1978 by Thailand. In 1978, royalties, export duties, and tributes accounted for approximately one-third to one-half of total production costs of tin mining in Thailand and Malaysia.[42] The burden is considerably lower in Indonesia, in which the overwhelming majority of tin-mining capacity is government-owned, and in Bolivia, where about two-thirds is government-owned and in which costs of mining are so high that the taxable capacity of the private mining sector is severely limited by world prices. Gillis calculated that in Bolivia, which levies no income tax on its private tin producers, total taxes amounted to 16.16 percent of the export value of tin in 1974.[43]

There is some suspicion that the synchronous tax increases of 1973 and 1974 were the result of overt agreement among the four principal tin producing countries. It is more probable that these increases resulted from the common but independent recognition of the opportunity created by the commodity boom and of the likelihood that the others would do the same. In either case, once the pattern was established, repetition of uniform but independent actions of this sort became more likely, particularly as all involved had reason to believe that the ITC would play its part in adjusting its price band and defending the floor.

The possibility that Thailand and Malaysia, at least, raised royalties to levels that were somewhat higher than optimal in terms of their own performance goals is indicated by later downward adjustments made by both countries at times when world tin prices were still firm or rising. The following comment on the Thai situation was made at a 1978 conference in Bangkok:

> As mentioned earlier, tin is the commodity whose price continued to climb in 1974–78, for the latter years following an excess demand for tin due to a mild economic recovery in industrialized countries. Like any government, the Thai government took advantage of the situation by increasing tin royalty. And, as expected, two things followed: A rise in tin smuggling to Singapore, and a fall in the opening of new mines.[44]

One of the most frequently made arguments for export control and the resulting capability of the ITC to assure defense of the floor price is that it will encourage new investment in tin mining. It should also be noted that if the major tin-producing nations are able to push royalties up to a level where virtually all of the additional profits earned in a period of price boom are appropriated by the government, the instability resulting from excessive private investment and a later outpouring of production from the new facilities will be curbed, since it is the prospect of higher profits rather than merely higher prices that attracts private capital. But a firm floor price cannot both encourage new investment and protect higher royalty collections unless these royalties are set at a level no higher than private user costs.

Those opposing export control have questioned whether it does in actuality encourage investment. Instead, the knowledge that they can be required to reduce output or shut down may deter investors, especially if the government bases quotas on several years' producing experience or otherwise discriminates against more recent entrants. The high shutdown costs of tin mining add strength to this argument. But tin miners can reduce output for some time by mining lower grades rather than by halting operations. Export control, then, provides an added incentive for the type of behavior noted as rational by Puey

Table 9.2. Estimated forgone exports of ITC members, 1957-76

Control period	Tonnage lost (metric tons)
December 1957-September 1960	111,400
September 1968-December 1969	14,600
January 1973-September 1973	18,300
April 1975-June 1976	39,600
Total	183,900

but not always followed by tin miners—the mining of high-grade deposits in times of high price and low-grade deposits when price is low. As long as this method of reducing output is available, investment will be deterred only if prospective investors think that the restriction is likely to lead to a proportionately greater decrease in their own output than in the price decline it prevents; and given that the ITC's restrictions now cover almost all of the world's marketed output of tin and that the short-run demand for tin is quite inelastic, this is a most unlikely event for them to anticipate. Long and severe export controls, however, such as occurred from December 1957 to September 1960, and which forced the closing of many mines, are another matter.

One knowledgeable commentator disputing the contention that the ITC has kept the price of tin above its long-run market trend level is Buffer Stock Manager de Koning, who argued in a 1977 talk that, to the contrary, the influence of the United States' strategic stockpile disposal program had been to keep the price below its natural level.[45] This is a difficult matter to assess, since expectations regarding stockpile disposals have affected prices and, very likely, levels of investment, as have expectations of ITC actions. But if one makes a crude estimate of the effects of the ITC's export controls on total output, the contention is a questionable one. A simple straight-line interpolation of ITC member nations' net exports, between the last full year before export control was imposed and the first full year after the control was ended, was used to estimate what exports would have been in the absence of control for each quarter during which control was in effect. Total actual quarterly exports of ITC members were subtracted from these figures to yield an estimate of forgone exports. The calculations are shown in table 9.2.[46]

Total disposals from the United States' stockpile, on the other hand, amounted to only 150,129 metric tons (shown as 147,758 long tons in table 3.5) from the inception of the program in 1962 through 1977. The comparison between export control and the stockpile program may be biased since the protracted and severe export controls from 1957 to 1960 were imposed in reaction to Soviet re-exports of Chinese tin. However, Soviet exports minus imports from nations other than China were less than the tonnage apparently eliminated by export control, amounting to only 72,124 metric tons between 1956 and 1961, after which the figure became negative. To compare more recent years: from 1968 through 1977 export control reduced output by an estimated 72,500 metric tons while stockpile disposals totaled 61,400 metric tons. Although the total of net Soviet exports plus United States' stockpile disposals through 1977 amounts to 222,253 metric tons against the 183,900 metric tons estimated as the effect of export control, if one compares only the export controls and stockpile disposals the conclusion seems inescapable that more tin was withheld to defend the ITC's floor prices than was sold by GSA to dampen price increases. The only significant qualification to be made is that there is no adjustment in the reduction attributed to export control to take account of smuggling; if such a correction could be made it might weaken the conclusion considerably.

The current disposal program, initiated in 1980, poses a different question, coming as it does at a time when the United States has asked that export control be eliminated from the sixth agreement, and when a group of the consuming nations has been able to restrict the use of such control. Undoubtedly these disposals have exerted a significant net downward pressure on the price of tin.

Alternatives to the ITC. Whatever the ITC's success may have been in raising tin prices—and the preponderance of evidence is that it has been quite successful—it is now one of the council's avowed and actual objectives. The equally firm and explicit avowal of the United States that it is opposed to international organizations with such a purpose may be incompatible with continued United States' membership in the ITC, as suggested by *Tin International* in its June 1980 issue.[47] But whether the continued existence of the ITC, with or without the participation of the United States, is in the interests of this country and other tin-consuming nations does not hinge on the extent to which the council has succeeded in raising the price of tin above free-market levels. Rather, the basic question is how different the industry's performance would be in this respect in the absence of the ITC.

The exercise of market power requires the ability to restrict output by some method in order to raise price. The hypothesized present division of function, stripped to its essence, is that the producing nations indicate the desired level of price through imposition of royalties; and the ITC, if consuming members holding two-thirds of the votes of that group are amenable, restricts output to the necessary degree through export control. If, as is true at the moment, consumer opposition prevents the ITC from playing its role in this process, several other options are open to the producing countries.

The producers might accept the ITC as a useful long-term price stabilizer, continue to support it, and otherwise allow the world tin market to function free from collective interference. It has been contended throughout this study that the structure of the tin industry is such that competition would then function efficiently, the one major restraint being the inability of a new smelter to enter the industry at a location of its own choosing and bid for ore supplies on a free market. This is the response of producers that is most favorable to the consuming nations, and perhaps the one sought by the delegations of the United States, Japan, and West Germany. But it is highly unlikely. If the consuming countries are able to eliminate the ITC's power to impose export control, the organization may not survive. Whether or not it does, the producers can turn to some other device for controlling output.

One alternative is that the tin producers themselves could exercise collective control of world output through a producers' cartel. Edwards has noted a number of features of the tin industry that would make such a cartel feasible: "Production is concentrated; the major consumers are almost wholly dependent on imports; the producers are countries which should find it reasonably easy to collaborate; by developing country standards they are relatively prosperous; the trade would be fairly easy for the producing countries to control (all Malaysia's tin being handled by two smelting companies, all Thailand's by one, and all Indonesia's by the state); demand is growing, albeit slowly; and demand appears fairly income elastic and, in the short run, not very price-elastic."[48] He concluded, however, that the prospects for substitution, the economic inability of some producing nations—Bolivia, certainly—to withhold sales for any length of time, and the threat of massive sales from the United States' stockpile made the long-run prospects for such a cartel unpromising.

Demand curves are thought typically to increase in elasticity as price increases. This is not the case for the long-run demand curve for tin. Edwards is correct in that if the price was raised substantially, the quantity of tin consumed in tinplate would drop off substantially enough to reduce total revenue in the long run, as can makers and others modified their production line equipment to handle substitutes. But there are no technically acceptable substitutes for tin in certain solder and bearing uses, and solder and bearings are normally insignificant cost components of their final products. Thus, still further increases in price would result in much smaller percentage reductions in purchases of tin. The long-run global profit-maximizing price of tin would probably be several times its present level with an associated output in the neighborhood of one-half the current rate. A cartel seeking such an objective would undoubtedly bring down the full force of the United States' stockpile, and in addition create severe unemployment and major dislocations in the member nations' domestic tin industries.

The discussion earlier in this chapter on the performance norms of the various producers is at variance with Edwards' opinion that "the producers are countries which should find it reasonably easy to collaborate." Even if it is not in the interest of any of the major producers to restrict output to the extent where joint profits are maximized, or even to push for individually maximizing prices, there is no reason to assume that the prices desired, in the interests of such national objectives as foreign exchange earnings and domestic employment, are the same for the several countries. Further, the costs of tin mining differ greatly among as well as within the producing nations.

For all of these reasons, the tin producers would find it difficult if not impossible to agree on the best price and then to allocate the required restrictions on output among themselves in a mutually satisfactory way. And the higher the cartel price, the greater the endemic problem of smuggling would become, as well as the need to include all present and potential minor producers.

Nevertheless, given the extent to which public ownership and control of domestic tin mining and smelting have grown in the independent producing countries, the governments of these nations cannot be expected to refrain from making some use of the economic power they have sought and obtained, in the event the ITC is abolished or stripped of its power to impose export control. The evident failure of the mystery buyer's attempt, in and of itself, should not be expected to discourage future efforts of a similar nature. The history of the modern tin industry is one of a succession of attempts by interested governments to counteract the effects of global competition that persistently frustrated private efforts to restrict production and raise prices. The international tin agreements are only the latest of these efforts.

While a formal tin cartel agreement is unlikely, it is equally unlikely that the producing countries would act completely independently of one another. A pattern similar to that which they appear to have adopted under the ITC might continue, with all of the major producers taking advantage of price booms to raise royalties, either automatically through a preexisting and possibly progressive ad valorem schedule or through changes in the rate, with each country firm in its expectation that all of the others would do likewise. Discussions among the producing nations on output, or on levels of price below which none would sell, could replace the ITC's export controls as devices for assuring that the new royalty rates could be maintained without imposing unacceptable losses on domestic miners. Prices would rise as the result of a ratchet effect until the level would be reached that was considered optimal by at least one major producer. Beyond that point, the others could not rely on that country for any further collaboration either in raising royalties or restricting production. Under present cost conditions, Thailand, as the lowest-cost producer, would probably be the first country to find the process no longer in its own best interest.

The adjustment to an equilibrium of this nature or the movement toward a new equilibrium in response to changes in demand or cost conditions would be smoother with only producers involved than under the ITC, in that it would be easier to agree on a price below which no one would sell or on appropriate levels of output if there were no need to justify the decision to consuming nations; but it would be difficult to design an enforcement mechanism as efficient as ITC export control. As long as the Penang market or a Kuala Lumpur commodities exchange set the day-to-day price for Malaysian tin, the producers would have to rely on output restrictions by the individual countries. The temptation to cheat on such restrictions has long been recognized as a major problem in maintaining collusive agreements, and a reliable method of detecting cheating is considered essential to the survival of effective oligopolistic coordination.

The efficacy of ITC export control is lessened by the ease with which tin concentrates can be smuggled and the structural flexibility that allows independent smelters outside of the producing countries, such as those in Singapore and Europe, to accomodate these illicit operations. The lack of cooperation from consuming nations, and the likelihood that at least some would actively encourage smuggling if they were not participants in or supporters of an international agreement, would heighten the problem. There would be no way of independently verifying the amounts of tin sold on long-term contracts or through individually negotiated sales of state trading companies. In all, unless producers can devise some more effective method of cooperation, they would appear to have little if anything to gain from substitution of a producers' agreement for the ITC as it was functioning until 1977.

Much would depend on the attitude of Malaysia. Without its full support, and probably leadership, no cooperative arrangement among producers could be workable. Since the Penang price now serves as a basis for the worldwide pricing of tin, it is conceivable that if Malaysia substituted a marketing board for the present arrangement, or posted predetermined prices to be paid to the smelters for tin metal, the other smelters, including the remaining private smelters outside of the producing countries, would recognize Malaysia as a price leader. But unless the Malaysian authorities were confident that the others would follow its prices, it would have nothing to gain by becoming a price leader rather than by allowing the Penang market or the Kuala Lumpur Commodities Exchange to operate subject to the government's power to set optimal royalty rates.

Malaysian officials with whom I discussed the matter in 1979 were unanimous in their views that direct control of tin output or prices would be impossible within the context of Malaysia's declared free-enterprise economic philosophy. Martin Rudner has identified a political aspect to this, noting, "Nor was the commitment to free enterprise based exclusively on ideology, for it incidentally enabled the Alliance to sublimate potentially embarrassing internal stresses over communal patterns of ownership and control, under the slogan 'human freedom.'"[49] In other words, the Alliance Party's 1957 bargain, which gave political power to the *bumiputras* was acceptable to the economically dominant ethnic Chinese only because of the protection provided for them by placing much of the nation's economic activity out of the reach of the government.

The bargain has come under increasing strain, however, with the socio-economic restructuring goals of the NEP; and the government's investment in the tin industry has grown with the establishment of MMC. Until recently, the Malaysian government has shown less interest in tin than in other commodities in which *bumiputra* participation is greater, other than restricting access to new mining sites. As both private *bumiputra* and government participation in the tin industry grow, however, this may change; and it is conceivable that Malaysia may in the future take a more active role in trying to further the economic interests of its tin industry, including establishing itself as a world price leader or, if the ITC remains in existence becoming as aggressive as Bolivia in its insistence on high price bands and heavy reliance on export control. Nevertheless, current plans for including trading in tin on the Kuala Lumpur Commodities Exchange indicate that the Malaysian government is not now interested in assuming any direct responsibility for setting the price of that metal, but rather is moving in a direction that will make it more difficult to do so in the future.

The ITC, at least as it was functioning until 1977, appears to have served the producing countries' interests about as well as any alternative that could have been established under present structural conditions and in light of the threat posed by the United States' stockpile. Recognition of this fact by Malaysia could provide an additional reason for that country's willingness to preserve the Penang market or substitute some other form of auction market-clearing pricing. ITC Executive Chairman Peter Lai stated recently that commodity agreements in general, and not just the tin agreement, "will and must operate within the free market."[50] Thus, Malaysia's abandonment of such a market could bring about the demise of the ITC even more surely than elimination of export control from its functions. If the Malaysian government were to set the price of the tin sold by that nation, with all of Indonesia's and most of Bolivia's tin sold by state trading organizations and the likelihood that Thailand would continue to simply follow Malaysia's price, there would be no role left for an international organization designating floor and ceiling prices and trying to defend them with a buffer stock, export control, or any other method.

Overall, despite the improbability of an effective cartel being formed, the ability of the producing governments to raise the price of tin seems to be roughly the same with or without reliance on the ITC's export control. The consuming countries, therefore, have nothing to gain by insistence on the elimination of such control; and by doing so they run the risk of destroying the council and thereby losing the benefits of stabilization.

The worst-case scenario from a global point of view is one in which the ITC ceases to function, the producers subsequently attempt to cartelize, fail both because of inability to reconcile differences among themselves and use of the United States' stockpile, and try again, perhaps after the stockpile is exhausted. Not only would economic performance deteriorate under the severe destabilization of price and investment that would accompany such a sequence of events, but also the sense of conflict associated with the inherent differences in economic interests between consumers and producers, and within the ranks of the producers, would be heightened to close to the maximum conceivable extent. This worst case strikes me as an improbable one, but it is worth noting both because it is not impossible and because it directs attention to an aspect of national self-interest that has been ignored up to this point. The political repercussions of an aggressive cartelization attempt countered by an equally aggressive use of the United States' stockpile might be worse than the economic effects.

While independent nations can be expected to pursue their self-interest, these interests extend beyond the economic. Malaysia's policies toward its tin industry are clearly the result of internal political concerns as well as economic interests. The Soviet Union's decision to join the ITC and China's decision to stay out were political. Similarly, the United States joined the ITC as a political response to pressures from the developing world, beginning with the establishment of UNCTAD and culminating in the Declaration of the NIEO and the Charter of Economic Rights and Duties of States, rather than as the result of a reassessment of the economic benefits and costs of membership.

An unfortunate feature of the major role of governments in the tin industry is that conflict over economic interests is escalated to the international political level. The structure of the industry, under which individual producers and consumers buy and sell on an impersonal competitive market, either directly or through an efficient network of dealers, minimizes the sense of economic conflict. The political advantages of the ITC seem overwhelming in similarly reducing the level of visible political conflict, provided all members recognize the value of and are concerned with this objective.

Some widely recognized and generally accepted political interests of the United States, including its desire for close and friendly relations with the ASEAN nations following military disengagement in Southeast Asia, and a long-standing concern for the economic and political health of Bolivia, have been damaged by the decision to withdraw from the ITC. But as long as the policy of the United States is based on adamant opposition to participation in any agreement that seeks to raise price above long-run free-market levels, constructive membership is probably impossible.

The final question: performance of the world tin industry. The final question is the extent to which the economic performance of the world tin industry diverges from the norm, or whether a norm can even be formulated.

Those who consider either producer interests or consumer interests as their sole criterion would not accept the following, but it seems a reasonable statement of a global pricing standard.

Producing countries should, ideally, retain a portion of the sales price of tin equal to the inherent in situ value of the metal removed plus the full cost of external diseconomies associated with mining operations. This may be regarded as social user cost. As in the competitive model, production should be carried out at the rate at which marginal cost equals price. In the case of tin, marginal cost is the sum of marginal extraction and smelting costs plus marginal social user cost.

If royalty charges are just high enough to capture social user cost, and if the metal is then sold on a competitive world market, the globally optimal equilibrium just described will result. The chief current source of misallocation of resources is the power of the producing nations to charge a royalty in excess of social user cost.

One of the most fundamental conclusions of this study is that the current structure of the world tin industry is highly conducive to competition. One of

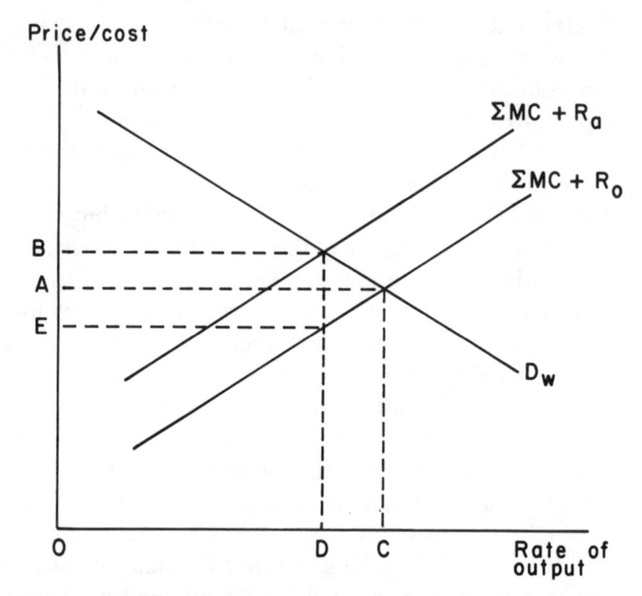

Figure 9.1. World prices and outputs in long-run equilibrium, under royalty charges equaling and exceeding the marginal social user costs of tin-producing nations

the principal features of structure underlying this aspect of performance is the basic nature of the metal itself, including high value relative to transportation costs, high content of metal in the ore as it leaves the mine, low costs and technical ease of smelting and refining most ores, and ease of storage. At the producing level, competition is facilitated by the large number of independent nonintegrated mines, coupled with ease of entry into smelting, the stage of production at which market power might otherwise be exerted. At Penang and on the LME there are well-functioning and informed competitive markets, linked to users throughout the world by an efficient network of dealers. Under the present structure of the industry, then, the world market price will tend toward a competitive equilibrium at which price equals long-run marginal cost, but with marginal cost including whatever royalties the governments of producing countries charge. The only significant qualification to this form of price determination is the pervasive downward pressure put on royalty charges and thus on prices by widespread smuggling of tin-in-concentrates.

Diagrammatically, the long-run equilibrium in such a situation, neglecting smuggling, is as shown in figure 9.1.

As before, D_w is the world demand for tin metal. The line $\Sigma MC + R_o$ represents the entire world industry's marginal extraction and smelting costs plus an optimal royalty set equal to social user cost. Under competitive conditions, $\Sigma MC + R_o$ is the industry's supply curve. The globally optimal price is OA and the associated output is *OC*. The line $\Sigma MC + R_a$ represents marginal extraction and smelting costs plus an actual royalty assumed to be higher than social user cost by the amount *BE*. The divergence from optimality is then indicated by the rise in price from *OA* to *OB* and the fall in output from *OC* to *OD*.

Only the roughest estimates can be made of the dimensions of this divergence. The magnitudes of the price and output differences are functions of the extent to which the actual royalty charge exceeds social user cost and of the elasticities of supply and demand. Neither social user cost nor the elasticities are known. A long-run elasticity of demand for tin of −1.5 and a long-run elasticity of supply of 1.0 are close to consensus figures on the basis of the econometric and other estimates reviewed in chapter 4. With these elasticities, a 10 percent shift upwards in the supply curve leads to a 4 percent increase in price and a 6 percent decrease in output. Suppose the optimal royalty equaled one-third of total cost. Then if the actual royalty were twice the optimal one, or half of the new total cost, price would be 12.2 percent above its optimal level and output would be reduced by 15 percent from the optimum.

The results are not extremely sensitive to plausible differences in assumed elasticities. A low elasticity estimate of −1.0 for demand and 0.5 for supply results in a 3.3 percent change in both price and output when the supply curve rises by 10 percent. A high estimate of −2.0 for demand and 1.5 for supply results in a 4.3 percent rise in price and an 8.6 percent decline in output.

Whether these estimates strike the reader as representing undue exploitation of consumers by producers will depend on personal value judgments. But there can be no question that in terms of pricing the industry's present performance is vastly superior to that which would result from private profit-maximizing monopoly.

There are major aspects of economic performance other than market determination of price levels. Stability of price and output has been a matter of common concern in international commodity markets. Given the short-run inelasticities of both supply and demand in the tin market, along with the technical production features of high fixed costs relative to variable and short-run irreversibilities of production, there is a widely recognized and accepted role for a price stabilization agency such as the ITC—entirely apart from what may be said for and against the use of such an organization to enhance the level of price. Evidence reviewed in this study indicates that the ITC has been markedly successful in dampening price fluctuations without undue exacerbation of short-term output instability; and that is probably as much as can be asked for in international control of a commodity market.

Long-range features of performance, such as the rate of resource depletion and the pace of technological advance, are extremely difficult to evaluate. Perhaps the most that can be said is that this study did not identify any major long-run problems or obvious changes that could be made at any level of the industry to improve its performance over future decades.

One of the best and most widely accepted definitions of workable competition is that of Jesse W. Markham, who wrote, "An industry may be judged to be workably competitive when, after the structural characteristics of its market and the dynamic forms that shaped them have been thoroughly examined, there is no clearly indicated change that can be effected through public policy measures that would result in greater social gains than social losses."[51]

By this criterion, I must conclude that the performance of the world tin industry is shaped by workable competition, notwithstanding sixty years of political pricing.

Notes

Introduction

1. *Business Times* (Malaysia), June 10, 1981. Reprinted in International Tin Council, *Notes on Tin*, July 1981.

2. The word *huge* is used advisedly. At the end of 1980, the United States' stockpile held 203,666 metric tons of tin metal, equal to 3.6 times that year's consumption of 56,362 tons, and 4.4 times imports of 45,982 tons. (Virtually all of the remainder of the nation's consumption was of tin metal recovered from scrap.) Total world output of tin for that year amounted to 248,104 tons, of which ITC members produced 170,964 tons.

3. *Comtel Reuter*, January 15, 1981. Reprinted in *Notes on Tin*, February 1981.

4. The ceiling price, expressed in pounds sterling, is approximate, as the ITC's price range is set in Malaysian ringgits. The conversions have been made at the average exchange rates for the years.

5. Failure to obtain the necessary signatures would not necessarily lead to the abandonment of any agreement. If less than 65 percent had not signed by April 30, those who had signed were to meet in June under auspices of the United Nations Conference for Trade and Development to see if they could frame an agreement of their own.

6. *Tin International*, November 1981, p. 431.

7. Texts of the stockpile agreements and quoted reactions to the renewed United States' sales taken from *Tin International*, June 1981, pp. 207 and 213.

8. *Tin International*, December 1981, p. 467.

9. *Business Times* (Malaysia), August 15, 1981. Reprinted in *Notes on Tin*, September 1981.

10. See, especially, "Prices of Tin Swing Widely," *New York Times*, November 27, 1981, p. D5, and "Tin Price Moves Jolt Market," *New York Times*, November 28, 1981, p. 29.

11. *New York Times*, November 27, 1981, p. D5.

12. *Business Times* (Malaysia), September 3, 1981. Reprinted in *Notes on Tin*, October 1981.

13. *Financial Times* (London), October 19, 1981. Reprinted in *Notes on Tin*, November 1981.

14. *Daily Telegraph* (London), November 16, 1981. Reprinted in *Tin International*, December 1981, p. 508.

15. *Observer* (London), December 6, 1981. Reprinted in *Tin International*, January 1982, p. 40.

16. *Far Eastern Economic Review*, December 11, 1981, p. 81.

17. *New York Times*, February 9, 1982, p. D23.

1. The Devil's Metal

1. C. L. Mantell, *Tin: Its Mining, Production, Technology, and Applications*, p. 3.

2. The geological, metallurgical, and engineering material in this chapter is drawn almost entirely from Mantell: J. L. Bray, *Non-Ferrous Production Metallurgy*; J. H. Cairns and P. T. Gilbert, *The Technology of Heavy Non-Ferrous Metals and Alloys*; and Kenneth Warren, *Mineral Resources*. Supplemental sources of information and insights into the pertinence of certain technical matters to the economics of the industry have been taken from William Fox, *Tin: The Working of a Commodity Agreement*, and from various issues of United States Bureau of Mines, *Minerals Yearbook*.

3. United States Central Intelligence Agency, *Soviet Tin Industry: Recent Developments and Prospects Through 1980*, Research Aid ER 77-1101 (January 1977).

4. Fox, p. 60.

5. Warren, pp. 165-67; and Robert J. Alexander, *The Bolivian National Revolution*, pp. 95-96.

6. Mantell, p. 71

7. Warren, p. 160.

8. Several references to such tactics are found in Wong Lin Ken, *The Malayan Tin Industry to 1914*.

9. Robert Gibson-Jarvie, *The London Metal Exchange: A Commodity Market*, p. 68.

10. United States Bureau of Mines, *Minerals Yearbook*, 1975, p. 1377.

11. Ibid., 1971, p. 1152.

12. Ibid., 1975, p. 1377–78.
13. Ibid., 1969, p. 1067.

2. Tin in The Economies of Producing Nations

1. U. S. Congress, Joint Economic Committee, *The United States Response to the New International Economic Order: The Economic Implications for Latin America and the United States*, pp. 15–16, 18–19.
2. Yip Yat Hoong, *The Development of the Tin Mining Industry of Malaya*, p. 110.
3. Wong Lin Ken, *The Malayan Tin Industry to 1914.*
4. L. L. Fermor, *Report upon the Mining Industry of Malaya*, pp. 21–22.
5. Wong Lin Ken, p. 17.
6. Ibid., p. 12.
7. H. S. Klein, *Parties and Political Change in Bolivia, 1880–1952*, pp. 31–32.
8. Most of the figures presented throughout this study on output of tin by various producing nations and on tin prices are taken from International Tin Council, *Statistical Yearbook, 1968* (London: 1970), *Tin Statistics, 1966–1976* and *1970–1980* (London: 1977 and 1981), and *Monthly Statistical Bulletin*, March 1982. Only sources other than these are hereafter cited individually.
9. C. H. Zondag, *The Bolivian Economy, 1952–65: The Revolution and its Aftermath*, pp. 18–19.
10. James M. Malloy, "Revolutionary Politics," in *Beyond the Revolution: Bolivia Since 1952*, eds. J. M. Malloy and R. S. Thorn, pp. 114–15.
11. Robert J. Alexander, *The Bolivian National Revolution*, p. 97.
12. James M. Malloy, *Bolivia: The Uncompleted Revolution*, p. 24.
13. Ibid., p. 25.
14. This point is made by both Alexander and Zondag.
15. R. S. Thorn, "The Economic Transformation," in Malloy and Thorn, p. 193.
16. Ibid., pp. 184–85.
17. *Metal Bulletin*, June 23, 1981. Reprinted in *Notes on Tin*, July 1981.
18. *Latin America Commodities Report*, August 1, 1980. Reprinted in *Notes on Tin*, August 1980.
19. *Tin International*, July 1981, p. 260.
20. John F. Cady, *The History of Post-war Southeast Asia*, p. 239.
21. Ibid.
22. Ibid., p. 263.
23. In addition to Cady, on this point see *inter alia* Hla Myint, *Southeast Asia's Economy*, and D. S. Paauw, "Southeast Asia's Economic Development in World Perspective," in *Southeast Asia, An Emerging Center of World Influence*, eds. W. Raymond and K. Mulliner.
24. Bank Negara Indonesia, *Report for the Financial Year 1968*, p. 215.
25. D. S. Paauw, "Southeast Asia's Economic Development in World Perspective," in Raymond and Mulliner, p. 47.
26. See, for example, "Bureaucratic Bottleneck in Indonesia," *Far Eastern Economic Review*, May 28, 1976, pp. 103–6, and in the same journal, "Suharto, A Decade of Deeds and Dilemmas," August 20, 1976, pp. 18–22.
27. *Far Eastern Economic Review*, May 22, 1969, p. 437.
28. Robert S. Milne and Diane K. Mauzy, *Politics and Government in Malaysia*, p. 129.
29. *Third Malaysia Plan, 1976–1980*, p. 7.
30. Ibid., p. 30.
31. Yip Yat Hoong, "Recent Changes in the Ownership and Control of Locally Incorporated Tin Dredging Companies in Malaya," pp. 70–88.
32. *Third Malaysia Plan*, p. 326. Number of mines taken from *Tin Statistics, 1966–1976*, p. 21.
33. Ibid., pp. 85, 88.
34. *Second Malaysia Plan, 1971–1975*, p. 172.
35. *Third Malaysia Plan*, p. 120. *New Straits Times*, March 20, 1979, p. 8.
36. States of Malaya Chamber of Mines, *Year Book* 1980, pp. 6–7.
37. *Asian Business and Industry*, p. 36.
38. *Financial Times*, February 24, 1978. Reprinted in *Notes on Tin*, February 1978.
39. States of Malaya Chamber of Mines, *Year Book* 1977, p. 3. *Far Eastern Economic Review*, September 23, 1977, p. 113.

40. John T. Thoburn, "Policies for Tin Exporters," p. 85.
41. *Wall Street Journal*, December 27, 1977, p. 16.
42. States of Malaya Chamber of Mines, *Year Book, 1977*, p. 47.
43. States of Malaya Chamber of Mines, *Year Book, 1976*, p. 13.
44. Bank Negara Malaysia. Reprinted in *Notes on Tin*, June 1979.
45. This episode aptly illustrates the tensions between the federal and state governments with respect to tin mining rights and revenues. Malaysia Mining Corporation (MMC) is 71.35 percent owned by an agency of the federal government holding shares for eventual distribution to individual *bumiputra* investors, with the remaining shares being held by a foreign concern, Charter Consolidated, Ltd. Berjuntai, in turn, is 37.4 percent owned by MMC. The federal government contended that Berjuntai's capital structure was within the NEP guidelines of 30 percent *bumiputra* ownership, and therefore the company was not subject to treatment as a foreign corporation by the state government of Selangor. Selangor, however, argued that the domestic interest in Berjuntai was only 71.35 percent of 37.4 percent, or 26.7 percent—well under the figure needed for exemption. *Far Eastern Economic Review*, February 2, 1979, p. 38. *Business Times* (Malaysia), March 10, 1979, p. 1.
46. *Business Times* (Malaysia), June 21, 1980. Reprinted in *Notes on Tin*, June 1980.
47. *Far Eastern Economic Review*, April 1, 1977, p. 33.
48. States of Malaya Chamber of Mines, *Year Book, 1977*, pp. 2–3.
49. Pernas is one of a number of public corporations established under the NEP and financed by a Bumiputra Investment Fund, designed to "selectively acquire," for the Malaysian government, "the reserved shares in enterprises with high growth potential for subsequent sale to Malays and other indigenous people." *Third Malaysia Plan*, p. 190.
50. *Tin International*, May 1979, p. 184.
51. *New Straits Times*, April 20, 1979, p. 10.
52. *Business Times* (Malaysia), August 7, 1981. Reprinted in *Notes on Tin*, September 1981.
53. *Far Eastern Economic Review*, September 23, 1977, p. 112.
54. *Mining Journal*, August 26, 1977. Reprinted in *Notes on Tin*, August 1977. *Tin International*, December 1977, p. 453.
55. *Tin International*, June 1978, p. 224.
56. *Tin International*, May 1979, p. 182.
57. *Tin International*, May 1978, p. 182.
58. *Wall Street Journal*, August 12, 1981, p. 26.
59. World Bank, *World Development Report 1981*, pp. viii, 134–35. An excellent treatment of the problems of comparing national income statistics of different nations, especially comparing national income figures of developed and developing nations, is found in Dan Usher, *The Price Mechanism and the Meaning of National Income Statistics* (London: Oxford University Press, 1967). See also, by the same author, "Income as a Measure of Productivity: Alternate Comparisons of Agricultural and Non-Agricultural Productivity in Thailand," *Econometrica* 33, no. 132 (November 1960): 430–41.

3. Tin in the Economies of Consuming Nations

1. Ceteris paribus, the less the elasticity of demand of consumers, and the more elastic the producers' supply, the greater will be the proportion of a cost increase that can be passed on. Market power will also affect the ability of a firm suffering an increase in cost to raise price. It has been argued that firms with enough discretionary market power, or under certain types of implicit oligopolistic understandings, may set price on the basis of cost plus a percentage markup and thus more than pass on any cost increase they incur to their customers; although it is an elementary proposition of microeconomic theory that a profit-maximizing firm acting independently and facing a downward-sloping but not perfectly inelastic demand curve for its product would never raise price by the full amount of a cost increase. Given a market demand curve with some elasticity, a cost increase would be reflected partly in an offsetting price increase and partly in a decrease in the industry's sales volume.
2. United States Bureau of Mines, *Minerals Yearbook 1943*, p. 731
3. Ibid., 1945, p. 724.
4. Ibid., 1957, p. 1162.
5. Mantell, pp. 485–501.
6. U. S. Department of Commerce, *Survey of Current Business* , various issues.
7. W. E. Hoare, "Tin and Its Competitors," in *Aspects of the Marketing of Tin: Papers Presented*

to the Committee on Development of the International Tin Council in 1973 and 1974, International Tin Council, pp. 30–31.

8. D. A. Robins, "Future Prospects for Tin Consumption," p. 301.
9. John E. Tilton, "Material Substitution and the Demand for Tin," p. 16.
10. Hoare, p. 48.
11. William Fox, *Tin: The Working of a Commodity Agreement*, p. 17
12. Ibid., p. 19.
13. *Minerals Yearbook, 1974*, pp. 1293–94. See also Tin Research Institute, *Tin Chemicals for Industry* (Perivale, England: Tin Research Institute, no date).
14. B. T. K. Barry and C. J. Thwaites, "Consumption Research Benefits the Tin Industry," paper no. 10, p. 26.
15. Tin Research Institute, *In Every Sphere*, pp. 17 and 21.
16. The above discussions of old and new uses for tin, prospects for substitution, and forecasts of future markets are drawn from a number of sources and several personal interviews. In addition to studies by Hoare, Robins, and the Tin Research Institute cited above, see the following: Ernest S. Hedges, *Tin in Social and Economic History*; International Tin Council and the Tin Research Institute, *Conference on Tin Consumption*; International Tin Research Council, *Annual Report, 1978*; and various issues of *Tin International* and *Minerals Yearbook*.
17. Fox, pp. 226–42, and Yip, *The Development of the Tin Mining Industry of Malaya*, pp. 250–51, 297–322.
18. Fox, p. 242.
19. These assurances and the legislation are reviewed in International Tin Council, *Annual Report, 1977–1978*, pp. 21–22.
20. United States Bureau of Mines, *Minerals Yearbook, 1962*, p. 1220.
21. United States Bureau of Mines, *Minerals Yearbook, 1963*, p. 1123.
22. United States Bureau of Mines, *Minerals Yearbook, 1967*, p. 1120.
23. United States Bureau of Mines, *Minerals Yearbook, 1968*, p. 1074.
24. International Tin Council, *The International Implications of United States Disposal of Stockpiled Tin*, p. 4.
25. *The Fifth International Tin Agreement*.
26. *Wall Street Journal*, March 11, 1980, p. 38.
27. International Tin Council, Press communique, March 13, 1980.
28. *American Metal Market*, June 27, 1980. Reprinted in *Notes on Tin*, July 1980.
29. *American Metal Market*, September 15, 1981. Reprinted in *Notes on Tin*, October 1981.
30. *Business Times* (Malaysia), November 26, 1981. Reprinted in *Notes on Tin*, December 1981.
31. *Tin International*, February 1982, p. 46.
32. States of Malaya Chamber of Mines, *Year Book*, 1980, p. 28.
33. *Tin International*, August 1981, p. 293.
34. Gordon W. Smith and George R. Schink, "The International Tin Agreement: A Reassessment," p. 721.
35. Ibid., p. 726.
36. For a cautionary note on such forecasts, however, consider the following observation of former ITC Buffer Stock Manager R. T. Adnan. "The published reports issued by dealers with a reputation of superior knowledge and foresight cannot at all times be correct in all respects, since they cannot afford to report the true situation if it would impair their own position," in "The Tin Agreement and the Market," p. 91.
37. *Wall Street Journal*, July 17, 1978, p. 20.
38. *Tin International*, April 1980, p. 138.
39. See chapter 5 for a discussion of the price elasticities of supply of and demand for tin. In light of the previous studies and evidence reviewed there, estimates of 0 to 0.3 for short-run elasticity of supply and of −0.3 to −0.5 for short-run elasticity of demand represent ranges of rough consensus.
40. Jindarah Phangmuangdee, "Analysis of United States Tin Disposals," pp. 68 and 72.
41. *New York Times*, October 23, 1979, p. A14.
42. *Business Times* (Malaysia), January 12, 1980. Reprinted in *Notes on Tin*, January 1980.

4. From Cartel to Cooperation

1. The account given here of the Bandoeng Agreement and Pool draws heavily on materials from the following sources: William Fox, *Tin: The Workng of a Commodity Agreement*; Klaus E. Knorr, *Tin Under Control*; Elizabeth S. May, "The International Tin Cartel," in William Y. Elliott et al., *International Control in the Non-ferrous Metals*, pp. 277–362; Puey Ungphakorn, "The Economics of Tin Control," Ph.D. thesis submitted to the University of London, 1949; and G. A. Roush, ed., *Mineral Industry*, various years; Yip Yat Hoong, *The Development of the Tin Mining Industry of Malaya.*
2. E. Baliol Scott, "Tin," 1921, p. 666.
3. Ibid., pp. 667, 669–72.
4. May, pp. 312–13.
5. Puey, pp. 293–96.
6. Yip, *The Development of the Tin Mining Industry of Malaya*, p. 156.
7. Knorr, p. 77.
8. Fox, pp. 116–17.
9. Knorr, p. 79.
10. Fox, p. 25.
11. Howeson was a leading figure in the Tin Producers' Association and the International Tin Committee as well as in the Anglo-Oriental Corporation. His career and role in the affairs of the tin industry were summarized in his obituary in the *New York Times* (the *London Times* did not carry any obituary item at the time of his death), reproduced in part below:

> John Howeson, "most important figure in the British tin industry," was sentenced to a year in jail in 1936 for issuing a fraudulent prospectus. He had headed a great pool in pepper, which collapsed.
> Howeson's name was once von Ernsthausen. He went to England from Calcutta, where he had been an obscure jute merchant. In 1921 he changed his name to Howeson, and embarked upon a career in tin. In eight years he rose to be head of the Anglo-Oriental Mining Corporation and a director of nineteen other companies, twelve of which he was chairman. [January 13, 1951, p. 15.]

12. See, for example, the listing in Kathleen M. Stahl, *The Metropolitan Organization of British Colonial Trade*, pp. 116–17.
13. This phenomenon is usually discussed in connection with the introduction of dredges. It should also be noted, however, that gravel-pump mining, while small scale, is highly capital intensive relative to older Chinese methods. The first gravel pump was introduced in Malaya in 1906, only six years prior to the first dredge. The technique of gravel-pump mining was adapted from Australian practice. (*Tin International*, February 1975, p. 45.) Hydraulic mining, utilizing a technology developed in the California gold fields, was introduced to Malaya in 1892. By 1900, monitors, steam engines, water pumps, and steam shovels were widely used in the Malayan tin fields. Osborne and Chappel played a major innovative role in introducing such Western devices and techniques; but the new methods were promptly adopted by Chinese miners, in some instances through takeovers of bankrupt European mines. Interesting accounts of this era are found in Wong Lin Ken, *The Malayan Tin Industry to 1914*, and G. C. Allen and Audrey G. Donnithorne, *Western Enterprise in Indonesia and Malaya.*
14. The organization that was established in 1956 to administer the postwar international tin agreements is known as the International Tin Council. Both the International Tin Committee and the International Tin Council are commonly referred to in the literature as the ITC, although there are major differences between the two bodies. Throughout this study, "ITC" will be used wherever the context is such that there can be no question as to which organization is meant. In all other cases the identification will be by full name, or the reference will be to the "committee" or the "council."
15. United States Bureau of Mines, *Minerals Yearbook, 1932–33*, p. 295.
16. J. K. Eastham, "Rationalisation in the Tin Industry," p. 13.
17. Yip, "Malaya Under the Pre-War International Tin Agreement," pp. 90–91.
18. May, p. 339.
19. John W. F. Rowe, *Primary Commodities in International Trade*, p. 141.
20. U. S. House of Representatives, Subcommittee of the House Committee on Foreign Affairs, *Tin Investigation: Report of the Subcommittee of the House Committee on Foreign Affairs.*
21. May, p. 346.
22. Knorr, p. 245.

23. Ibid.
24. Puey, p. 361.
25. May, p. 332.
26. Knorr, p. 141.
27. Puey, p. 389.
28. Fox, p. 168.
29. Knorr, p. 225.
30. Puey, pp. 419–21.
31. Yip, *The Development of the Tin Mining Industry of Malaya*, p. 284.
32. The expression "burdensome surplus" is one which was frequently used in position papers and international agreements during World War II and the early postwar years. It became a term of art, referring to a short-run level of production which resulted in an unacceptably rapid or deep decline in price or necessitated a temporary cutback in output which unduly disrupted the industry. A "burdensome surplus" was regarded as one which normal market forces could not correct without leading to excessive losses to either producers or consumers, or excessive suffering by workers, and which therefore justified governmental intervention to control price and output. To my knowledge, the expression was never defined more precisely.
33. This account draws heavily on Fox, pp. 196–98.
34. The Havana Charter is published in its entirety as an appendix to Clair Wilcox, *A Charter for World Trade*.
35. Karin Kock, *International Trade Policy and the Gatt, 1947–1967*, p. 65.
36. *United Nations Tin Conference 1950 and 1953: Summary of Proceedings*.
37. Fox, pp. 264–65.
38. Yip, *The Development of the Tin Mining Industry of Malaya*, p. 327.
39. *United Nations Tin Conference, 1960: Summary of Proceedings*, p. 270.
40. Lim Chong Yah, "A Reappraisal of the 1953 International Tin Agreement," pp. 13–24.
41. Ibid., p. 14.
42. Ibid., p. 16.
43. Rowe, pp. 84–88.
44. For a description of the extent of British capital in the Malayan tin industry and the degree of concentration of the industry in the early 1950s, see Siew Nim Chee, "The International Tin Agreement of 1953," pp. 35–53.
45. Yip, *The Development of the Tin Mining Industry of Malaya*, p. 321.
46. Fox, p. 253.
47. Ibid., pp. 252–3.
48. *First Annual Report of the International Tin Council, 1956–57*, p. 6.
49. United States Bureau of Mines, *Minerals Yearbook, 1954*, p. 1186.
50. *First Annual Report of the International Tin Council*, p. 17.
51. Fox, p. 260.
52. Ibid.
53. Ibid., p. 250.
54. Ibid., p. 285.
55. Ibid., p. 323.
56. The expression is taken from James C. Ingram, *International Economic Problems*, p. 95.
57. *United Nations Tin Conference, 1965: Summary of Proceedings*, p. 1.
58. Fox, p. 247.
59. Ibid., p. 359.
60. *United Nations Tin Conference, 1970: Summary and Proceedings*.
61. R. T. Adnan, "The Tin Agreement and the Market," p. 85.
62. Fox, p. 258.
63. Adnan, p. 89.
64. *United Nations Tin Conference, 1975*, p. 3.
65. Ibid.
66. United Nations General Assembly, *Resolutions Adopted by the General Assembly During its Twenty-Ninth Session, 17 September – 18 December 1974*, p. 52.
67. *Tin International*, November 1975, p. 388.
68. U. S. Senate, Committee on Foreign Relations, *Fifth Annual Tin Agreement: Report*.
69. Ibid., pp. 16 and 17.

70. *Comtel Reuter*, June 8, 1977. Reprinted in *Notes on Tin*, June 1977.
71. U. S. Congress, Joint Economic Committee, *Issues at the Summit*, p. 29.
72. *Mining Journal*, November 17, 1978. Reprinted in *Notes on Tin*, November 1978.
73. C. L. Gilbert, "The Post-war Tin Agreements: An Assessment," pp. 108–17.
74. Ibid., p. 109.
75. Ibid., p. 113.
76. *Tin International*, January 1977, p. 3.
77. *Far Eastern Economic Review*, May 14, 1976, p. 41.
78. *Tin International*, March 1977, p. 83. See also *Far Eastern Economic Review*, March 18, 1977, pp. 95–97.
79. *Tin International*, April 1977, p. 124.
80. International Tin Council, *Annual Report 1977–78*, p. 7.
81. Ibid., *1979–80*, p. 17.
82. International Tin Council, *The Fifth International Tin Agreement, Rules of Procedure* (no date), rule 18 (d).
83. William Page, *One Attempt at Taking a Long-Term View to Assist Decision-Makers in the World Tin Industry*, p. 22.2.
84. International Tin Council and the Tin Research Institute, *Conference on Tin Consumption*, p. 437.
85. P. A. A. de Koning, "The Achievements of the International Tin Council," p. 5.

5. Supply and Demand

1. Yip Yat Hoong, *The Development of the Tin Mining Industry of Malaya*, p. 10.
2. Meghnad Desai, "An Econometric Model of the World Tin Economy, 1948–1961."
3. Ferdinand E. Banks, "An Econometric Model of the World Tin Economy: A Comment," pp. 749–52.
4. Meghnad Desai, "An Econometric Model of the World Tin Economy: Reply to a Comment by F. E. Banks," p. 753.
5. K. A. Mohamed Ariff, *Export Trade and the West Malaysian Economy*.
6. Ibid., p. 52.
7. Ibid., pp. 52–53.
8. Ibid., p. 57.
9. Jasbir Chhabra, Enzo Grilli, and Peter Pollak, "The World Tin Economy: An Econometric Analysis," World Bank Staff Commodity Paper No. 1, June 1978.
10. Ibid., p. 16.
11. Ibid.
12. Ariff, p. 51.
13. Chhabra, Grilli, and Pollak, p. 23.
14. David Lin Shu Lim, *Supply Responses of Primary Producers*, and Jere R. Behrman, *Development, the International Economic Order, and Commodity Agreements*.
15. Puey Ungphakorn, "The Economics of Tin Control," pp. 225–32.
16. William Fox, *Tin: The Working of a Commodity Agreement*, pp. 101–3.
17. Ibid., p. 105.
18. William Robertson, *Report on the World Tin Position, with Projections for 1965 and 1970*.
19. Ibid., p. 3.
20. Ibid., p. 58.
21. Ibid., p. 152.
22. Ibid., p. 150.
23. See Joan Robinson, *Economics of Imperfect Competition*, especially chap. 25, "Monopolistic Exploitation of Labour," and chap. 26, "Monopsonistic Exploitation of Labour," for development and formal application of this concept of exploitation.
24. These results depend upon three assumptions made to approximate markets for tin-in-concentrates: first, that the supply curve is upward-sloping but not perfectly inelastic; second, that the smelter's demand curve is perfectly elastic; and third, that tin concentrates are used in constant proportion to other inputs in the production of tin metal. There is also an implicit assumption that the process of integration is costless, so that only the present value of the expected future income stream

need be paid to acquire a mine. For analysis of a more general case, with these restrictive assumptions relaxed, see Martin K. Perry, "Vertical Integration: the Monopsony Case."

25. The precise relationship between the supply and demand curves and the relative sizes of *BmpL* and *mgn* can be shown in a simple expression if D_w, MC_m, and MC_s are treated as straight lines throughout the pertinent ranges of price and output.

Let $AC_s = a + bQ$.

Then $MC_s = a + 2bQ$.

Under competitive purchasing of tin-in-concentrates, where $P = AC_s$,

(1) $OD = (OB - a)/b.$

And under monopsonistic purchasing, where $P = MC_s$,

(2) $OQ = (OL - a)/2b.$

Let world demand, D_w, be given by the expression $P = \alpha - \beta Q$.
Let n = the proportion of world output accounted for by the "representative" country. Then,

$$P = \alpha - (\beta/n)Q,$$

for the output of the representative country, on the assumption that all producers share proportionately in any increase or decrease in quantities of tin metal sold.

To simplify the notation, let $(\beta/n) = \gamma$, and

(3) $P = \alpha - \gamma Q.$

From equations (1) and (3), the equilibrium quantity under competition can be expressed as

$$OD = (\alpha - \gamma OD - a)/b.$$

Rearranging,

(4) $OD = (\alpha - a)/(b + \gamma).$

Under monopsony,

(5) $OQ = (\alpha - a)/(2b + \gamma).$

The competitive price is $OB = a + b \cdot OD$ or, substituting (4) for OD,

(6) $OB = a + b \cdot (\alpha - a)/(b + \gamma).$

The monopsonistic price, by similar derivation, is

(7) $OL = a + 2b \cdot (\alpha - a)/(2b + \gamma).$

The area *BmpL* is equal to $(OL - OB) \cdot OQ$. Inserting the expressions in (5), (6), and (7), and rearranging, we obtain

(8) $BmpL = \{[\gamma(\alpha - a)b]/[(2b + \gamma)(b + \gamma)]\} \cdot (\alpha - a)/(2b + \gamma).$

The area *mng* is equal to $\frac{1}{2}(nm \cdot mg)$. The vertical distance from Q to n, or OM, may be treated as if it were the competitive equilibrium price at the output OQ. Thus,

(9) $OM = a + b \cdot (\alpha - a)/(2b + \gamma).$

$OB - OM = nm$. From (6) and (9), and rearranging,

(10) $nm = b^2(\alpha - a)/[(b + \gamma)(2b + \gamma)].$

$OD - OQ = mg$. From (4) and (5), and rearranging,

(11) $mg = b(\alpha - a)/[(b + \gamma)(2b + \gamma)].$

Therefore,

(12) $mng = \frac{1}{2} \cdot \{b^2 (\alpha - a)/[(b + \gamma)(2b + \gamma)]\} \cdot \{[b(\alpha - a)]/[(b + \gamma)(2b + \gamma)]\}.$

Comparing (8) and (12), $BmpL > mng$ when

$$\gamma > \frac{1}{2}b^2/(b + \gamma),$$

or when,

$$(13) \qquad \gamma^2 + \gamma b - \tfrac{1}{2}b^2 > 0.$$

Since γ is a multiple of β, the slope of the world demand curve, as γ rises so does β. Thus, as the slope of the world demand curve increases, or as its elasticity at any given point decreases, the size of *BmpL* relative to that of *mng* increases, or monopsonization becomes more attractive. As noted for supply elasticity in general, the effect of a change in b, the slope of the supply curve AC_s, is ambiguous. The derivative of (13) with respect to b is simply $\gamma - b$, implying that as long as $b < \gamma$, an increase in b, corresponding to a decrease in elasticity of supply, will lead to an increase in the size of *BmpL* relative to *mng*. But when $b > \gamma$, a further increase in the size of b, and hence a further decrease in elasticity of supply, will cause *mng* to increase relative to *BmpL*.

26. See, among others: F. V. Waugh, "Does the Consumer Benefit from Price Instability?" pp. 602–14; Walter Oi, "The Desirability of Price Instability Under Perfect Competition," pp. 58–64; Benton F. Massell, "Price Stabilization and Welfare," pp. 284–98; C. P. Brown, "Some Implications of Tin Price Stabilisation," pp. 99–118; Harry G. Johnson, "The Elementary Geometry of Buffer Stock Price Stabilization," pp. 1–9 and "Commodities: Less Developed Countries' Demands and Developed Countries' Responses, " in *The New International Economic Order*, ed. J. N. Bhagwati, especially "Appendix: The Elementary Algebra of Buffer Stocks," pp. 249–51; Jere R. Behrman, *Development, the International Economic Order, and Commodity Agreements*, pp. 29–38; Stephen J. Turnovsky, "The Distribution of Welfare Gains from Price Stabilization: A Survey of Some Theoretical Issues," in *Stabilizing World Commodity Markets*, eds. F. G. Adams and S. A. Klein, pp. 119–47; and D. M. G. Newbery and J. E. Stiglitz, "The Theory of Commodity Price Stabilisation Rules: Welfare Impacts and Supply Responses," pp. 799–817.

6. Price and Output

1. Yip Yat Hoong, *The Development of the Tin Mining Industry of Malaya*, p. 2.
2. Harold W. Allen, "The International Tin Agreement: Why It Works," p. 2.
3. William Fox, *Tin: The Working of a Commodity Agreement*, p. 382.
4. These events have been discussed and interpreted by Fox, pp. 384–86, P. A. A. de Koning, "The Achievements of the International Tin Council," and C. A. J. Herkstroeter, "Some Aspects of the Marketing of Tin, the Tin Price, and the Role of the London Metal Exchange, the Penang Market and the United States Market in Establishing World Market Prices for Tin," pp. 59–81.
5. I have calculated another set of correlation coefficients for the actual values of price and output, as well as two other series (for both actual data and their logarithms) using the standard error of the estimate normalized at an initial value of 1 as the index of instability. The latter index has been used by Jere R. Behrman, *International Commodity Agreements: An Evaluation of the UNCTAD Integrated Commodity Programme*. The general patterns of statistical comparisons are similar regardless of how stability is measured; and all of the statistically significant differences reported here are corroborated by the alternative calculations. I have preferred to use correlation coefficients because there is an established procedure for measuring the significance of differences between any two of them.
6. Fox, pp. 323–53.
7. The interpretation of this confidence level is that if two samples of twenty-one observations each were drawn at random from a normally distributed universe, their correlation coefficients would differ by less than they actually did in at least 95 percent of the drawings.
8. Joseph D. Coppock, *International Economic Instability*. The Coppock Index is calculated from the following expression:

$$V_{\log} = \Sigma\{[\log(X_t + 1)/\log X_t] - m\}^2/(N - 1)$$

where X = the fluctuating variable, N = the number of years, and m = the arithmetic mean of the year-to-year differences between the logarithms of X. The Index is

$$I - I = 100\ (\text{antilog}\ \sqrt{V_{\log}} - 1)$$

9. Ibid., p. 45.
10. K. A. Mohamed Ariff, *Export Trade and the West Malaysian Economy*, p. 19.
11. Reproduced in L. N. Rangarajan, *Commodity Conflict*, p. 191.
12. "Moreover, the ability of the London tin market to withstand shocks is said to be declining, because tin price stabilization is causing speculators to transfer their activities to other markets."

United States Bureau of Mines, *Minerals Yearbook, 1935*, p. 512.

13. Fox, p. 400.

14. See J. M. Keynes, *A Treatise on Money*, 2:142–47 and *The General Theory of Employment Interest and Money*, pp. 147–64; and John R. Hicks, *Value and Capital*, pp. 135–39. For a clear summary of the "Keynes-Hicks view" and the alternative posed by Holbrook Working, see Jack Hirshleifer, "The Theory of Speculation Under Alternative Regimes of Markets," pp. 975–76 and "Speculation and Equilibrium: Information, Risk, and Markets," pp. 519–20.

15. Keynes, *A Treatise on Money*, especially pp. 142–44.

16. Hicks, p. 137.

17. Keynes, *A Treatise on Money*, p. 144.

18. Holbrook Working, "Futures Trading and Hedging," p. 320.

19. Herkstroeter, p. 60 (italics in original).

20. D. Lorenz-Meyer, "The Marketing of Tin in Europe," pp. 21–22.

21. Leland L. Johnson, on the basis of interviews with twenty firms in the New York coffee trade, came to a similar conclusion, that "hedging activities get mixed in very closely with speculative operations in the accounts of the individual trader." See Leland L. Johnson, "The Theory of Hedging and Speculation in Commodity Futures," in *The Economics of Futures Trading*, eds. B. A. Goss and B. S. Yamey, p. 87.

22. Working, p. 342.

23. Nicholas Kaldor, "Speculation and Economic Stability," p. 1.

24. Milton Friedman, *Essays in Positive Economics*, p. 175.

25. William J. Baumol, "Speculation, Profitability, and Stability," p. 263.

26. Keynes, *General Theory*, p. 159.

27. Ibid., p. 155.

28. John W. F. Rowe, *Primary Commodities in International Trade*, pp. 56–62.

29. R. T. Adnan, "The Tin Agreement and the Market," pp. 90–91.

30. Rowe, p. 54.

31. *Wall Street Journal*, September 26, 1978, p. 38.

32. *Tin International*, November 1979, p. 448.

33. *New York Times*, February 9, 1982, p. D23.

34. *Wall Street Journal*, February 26, 1982, p. 40.

35. See, for example, D. R. Williamson, "The Use of the London Metal Exchange for Hedging Purposes," p. 4. Williamson states that a contango is "normal" since the higher forward price reflects interest, warehouse storage, and insurance costs incurred in carrying tin. Thus, an actual contango is consistent with Keynes's "normal backwardation" if these carrying charges exceed the subjective value of the insurance provided to hedgers. A backwardation, Williamson continues, is an "unusual market," indicating a period of shortage of prompt tin. Backwardation is a signal to the market tin is needed immediately and serves to attract metal for prompt sale.

36. The description of hedging on the LME is based on Robert Gibson-Jarvie, *The London Metal Exchange: A Commodity Market*, pp. 119–26.

37. C. P. Brown, *Primary Commodity Control*, p. 6, noted: "The period used for measuring instability will depend on the problem under consideration. If you are interested in controlling short run speculation on an organized commodity market, comparison of average day-to-day prices may be made."

38. Peter J. W. N. Bird, "An Investigation of the Role of Speculation in the 1972 to 1975 Commodity Price Boom," has analyzed the effect of the commodity boom on the day-to-day price stability of tin on the LME, and found it to be destabilizing. The contrast between Bird's findings for the 1972–75 period and mine for 1977–79 should not be regarded as mutually contradictory. Rather, these two together appear to bear out the views of Keynes and Rowe that informed professional speculation is stabilizing under normal market conditions, but that when the markets attract uninformed speculators—as occurred during the commodity boom when people entered the commodities markets in an effort to protect themselves from erosion in the real value of money holdings—and the professional speculators are forced to consider the psychological attitudes and lack of knowledge of the newcomers, speculation can become destabilizing.

7. Industry Structure

1. E. Baliol Scott, p. 625.
2. Ibid., p. 626.
3. The Pulau Brani smelter was put on standby basis in 1955 but not finally closed until 1972. United States Bureau of Mines, *Minerals Yearbook, 1972*, p. 1223.
4. J. J. Puthucheary, *Ownership and Control in the Malayan Economy*, p. 93, discusses shareholdings in the Straits Trading Company and patterns of the tin trade in Malaya. See also K. G. Tregonning, *Straits Tin: A Brief Account of the First Seventy-five Years of the Straits Trading Company, Limited, 1887-1962*, pp. 1–25 for a discussion of the founding of the Company, including the role of the Chartered Bank in providing essential financial support.
5. *Minerals Yearbook, 1934*, p. 462.
6. For an excellent brief account of the growth of Patiño's mining enterprises in Bolivia and his expansion into smelting in an initial partnership with the National Lead Company of the United States, see Herbert S. Klein, "The Creation of the Patiño Tin Empire," pp. 3–23.
7. Roberto Arce, "Influencia de las Empresas Transnacionales en la Mineria del Estaño: el Caso de Bolivia," p. 30.
8. M. Gratacap, "L'industrie Minière en Bolivie," *Mémoire de la Société des Ingénieurs Civils de France*, July-August 1958. Cited and discussed by William Robertson, *Report on the World Tin Position with Projections for 1965 and 1970*, p. 81.
9. Ibid.
10. Arce, p. 29.
11. Peter Kilby, *Industrialization in an Open Economy: Nigeria 1945–1966*, pp. 174–77. Embel utilized an electrometallurgical process and required only inputs such as wood, coal, and local limestone that could be obtained from domestic sources. Ironically, its design was in many ways better suited to the isolated location at Jos than was the more traditional design of Makere, both from its own point of view and that of the developing nation. Embel failed because it was unable to obtain sufficient and reliable electricity.
12. *Metal Bulletin*, January 26, 1982. Reprinted in *Notes on Tin*, February 1982.
13. *Mining Magazine*, December 1977, reprinted in *Notes on Tin*, December 1977. *Mining Journal*, September 22, 1978, reprinted in *Notes on Tin*, September 1978.
14. *Tin International*, January 1980, p. 12.
15. United Nations Industrial Development Organization, *Non-Ferrous Metals Industry*, p. 68.
16. *Tin International*, December 1980, p. 520.
17. *Metal Bulletin*, August 4, 1981. Reprinted in *Notes on Tin*, September 1981.
18. International Tin Council, *Monthly Statistical Bulletin*, March 1982. The number of operating mines reflects the depressed state of the world market in 1981. Comparable figures for 1978 were 898 mines, the sum of 803 gravel pumps, 16 open-cast, 53 dredges, 23 underground, and 3 other.
19. Wong Lin Ken, *The Malayan Tin Industry to 1914*, and Yip Yat Hoong, *The Development of the Tin Mining Industry of Malaya*, pp. 99–105, have excellent discussions of the reasons for the failure of British and other foreign mining companies to succeed in gravel-pump mining. A striking contemporary account is given in Frank A. Swettenham, *About Perak*, pp. 33–35, as follows.

> The European first bores. I have said it is not an altogether reliable plan but it may, if carefully done, be almost as successful as complete trust in the Malay Pâwang [medicine man]. Then, usually, European mining is done by companies, and company's money is almost like Government money. It is not of too much account because it seems to belong to no one in particular, and is given by Providence for the support of deserving expert and often travelled individuals. Several of these are necessary to fairly start a European mining venture, and they are mostly engaged long before they are wanted. There is the manager and the sub-manager, the accountant, the engineer, the smelter—but do we not all know the oft-told tale that never seems to point to any moral at all. Machinery is bought, houses are built, in fact the capital of the company is spent—no doubt that is what it was subscribed for, and the share-holders shall not be disappointed if the management, the experts and the employés can help it. And then—if ever things get so far—some Chinese are employed on wages or contract, the former for choice, to remove the overburden. After possibly a series of great hardships to the staff and disasters to the company, it is found that the tin raised is infinitesimal in value when compared with the rate of expenditure and that longer the work goes on

the greater will be the losses. This is usually discovered when the paid up capital is all but exhausted. The company is wound up and the State [Perak] gets a bad name with investors, and the only people who really enjoy themselves are the neighboring Chinese miners who buy the mine and plant for an old song and make several large fortunes out of working on their own ridiculous and primitive methods.

20. Puthucheary, pp. 84–85.

21. Yip, *The Development of the Tin Mining Industry of Malaya*, pp. 138–48, 160–61.

22. D. R. Williamson, "Tin Shares—Some of the Problems of Marketability and Company Structure Examined," pp. 138–49.

23. Kathleen M. Stahl, *The Metropolitan Organization of British Colonial Trade*, p. 104.

24. John W. F. Rowe, *Primary Commodities in International Trade*, p. 16.

25. For a description of Pernas, see chap. 2, note 49.

26. *Tin International*, February 1978, p. 51, and *Far Eastern Economic Review*, April 1, 1977, p. 38.

27. Kuala Lumpur Stock Exchange, *Annual Companies Handbook*.

28. The information on interlocking directorates between the Malaysian smelters and mines is taken from Williamson, the *Financial Times* (London), and *Mining International Yearbook, 1979*.

29. International Tin Council, *Monthly Statistical Bulletin*, March 1982. The quarterly output figures are misleading in that the suction boats and offshore dredges cannot operate during the monsoon season in the second and third quarters. For the entire year 1980, percentages of production were dredges, 12.9; suction boats, 42.6; gravel pumping and hydraulicking, 29.6; dulang washing, 2.9; and other, 12.0.

30. Puey, "The Economics of Tin Control," p. 155.

31. For an account of the development of dredge mining in Thailand since 1907, see Wit Sakyarakwit, "Tin: A Comparison of Gravel Pump and Dredging Mining." See also International Tin Council, *Statistical Yearbook, 1968*, pp. 128–29.

32. *The Nation* (Bangkok), July 2, 1978, special supplement, p. 10.

33. A few mining engineers with whom I discussed this question doubted the claim. In their opinions, dredges can rework areas previously mined by the small boats efficiently enough so that the total tin ore recovered will be just about equal to the amount that would have been recovered by dredging alone, but admittedly at the cost of moving quite a bit more sand and silt. A similar view is expressed in a recent study: John Thoburn, *Multinationals, Mining, and Development: A Study of the Tin Industry*, pp. 85 and 137.

34. *Business Review*, June 1976, pp. 204–5 (italics in original). See also *Wall Street Journal*, January 3, 1977, p. 8, and *Business in Thailand*, September 1978, pp. 29–35.

35. *Business Review*, November 1975, p. 527.

36. *Investor* (Bangkok), March 1976, p. 108.

37. *Bangkok Post*, February 12, 1979, sec. 2, p. 1.

38. *Tin International*, February 1969, p. 59, and May 1979, p. 182.

39. Thailand, Department of Mineral Resources, *The Mining Industry of Thailand*, September 1978, insert.

40. R. Olumfeni, *An Economic History of Nigeria 1860–1960*, pp. 176–81.

41. Gold Base and Metal Mines of Nigeria Ltd., *Annual Report for 1977*, and Ex-Lands Nigeria Ltd., 1977 *Annual Report*, both reprinted in *Notes on Tin*, November 1977.

42. Nigerian Mining Corporation, *Annual Report and Accounts for the Year Ended 31st December 1977*, p. 1.

43. As a result of the nearly simultaneous takeovers of Anglo-Oriental by MMC and of Amalgamated Tin Mines of Nigeria by NMC, the capital structure of Amalgamated became quite complex. Amalgamated Tin Mines of Nigeria (Holding) Ltd. transferred 60 percent of its equity in Amalgamated Tin Mines of Nigeria Ltd. to NMC and to a pool for distribution to its Nigerian employees. Anglo-Oriental retained a 31.5 percent ownership interest in the parent holding company, giving it a 12.6 percent interest in the operating company. MMC subsequently assumed this interest. Makeri Smelting Company Ltd., itself 40 percent owned by NMC, holds 9.6 percent of the shares of Amalgamated (Holding).

44. Still earlier, from 1722 until the British occupied the area in 1795, the Dutch East India Company held a monopoly on the trade of Banka and Belitung tin, granted by the Sultan of Palembang in return for assistance in retaining his throne. Smuggling was at least as common a response to restraints on tin trading then as it is now. Contemporary observers estimated that during the 1770s and

'80s less than half of the islands' tin output was delivered to the Dutch factory at Palembang, despite the fact that Dutch warships patrolled the area to enforce the Sultan's decree that all ships leaving Banka and Belitung could proceed only to Palembang. Smuggling was assisted by corrupt local Dutch officials, and the Sultan himself was involved in the illegal trade. James C. Jackson, "Mining in 18th Century Bangka: the Pre-European Exploitation of a Tin Island," pp. 40–41.

See also George C. Allen and Audrey G. Donnithorne, *Western Enterprise in Indonesia and Malaya*, pp. 22–29 and 168–70, for accounts of the development of the Netherlands Indies tin mining and smelting industries in the nineteenth century under Prince Henry of Holland, and the initiative taken by Prince Henry in forming the Billiton Company.

45. Thoburn, *Multinationals, Mining, and Development*, p. 80.

46. *Financial Times* (London), August 21, 1980. Reprinted in *Notes on Tin*, September 1980.

47. Malcolm Gillis et al., *Taxation and Mining: Nonfuel Minerals in Bolivia and Other Countries*, p. 51.

48. William Fox, *Tin: The Working of a Commodity Agreement*, pp. 66–67.

49. *Notes on Tin*, June and August 1977.

50. *Tin International*, December 1975, p. 429.

51. *Tin International*, April 1978, p. 142, and August 1979, p. 300.

52. *Minerals Yearbook, 1971*, p. 1161.

53. Ibid., p. 1163.

54. Ibid., *1972*, p. 1219.

55. *Tin International*, January 1975, pp. 7 and 40.

56. J. S. Bettencourt et al., "Brazilian Tin Deposits and Potential."

57. *American Metal Market*, October 21, 1977. Reprinted in *Notes on Tin*, October 1977.

58. *Tin International*, August 1978, p. 300.

59. Kung-Pin Wang, *Mineral Resources and Basic Industries in the People's Republic of China*, pp. 153–55.

60. *Far Eastern Economic Review*, June 6, 1980, p. 58.

61. *Tin International*, September 1980, p. 364. For a recent survey of Chinese tin smelting facilities and plans, see Thomas S. Mackey, "Tin Smelting and Refining in China," pp. 129–32.

62. Puey, pp. 196–202, Yip, *The Development of the Tin Mining Industry of Malaya*, Puthucheary, and Siew, also discuss these groups. In addition, see United States Congress, Senate, *Sixth Report of the Preparedness Subcommittee of the Committee on Armed Services, Tin, 1951*.

8. Mining Costs and Market Pricing

1. John T. Thoburn, "Commodity Prices and Appropriate Technology: Some Lessons from Tin Mining," pp. 35–52.

2. John T. Thoburn, *Multinationals, Mining, and Development: A Study of the Tin Industry*, p. 133.

3. Bernard C. Engel and Harold W. Allen, *Tin Production and Investment*.

4. Yip Yat Hoong, "The Domestic Implementation of the 1953 International Tin Agreement in Malaya," p. 60.

5. Puey, "The Economics of Tin Control," pp. 332–41.

6. Engel and Allen, p. 76.

7. Abdullah Hasbi Hassan, "A Regional Approach to Research on Tin Production," p. 4.

8. Percy P. Courtenay, "International Tin Restriction and its Effects on the Malayan Tin Mining Industry," pp. 223–31.

9. Tan Teong Hean, "A Study of the Sources and Nature of Credit in Chinese Tin Mining in Malaya," p. 25.

10. Estimates made available from internal records of of the International Tin Council.

11. States of Malaya Chamber of Mines, *Year Book, 1980*, pp. 34–37.

12. Sirman Widiatmo, "Vertical Integration in Indonesian Tin Mining," pp. 11–15.

13. The Penang market is described in the following: Ewen Fergusson, "The Marketing of Tin in Malaysia," pp. 1–5; Ahmad Zubeir Noordin, "The Penang Tin Market," pp. 47–52; Datuk Keramat Smelting Sdn. Berhad, "Guide to Straits Tin Market: the Operations of Straits Tin Market"; Straits Trading Company Limited, "Penang Tin Market," and "Tin Sales and Conditions Relating to Tin Bids."

14. Straits Trading Company Limited, "Penang Tin Market," p. 1.

15. For an exchange of views on the extent of this practice see Noordin, presentation and discussion, pp. 54–55.
16. Kathleen M. Stahl, *The Metropolitan Organization of British Colonial Trade*, p. 114.
17. Yip Yat Hoong, "The Marketing of Tin Ore in Kampar," pp. 45–55.
18. Fergusson, p. 5.
19. K. G. Tregonning, *Straits Tin: A Brief Account of the First Seventy-five Years of the Straits Trading Company, Limited, 1887-1962*, pp. 40–42.
20. *Tin International*, March 1980, p. 94. See also, for early discussions of the likely effects of the KLCE: J. W. Landon, "Prospects for a Kuala Lumpur Commodity Exchange."
21. Commodities Research Unit Ltd., *Establishment of a Tin Exchange in Malaysia: Draft Report Submitted to the Ministry of Primary Industries, Federation of Malaysia*, August 1977, p. 31.
22. Ibid., p. 54.
23. Ibid., p. 56.
24. *Metal Bulletin*, July 24, 1981. Reprinted in *Notes on Tin*, August 1981.
25. Ismail Ahmad and H. J. Wilson, "Establishment of a Tin Futures Exchange in Malaysia."
26. For an excellent discussion of the advantages and disadvantages of the Penang market from the point of view of the tin trader, see C. A. J. Herkstroeter,"Some Aspects of the Marketing of Tin, the Tin Price, and the Role of the London Metal Exchange, the Penang Market and the United States Market in Establishing World Market Prices for Tin," pp. 70–78.
27. *Comtel Reuter*, October 21, 1980. Reprinted in *Notes on Tin*, October 1980.
28. *Business Times* (Malaysia), November 24, 1980. Reprinted in *Notes on Tin*, December 1980.
29. Ahmad and Wilson, p. 4.
30. *Tin International*, July 1979, p. 272.
31. R. T. Adnan, "The Tin Agreement and the Market," p. 89.
32. *New York Times*, April 1, 1980, p. D5, and April 6, 1980, pp. F1, F9.
33. "Workably" competitive is used in the sense of the concept first formulated in John M. Clark, "Toward a Concept of Workable Competition," pp. 241–56. Clark and others have noted that in some markets all of the structural conditions prescribed in the model of pure and perfect competition cannot possibly exist, but that if the incentives for competition could not be improved upon by any practicable change in conditions the market should be viewed as "workably" competitive. See also Jesse W. Markham, "An Alternative Approach to the Concept of Workable Competition," pp. 349–61, and Stephen Sosnick, "A Critique of Concepts of Workable Competition," pp. 380–423.
34. Herkstroeter, p. 61.
35. *Tin International*, January 1979, pp. 6–7.
36. Herkstroeter, p. 62.
37. *American Metal Market*, November 16, 1964, p. 1459.

9. Economic Performance and Public Policy

1. The discussion of the economics of the mine, throughout this section, draws heavily on the following: Harold Hotelling, "The Economics of Exhaustible Resources," pp. 137–75; Orris C. Herfindahl, "Some Fundamentals of Mineral Economics," pp. 131–38; Orris C. Herfindahl and Allen V. Kneese, *Economic Theory of Natural Resources*, pp. 114–84, especially pp. 114–38; Anthony C. Fisher and John V. Krutilla, "Conservation, Environment, and the Rate of Discount," pp. 358–70; D. W. Pearce and J. Rose, eds., *The Economics of Natural Resource Depletion*; Frederick M. Peterson and Anthony C. Fisher, "The Exploitation of Extractive Resources: A Survey," pp. 681–721; Robert S. Pindyck, "The Optimal Exploration and Production of Nonrenewable Resources," pp. 841–61. Milton C. Weinstein and Richard J. Zeckhauser, "Optimal Consumption of Depletable Resources," pp. 371–92; and M. L. Cropper, Milton C. Weinstein, and Richard J. Zeckhauser, "The Optimal Consumption of Depletable Natural Resources: An Elaboration, Correction, and Extension," pp. 337–44.
2. Peterson and Fisher, p. 702.
3. Pearce and Rose, pp. 118–39.
4. Peterson and Fisher, p. 711.
5. A monopolist, under certain exceptional circumstances, may underconserve rather than overconserve. See Weinstein and Zeckhauser, pp. 387–89.
6. John W. F. Rowe, *Primary Commodities in International Trade*, pp. 184–88. See also, Klaus E. Knorr, *Tin Under Control*, pp. 246–47, and Christopher D. Rogers, "Consumer Participation in the International Tin Agreements," pp. 113–29.

7. Anthony Edwards, *The Potential for New Commodity Cartels*, p. 2.

8. Joseph A. Schumpeter, *The Theory of Economic Development*, and *Capitalism, Socialism, and Democracy*.

9. International Tin Council and the Tin Research Institute, *Conference on Tin Consumption*, p. 412.

10. Rowe, p. 128.

11. Ibid., p. 198.

12. International Tin Council, *Annual Report, 1978*. Tin Research Institute, *In Every Sphere*.

13. Mines Research Institute Malaysia, *Institut Penyeldikan Galian Malaysia* (*The Mines Research Institute Malaysia*).

14. Yip Yat Hoong, *The Development of the Tin Mining Industry of Malaya*, p. 284.

15. Ibid., p. 273.

16. In addition to Joseph D. Coppock, *International Economic Instability*, see also: John R. Hanson II, "Export Instability in Historical Perspective," pp. 293–310; Robert G. Hawkins, José Epstein, and Joaquin Gonzales, *Stabilization of Export Receipts and Economic Development*; Alasdair I. McBean, *Export Instability and Economic Development*; David L. McNicol, *Commodity Agreements and Price Stabilization*; and D. T. Nguyen, "The Implications of Price Stabilization for the Short-term Instability and Long-term Level of LDCs' Export Earnings." pp. 149–54.

17. Ralph G. M. Sultan, *Pricing in the Electrical Oligopoly*, vol. 1, p. 322.

18. Roland Brown and Mike Faber, *Some Policy and Legal Issues Affecting Mining Legislation in African Commonwealth Countries*, p. 3.

19. The major tax on tin in Bolivia is the "Regalia," which is nominally a profits tax. It is based on the current LME price of tin less "presumptive cost." The presumptive cost is so low, however, that the Regalia is actually a progressive royalty charge.

20. David N. Smith and Louis T. Wells, *Negotiating Third-World Mineral Agreements*, pp. 34–35.

21. Malcolm Gillis et al., *Taxation and Mining: Nonfuel Minerals in Bolivia and Other Countries*, pp. 122–23.

22. States of Malaya Chamber of Mines, *Year Book, 1980*, pp. 6–7.

23. F. J. Blok, "Considerations Related to Investments in Tin Mining," p. 6.

24. John T. Thoburn, *Multinationals, Mining, and Development*, pp. 114, 158–59.

25. Robert F. Conrad and R. Bryce Hool, *Taxation of Mineral Resources*, pp. 35–40.

26. *Tin International*, June 1980, p. 223.

27. United Nations Conference on Trade and Development, *Sixth International Tin Agreement*.

28. *Business Times* (Malaysia), July 13, 1981. Reprinted in *Notes on Tin*, August 1981.

29. For a discussion of slight differences in purity of tin, even of the same grade, from different sources that might be important to some users, and the extent of reliance on brand names in the trade, see N. J. B. Pocock, "The Qualities and Forms of Tin and the Requirements of the Consumer," pp. 63–78.

30. McNicol, pp. 45–61, using an approach similar to that illustrated in figure 4.4 above and discussed in chapter 4, has estimated net gains in producers' minus consumers' surplus when supply varies, and net gains in consumers' minus producers' surplus when demand varies, for a number of commodities including tin, assuming that variations of ±10 percent in output are curbed by buffer stock stabilization.

31. R. T. Adnan, "The Tin Agreement and the Market," p. 85.

32. Harold W. Allen, "How the Tin Agreement Works," p. 452.

33. Bernard C. Engel, "International Tin Agreements," in *A New International Commodity Regime*, eds. Geoffrey Goodwin and James Mayall, pp. 87–88.

34. Adnan, p. 85.

35. Puey, "The Economics of Tin Control," p. 17.

36. Rowe, p. 220.

37. *The Fifth International Tin Agreement, as adopted 21 June 1975*, p. 20.

38. Yip, *The Development of the Tin Mining Industry of Malaya*, pp. 328–29.

39. Siew Nim Chee, "The International Tin Agreement of 1953," p. 52.

40. United States Congress, Joint Economic Committee, *Issues at the Summit*, p. 29.

41. American Metal Market, *Metal Statistics*, various issues.

42. Estimates based on confidential information.

43. Gillis et al., p. 203.

44. Narongchai Akrasanee and Vinyu Vichit-Vadakan, eds., *ASEAN Co-operation in Selected Primary Commodities*, vol. 2, p. 215.

See also, for opinions that the Thai government had raised royalties too high by 1977, "Royalties a Key: Prospects for Tin," *Business in Thailand*, September 1977, pp. 44–49, and "Tin Mining in Crisis," *The Investor* (Thailand), November 1977, pp. 36–40.

45. *Tin International*, December 1977, p. 462.

46. International Tin Council, *Annual Report*, several issues; *Statistical Yearbook, 1968*; *Tin Statistics*, 1966–1976; *Monthly Statistical Bulletin*, June 1979.

47. *Tin International*, June 1980, p. 223.

48. Edwards, p. 83.

49. Martin Rudner, *Nationalism, Planning, and Economic Modernization in Malaysia*, p. 25.

50. *American Metal Market*, October 21, 1977. Reprinted in *Notes on Tin*, October 1977.

51. Jesse W. Markham, "An Alternative Approach to the Concept of Workable Competition," p. 361.

Bibliography

Books and pamphlets

Adams, F. Gerard, and Sonia A. Klein, eds. *Stabilizing World Commodity Markets.* Lexington, Mass.: D. C. Heath, 1978.

Alexander, Robert J. *The Bolivian National Revolution.* New Brunswick, N.J.: Rutgers University Press, 1958.

Allen, George C. and Audrey G. Donnithorne. *Western Enterprise in Indonesia and Malaya.* New York: Macmillan, 1957.

Ariff, K. A. Mohamed. *Export Trade and the West Malaysian Economy.* Kuala Lumpur: University of Malaya, 1972.

Behrman, Jere R. *Development, the International Economic Order, and Commodity Agreements.* Reading, Mass.: Addison-Wesley, 1978.

______. *International Commodity Agreements: An Evaluation of the UNCTAD Integrated Commodity Programme.* Washington: Overseas Development Council, 1977.

Bhagwati, Jagdish N., ed. *The New International Economic Order.* Cambridge: MIT Press, 1977.

Bray, John L. *Non-Ferrous Production Metallurgy.* New York: Wiley, 1941.

Brown, C. P. *Primary Commodity Control.* Kuala Lumpur: Oxford University Press, 1975.

Brown, Roland, and Mike Faber. *Some Policy and Legal Issues Affecting Mining Legislation in African Commonwealth Countries.* London: Commonwealth Secretariat, 1977.

Cady, John F. *The History of Post-war Southeast Asia.* Athens: Ohio University Press, 1974.

Cairns, John H., and Peter T. Gilbert. *The Technology of Heavy Non-Ferrous Metals and Alloys.* London: George Newnes Ltd., 1967.

Conrad, Robert F., and R. Bryce Hool. *Taxation of Mineral Resources.* Lexington, Mass.: D. C. Heath, 1980.

Coppock, Joseph D. *International Economic Instability.* New York: McGraw-Hill, 1962.

Edwards, Anthony. *The Potential for New Commodity Cartels.* London: Economist Intelligence Unit Ltd., 1975.

Elliott, William Y., Elizabeth S. May, John W. F. Rowe, Alex Skelton, and Donald H. Wallace. *International Control in the Non-Ferrous Metals.* New York: Macmillan, 1937.

Engel, Bernard C. and Harold W. Allen. *Tin Production and Investment.* London: International Tin Council, 1979.

Fermor, Lewis L. *Report upon the Mining Industry of Malaya.* Kuala Lumpur: Federated Malay States Government Press, 1940.

Fox, William. *Tin: The Working of a Commodity Agreement.* London: Mining Journal Books, 1974.

Friedman, Milton. *Essays in Positive Economics.* Chicago: University of Chicago Press, 1953.

Gibson-Jarvie, Robert. *The London Metal Exchange: A Commodity Market.* London: Woodland-Faulkner, 1976.

Gillis, Malcolm, Meyer W. Bucovetsky, Glenn P. Jenkins, Ulrich Peterson, Louis T. Wells,

Jr., and Brian D. Wright. *Taxation and Mining: Nonfuel Minerals in Bolivia and Other Countries*. Cambridge, Mass.: Ballinger, 1978.

Goodwin, Geoffrey, and James Mayall, eds. *A New International Commodity Regime*. New York: St. Martin's Press, 1980.

Goss, B. A., and Basil S. Yamey. *The Economics of Futures Trading*. New York: Wiley, 1976.

Hawkins, Robert G., José Epstein, and Joaquin Gonzales. *Stabilization of Export Receipts and Economic Development*. New York: New York University Graduate School of Business Administration, 1966.

Hedges, Ernest S. *Tin in Social and Economic History*. London: Edward Arnold Ltd., 1964.

Herfindahl, Orris C., and Allen V. Kneese. *Economic Theory of Natural Resources*. Columbus: C. E. Merrill, 1974.

Hicks, John R. *Value and Capital*, 2d ed. Oxford: Clarendon Press, 1946.

Ingram, James C. *International Economic Problems*. New York: Wiley, 1966.

International Tin Council. *Aspects of the Marketing of Tin: Papers Presented to the Committee on Development of the International Tin Council in 1973 and 1974*. London: International Tin Council, no date.

———. *Fourth World Conference on Tin: Kuala Lumpur, 1974*, vol. 4, *Marketing and Consumption*. London: International Tin Council, 1975.

———. *The International Implications of United States Disposal of Stockpiled Tin*. London: International Tin Council, 1973.

———. *Trade in Tin 1960–1974*. London: International Tin Council, no date.

———, and the Tin Research Institute. *Conference on Tin Consumption*. London: International Tin Council, 1972.

Keynes, John Maynard. *The General Theory of Employment Interest and Money*. London: Macmillan, 1936.

———. *A Treatise on Money*, vol. 2. New York: Harcourt, Brace, 1930.

Kilby, Peter. *Industrialization in an Open Economy: Nigeria 1945–1966*. Cambridge: Cambridge University Press, 1960.

Klein, Herbert S. *Parties and Political Change in Bolivia 1880–1952*. Cambridge: Cambridge University Press, 1969.

Knorr, Klaus E. *Tin Under Control*. Stanford: Food Research Institute, 1945.

Kock, Karin. *International Trade Policy and the Gatt, 1947–1967*. Stockholm: Almqvist & Wiksell, 1969.

Lim, David L. S. *Supply Responses of Primary Producers*. Kuala Lumpur: Penerbit University Malaya, 1975.

McBean, Alasdair I. *Export Instability and Economic Development*. Cambridge: Harvard University Press, 1966.

McNicol, David L. *Commodity Agreements and Price Stabilization*. Lexington, Mass.: D. C. Heath, 1978.

Malloy, James M. *Bolivia: The Uncompleted Revolution*. Pittsburgh: Pittsburgh University Press, 1970.

———, and Richard S. Thorn, eds. *Beyond the Revolution: Bolivia Since 1952*. Pittsburgh: Pittsburgh University Press, 1971.

Mantell, Charles L. *Tin: Its Mining, Production, Technology, and Applications*. 2d. ed. New York: Hafner, 1970.

Milne, Robert S., and Diane K. Mauzy. *Politics and Government in Malaysia*. Singapore: Federal Publications, 1978.

Ministry of Primary Industries of the Government of Malaysia and International Tin Council. *Fifth World Conference on Tin, Kuala Lumpur, 19–23 October 1981.* London: International Tin Council, forthcoming.

Myint, Hla. *Southeast Asia's Economy*. New York: Praeger, 1972.

Narongchai Akrasanee and Vinyu Vichit-Vadakan, eds. *ASEAN Co-operation in Selected Primary Commodities*, 2 vols. Bangkok: United Nations Asian and Pacific Development Institute, August 1978.

Olumfeni, R. *An Economic History of Nigeria 1860–1960*. New York: Africana Publishing Company. 1973.

Pearce, David W., and James Rose, eds. *The Economics of Natural Resource Depletion.* London: Macmillan, 1975.

Puthucheary, James J. *Ownership and Control in the Malayan Economy*. Singapore: Eastern Universities Press, 1960.

Rangarajan, L. N. *Commodity Conflict*. Ithaca, N. Y.: Cornell University Press, 1978.

Raymond, Wayne, and K. Mulliner, eds. *Southeast Asia, An Emerging Center of World Influence*. Athens: Ohio University Center for International Studies, 1977.

Robertson, William. *Report on the World Tin Position, with Projections for 1965 and 1970*. London: International Tin Council, 1965.

Robinson, Joan. *Economics of Imperfect Competition*. London: Macmillan, 1933.

Rowe, John W. F. *Primary Commodities in International Trade*. Cambridge: Cambridge University Press, 1965.

Rudner, Martin. *Nationalism, Planning, and Economic Modernization in Malaysia.* Beverly Hills, Calif.: Sage Publications, 1975.

Schumpeter, Joseph A. *Capitalism, Socialism, and Democracy*. New York: Harper & Brothers, 1942.

______. *The Theory of Economic Development*. Translated by Redvers Opie. Cambridge: Harvard University Press, 1934.

Smith, David N., and Louis T. Wells, Jr. *Negotiating Third-World Mineral Agreements.* Cambridge, Mass.: Ballinger, 1975.

Stahl, Kathleen M. *The Metropolitan Organization of British Colonial Trade*. London: Faber & Faber, 1951.

Sultan, Ralph G. M. *Pricing in the Electrical Oligopoly*, 2 vols. Cambridge: Harvard University Press, 1974.

Swettenham, Frank A. *About Perak*. Singapore: Straits Times Press, 1893.

Thoburn, John T. *Multinationals, Mining, and Development: A Study of the Tin Industry.* Westmead, England: Gower Publishing Co., 1981.

Tin Research Institute. *In Every Sphere*. Perivale, England: Tin Research Institute, no date.

______. *Tin Chemicals for Industry*. Perivale, England: Tin Research Institute, no date.

Tregonning, K. G. *Straits Tin: A Brief Account of the First Seventy-five Years of the Straits Trading Company, Limited, 1887–1962*. Singapore: Straits Times Press, 1962.

Usher, Dan. *The Price Mechanism and the Meaning of National Income Statistics*. London: Oxford University Press, 1968.

Wang, Kung-Ping. *Mineral Resources and Basic Industries in the People's Republic of China*. Boulder: Westview Press, 1977.

Warren, Kenneth. *Mineral Resources*. New York: Wiley, 1973.

Wilcox, Clair. *A Charter for World Trade*. New York: Macmillan, 1949.

Wong Lin Ken. *The Malayan Tin Industry to 1914*. Tucson: University of Arizona Press, 1965.

Yip Yat Hoong. *The Development of the Tin Mining Industry of Malaya*. Kuala Lumpur: University of Malaya Press, 1969.

Zondag, Cornelius H. *The Bolivian Economy, 1952–65: The Revolution and its Aftermath*. New York: Praeger, 1966.

Articles

Adnan, R. T. "The Tin Agreement and the Market." International Tin Council and Tin Research Institute. *Conference on Tin Consumption* (London: International Tin Council, 1972), pp. 85–95.

Ahmad, Ismail, and H. J. Wilson. "Establishment of a Tin Futures Exchange in Malaysia." Paper no. 15. Ministry of Primary Industries of the Government of Malaysia and International Tin Council. *Fifth World Conference on Tin, Kuala Lumpur, 19–23 October 1981* (London: International Tin Council, forthcoming).

Allen, Harold W. "How the Tin Agreement Works." International Tin Council and the Tin Research Institute. *Conference on Tin Consumption* (London: International Tin Council, 1972), pp. 451–70.

———. "The International Tin Agreement: Why It Works." Supplement to *Tin International*, December 1975.

Banks, Ferdinand E. "An Econometric Model of the World Tin Economy: A Comment." *Econometrica* 40, 4 (July 1972): 749–52.

Barry, B. K. T. and C. J. Thwaites. "Consumption Research Benefits the Tin Industry." Paper no. 10. Ministry of Primary Industries of the Government of Malaysia and International Tin Council. *Fifth World Conference on Tin, Kuala Lumpur, 19-23 October 1981* (London: International Tin Council, forthcoming).

Baumol, William J. "Speculation, Profitability, and Stability." *Review of Economics and Statistics* 39, 3 (August 1957): 263–71.

Bettencourt, J. S., E. C. Damasceno, W. S. Fontanelli, J. R. M. Franco, and N. M. Pereira. "Brazilian Tin Deposits and Potential." Paper no. 3(ii). Ministry of Primary Industries of the Government of Malaysia and International Tin Council. *Fifth World Conference on Tin, Kuala Lumpur, 19–23 October 1981* (London: International Tin Council, forthcoming).

Blok, F. J. "Considerations Related to Investments in Tin Mining." Paper no. 18(iii). Ministry of Primary Industries of the Government of Malaysia and International Tin Council. *Fifth World Conference on Tin, Kuala Lumpur, 19–23 October 1981* (London: International Tin Council, forthcoming).

Brown, C. P. "Some Implications of Tin Price Stabilisation." *Malayan Economic Review* 17, 1 (April 1972): 99–118.

Clark, John M. "Toward A Concept of Workable Competition." *American Economic Review* 30, 2, part 1 (June 1940): 241–56.

Courtenay, Percy P. "International Tin Restriction and its Effects on the Malayan Tin Mining Industry." *Geography* 46, part 3 (July 1961): 223–31.

Cropper, M. L., Milton C. Weinstein, and Richard J. Zeckhauser. "The Optimal Consumption of Depletable Natural Resources: An Elaboration, Correction, and Extension." *Quarterly Journal of Economics* 92, 2 (May 1978): 337–44.

Desai, Meghnad. "An Econometric Model of the World Tin Economy, 1948–1961." *Econometrica* 34, 1 (January 1966): 105–34.

———. "An Econometric Model of the World Tin Economy: Reply to a Comment by F. E. Banks." *Econometrica* 40, 4 (July 1972): 753–55.

Eastham, J. K. "Rationalisation in the Tin Industry." *Review of Economic Studies* 4, 1 (1936): 13–32.

Engel, Bernard C. "International Tin Agreements." In *A New International Commodity Regime*, edited by Geoffrey Goodwin and James Mayall. (New York: St. Martin's Press, 1980), pp. 86–94.

Fergusson, Ewen. "The Marketing of Tin in Malaysia." International Tin Council. *Aspects of the Marketing of Tin: Papers Presented to the Committee on Development of the International Tin Council in 1973 and 1974*. (London: International Tin Council, no date), pp. 1–5.

Fisher, Anthony C., and John V. Krutilla. "Conservation, Environment, and the Rate of Discount." *Quarterly Journal of Economics* 89, 3 (August 1975): 358–70.

Gilbert, C. L. "The Post-war Tin Agreements: An Assessment." *Resources Policy* 3, 2 (June 1977): 108–17.

Hanson, John R., II. "Export Instability in Historical Perspective." *Explorations in Economic History* 14, 4 (October 1977): 293–310.

Hasbi Hassan, Abdullah. "A Regional Approach to Research on Tin Production." Paper 5(i). Ministry of Primary Industries of the Government of Malaysia and International Tin Council. *Fifth World Conference on Tin, Kuala Lumpur, 19–23 October 1981* (London: International Tin Council, forthcoming).

Heal, Geoffrey. "Economic Aspects of Natural Resource Depletion." In *The Economics of Natural Resource Depletion*, edited by D. W. Pearce and J. Rose. London: Macmillan, 1975.

Herfindahl, Orris C. "Some Fundamentals of Mineral Economics." *Land Economics* 31, 2 (May 1955): 131–38.

Herkstroeter, C. A. J. "Some Aspects of the Marketing of Tin, the Tin Price, and the Role of the London Metal Exchange, the Penang Market and the United States Market in Establishing World Market Prices for Tin." *Fourth World Conference on Tin: Kuala Lumpur, 1974*, vol. 4, *Marketing and Consumption*. (London: International Tin Council, 1975), pp. 59–81.

Hirshleifer, Jack. "Speculation and Equilibrium: Information, Risk, and Markets." *Quarterly Journal of Economics* 89, 4 (November 1975): 519–42.

______. "The Theory of Speculation Under Alternative Regimes of Markets." *Journal of Finance* 32, 4 (September 1977): 975–99.

Hoare, W. E. "Tin and Its Competitors." International Tin Council. *Aspects of the Marketing of Tin: Papers Presented to the Committee on Development of the International Tin Council in 1973 and 1974*. (London: International Tin Council, no date), pp. 29–59.

Hotelling, Harold. "The Economics of Exhaustible Resources." *Journal of Political Economy* 39, 2 (April 1931): 137–75.

Jackson, James C. "Mining in 18th Century Banka: the Pre-European Exploitation of a Tin Island." *Pacific Viewpoint* 10, 2 (September 1969): 28–54.

Johnson, Harry G. "Commodities: Less Developed Countries' Demands and Developed Countries' Responses." In *The New International Economic Order*, edited by J. N. Bhagwati. Cambridge: MIT Press, 1977.

______. "The Elementary Geometry of Buffer Stock Price Stabilization." *Malayan Economic Review* 22, 1 (April 1977): 1–9.

Johnson, Leland L. "The Theory of Hedging and Speculation in Commodity Futures." In *The Economics of Futures Trading*, edited by B. A. Goss and Basil A. Yamey. (New York: Wiley, 1976), pp. 83–99.

Kaldor, Nicholas. "Speculation and Economic Stability." *Review of Economic Studies* 7 (1939): 1–27.

Klein, Herbert S. "The Creation of the Patiño Tin Empire." *Inter-American Economic Affairs* 19, 2 (August 1965): 3–23.

Lim Chong Yah. "A Reappraisal of the 1953 Tin Agreement." *Malayan Economic Review* 5, 1 (April 1960): 13–24.

Lorenz-Meyer, D. "The Marketing of Tin in Europe." International Tin Council. *Aspects of the Marketing of Tin: Papers Presented to the Committee on Development of the International Tin Council in 1973 and 1974.* (London: International Tin Council, no date), pp. 21–26.

Mackey, Thomas S. "Tin Smelting and Refining in China." *Tin International* (April 1981): 129–32.

Malloy, James M. "Revolutionary Politics." In *Beyond the Revolution: Bolivia Since 1952*, edited by James M. Malloy and Richard S. Thorn. (Pittsburgh: University of Pittsburgh Press, 1971): 111–56.

Markham, Jesse W. "An Alternative Approach to the Concept of Workable Competition." *American Economic Review* 40, 2 (June 1950): 349–61.

Massell, Benton F. "Price Stabilization and Welfare." *Quarterly Journal of Economics* 83, 2 (May 1969): 284–98.

May, Elizabeth S. "The International Tin Cartel." In *International Control in the Non-Ferrous Metals*, William Y. Elliott, Elizabeth S. May, John W. F. Rowe, Alex Skelton, and Donald H. Wallace. (New York: Macmillan, 1937), pp. 277–362.

Newbery, D. M. G., and Joseph E. Stiglitz. "The Theory of Commodity Price Stabilisation Rules: Welfare Impacts and Supply Responses." *Economic Journal* 89, 356 (December 1979): 799–817.

Nguyen, D. T. "The Implications of Price Stabilization for the Short-term Instability and Long-term Level of LDCs' Export Earnings." *Quarterly Journal of Economics* 92, 1 (February 1979): 149–54.

Noordin, Ahmad Zubeir. "The Penang Tin Market." *Fourth World Conference on Tin: Kuala Lumpur, 1974*, vol. 4, *Marketing and Consumption.* (London: International Tin Council, 1975), pp. 47–52.

Oi, Walter, "The Desirability of Price Instability Under Perfect Competition." *Econometrica* 29, 1 (January 1961): 58–64.

Paauw, Douglas S. "Southeast Asia's Economic Development in World Perspective." In *Southeast Asia, An Emerging Center of World Influence*, edited by Wayne Raymond and K. Mulliner. Athens: University Center for International Studies, 1977.

Perry, Martin K. "Vertical Integration: the Monopsony Case." *American Economic Review* 68, 4 (September 1978): 561–70.

Peterson, Frederick N., and Anthony C. Fisher. "The Exploitation of Extractive Resources: A Survey." *Economic Journal* 86, 348 (December 1977): 681–721.

Pindyck, Robert S. "The Optimal Exploration and Production of Nonrenewable Resources." *Journal of Political Economy* 86, 5 (October 1978): 841–61.

Pocock, N. J. B. "The Qualities and Forms of Tin and the Requirements of the Consumer." International Tin Council and the Tin Research Institute. *Conference on Tin Consumption.* (London: International Tin Council, 1972), pp. 63–78.

Robins, D. A. "Future Prospects for Tin Consumption." *Metals and Materials* 5, 9 (1971): 298–302.

Rogers, Christopher D. "Consumer Participation in the International Tin Agreements." *Malayan Economic Review* 14, 2 (October 1969): 113–29.

Scott, E. Baliol. "Tin." *Mineral Industry*. Several issues.

Siew Nim Chee. "The International Tin Agreement of 1953." *Malayan Economic Review* 2, 1 (April 1957): 35–53.

Smith, Gordon W. and George R. Schink. "The International Tin Agreement: A Reassessment." *Economic Journal* 86, 344 (December 1976): 715–28.

Sosnick, Stephen. "A Critique of Concepts of Workable Competition." *Quarterly Journal of Economics* 72, 3 (August 1958): 380–423.

Tan, Lee, and Yao Chi-Lung. "Abundance of Chemical Elements in the Earth's Crust and its Major Tectonic Units." *International Geology Review* 12, 7 (July 1970): 778–86.

Thoburn, John T. "Commodity Prices and Appropriate Technology: Some Lessons from Tin Mining." *Journal of Development Studies* 14, 1 (October 1977): 35–52.

______. "Policies for Tin Exporters." *Resources Policy* 7, 2 (June 1981): 74–86.

Thorn, Richard S. "The Economic Transformation." In *Beyond the Revolution: Bolivia Since 1952*, edited by James M. Malloy and Richard S. Thorn. (Pittsburgh: University of Pittsburgh Press, 1971), pp. 157–216.

Tilton, John E. "Material Substitution and the Demand for Tin." Paper no. 9. Ministry of Primary Industries of the Government of Malaysia and International Tin Council. *Fifth World Conference on Tin, Kuala Lumpur, 19–23 October 1981* (London: International Tin Council, forthcoming).

Turnovsky, Stephen J. "The Distribution of Welfare Gains from Price Stabilization: A Survey of Some Theoretical Issues." In *Stabilizing World Commodity Markets*, edited by F. Gerard Adams and Sonia A. Klein. (Lexington: D. C. Heath, 1978), pp. 119–47.

Usher, Dan. "Income as a Measure of Productivity: Alternate Comparisons of Agricultural and Non-Agricultural Productivity in Thailand." *Econometrica* 33, 132 (November 1960): 430–41.

Waugh, Frederick V. "Does the Consumer Benefit from Price Instability?" *Quarterly Journal of Economics* 58, 3 (August 1944): 602–14.

Weinstein, Milton C., and Richard J. Zeckhauser. "Optimal Consumption of Depletable Resources." *Quarterly Journal of Economics* 89, 3 (August 1975): 371–92.

Widiatmo, Sirman. "Vertical Integration in Indonesian Tin Mining." International Tin Council. *Aspects of the Marketing of Tin: Papers Presented to the Committee on Development of the International Tin Council in 1973 and 1974*. (London: International Tin Council, no date), pp. 11–15.

Williamson, D. R. "Tin Shares—Some of the Problems of Marketability and Company Structure Examined." *Fourth World Conference on Tin: Kuala Lumpur, 1974*, vol. 4, *Marketing and Consumption*. (London: International Tin Council, 1975) pp. 138–49.

______. "The Use of the London Metal Exchange for Hedging Purposes." Paper no. 16. Ministry of Primary Industries of the Government of Malaysia and International Tin Council. *Fifth World Conference on Tin, Kuala Lumpur, 19–23 October 1981* (London: International Tin Council, forthcoming).

Working, Holbrook. "Futures Trading and Hedging." *American Economic Review* 43, 3 (June 1953): 314–43.

Yip Yat Hoong. "The Domestic Implementation of the 1953 International Tin Agreement in Malaya." *Malayan Economic Review* 5, 20 (October 1960): 59–65.

______. "Malaya Under the Pre-War International Tin Agreement." *Malayan Economic Review* 8, 1 (April 1963): 81–97.

"The Marketing of Tin Ore in Kampar." *Malayan Economic Review* 4, 2 (October 1959): 45–55.

———. "Recent Changes in the Ownership and Control of Locally Incorporated Tin Dredging Companies in Malaya." *Malayan Economic Review*, 13, 1 (April 1968): 70–88.

Government and public agency documents

International Tin Council, *The Fifth International Tin Agreement as adopted 21 June 1975*. London: International Tin Council, 1976.

International Tin Council. *The Fifth International Tin Agreement, Rules of Procedure*. No date.

Mines Research Institute Malaysia. *Institut Penyeldikan Galian Malaysia (The Mines Research Institute Malaysia)*. January 1978.

Second Malaysia Plan, 1971–1975. Kuala Lumpur: Government Press, 1971.

Thailand, Department of Mineral Resources. *The Mining Industry of Thailand*. Bangkok: Cabinet Printing Office, September 1978.

Third Malaysia Plan, 1976–1980. Kuala Lumpur: Government Press, 1976.

United Nations Conference on Trade and Development. *Sixth International Tin Agreement*. New York: United Nations, 1981.

United Nations General Assembly. *Resolutions Adopted by the General Assembly During its Twenty-ninth Session, 17 September-18 December 1974*. New York: United Nations, 1975.

United Nations Industrial Development Organization. *Non-Ferrous Metals Industry*. UNIDO Monographs on Industrial Development, no. 1. New York: United Nations, 1969.

United Nations Tin Conference, 1950 and 1953. *Summary of Proceedings*. New York: United Nations, 1954.

United Nations Tin Conference, 1960. *Summary of Proceedings*. New York: United Nations, 1961.

United Nations Tin Conference, 1965. *Summary of Proceedings*. New York: United Nations, 1965.

United Nations Tin Conference, 1970. *Summary of Proceedings*. New York: United Nations, 1970.

United Nations Tin Conference, 1975. *United Nations Tin Conference, Geneva, 1975*. New York: United Nations, 1976.

United States Central Intelligence Agency. *Soviet Tin Industry: Recent Developments and Prospects Through 1980*. Research Aid ER 77–1101 (January 1977).

United States Congress, Joint Economic Committee. *Issues at the Summit*. Washington: Government Printing Office, 1977.

———. *The United States Response to the New International Economic Order: The Economic Implications for Latin America and the United States*. Washington: Government Printing Office, 1977.

United States Congress, House of Representatives, Subcommittee of the House Committee on Foreign Affairs. *Tin Investigation: Report of the Subcommittee of the House Committee on Foreign Affairs*. Washington: Government Printing Office, 1935.

United States Congress, Senate, Committee on Foreign Relations. *Fifth Annual Tin Agreement: Report*. Washington: Government Printing Office, 1976.

United States Congress, Senate, Preparedness Subcommittee of the Committee on Armed Services. *Sixth Report of the Preparedness Subcommittee of the Committee on Armed Services: Tin, 1951*. Washington: Government Printing Office, 1951.

Magazines and newspapers

American Metal Market
Asian Business and Industry
Bangkok Post
Business in Thailand
Business Review (Thailand)
Business Times (Malaysia)
Comtel Reuter
Engineering and Mining Journal
Far Eastern Economic Review
Financial Times (London)
Fortune
The Investor (Thailand)
Latin America Commodities Report
Metal Bulletin
Metals and Materials
Mining Journal
Mining Magazine
The Nation (Thailand)
New Straits Times
New York Times
Tin International
Wall Street Journal

Statistical and reference series

American Bureau of Metal Statistics. *Non-Ferrous Metal Data*. New York: American Bureau of Metal Statistics, Inc.

American Bureau of Metal Statistics. *Yearbook*. New York: American Bureau of Metal Statistics, Inc.

American Metal Market. *Metal Statistics*. New York: Fairchild Publications.

Bank Negara Indonesia. *Report*. Jakarta, Bank Negara Indonesia.

Bank Negara Malaysia. *Annual Report*. Kuala Lumpur: Government Press.

Economist Intelligence Unit. *Quarterly Economic Review of Nigeria*. London: Economist Intelligence Unit Ltd.

Engineering and Mining Journal. *E/MJ International Directory of Mining and Mineral Processing Operations*. New York: McGraw-Hill.

International Monetary Fund. *International Financial Statistics*. Washington: International Monetary Fund.

International Tin Council. *Annual Report*. London: International Tin Council.

______. *Monthly Statistical Bulletin*. London: International Tin Council.

______. *Statistical Yearbook*. London: International Tin Council.

______. *Tin Statistics*. London: International Tin Council.

International Tin Research Council. *Annual Report*. Greenford: International Tin Research Council.

Kuala Lumpur Stock Exchange. *Annual Companies Handbook*. Kuala Lumpur: Kuala Lumpur Stock Exchange.

Metal Bulletin Books. *Non-Ferrous Metal Works of the World*. Worcester Park: Metal Bulletin Books Ltd.

Mineral Industry. New York: McGraw-Hill.

Mining International Yearbook. London: Financial Times Ltd.

National Institute of Economic and Social Research. *National Institute Economic Review*. London: National Institute of Economic and Social Research.

States of Malaya Chamber of Mines. *Year Book*. Kuala Lumpur: United Selangor Press.

Thailand, Department of Mineral Resources. *Mineral Statistics of Thailand*. Bangkok: Cabinet Printing Office.

United Nations. *Statistical Yearbook*. New York: United Nations.

———. *Yearbook of International Trade Statistics*. New York: United Nations.

United States Bureau of Mines. *Minerals Yearbook*. Washington: Government Printing Office.

United States Council of Economic Advisers. *Annual Report*. Washington: Government Printing Office.

United States Department of Commerce. *Survey of Current Business*. Washington: Government Printing Office.

World Bank. *World Bank Atlas*. Washington: World Bank.

———. *World Development Report*. New York: Oxford University Press.

Unpublished materials

Arce, Roberto. "Influencia de las Empresas Transnacionales en la Mineria del Estaño: el Caso de Bolivia." Mimeographed. United Nations Economic Commission for Latin America, July 1977.

Bird, Peter J. W. N. "An Investigation of the Role of Speculation in the 1972 to 1975 Commodity Price Boom." Ph.D. dissertation, Stirling University, 1981.

Chhabra, Jasbir, Enzo Grilli, and Peter Pollak. "The World Tin Economy: An Econometric Analysis." World Bank Staff Commodity Paper no. 1, June 1978.

Chong Pit Loo. "The Tin Smelting Industry of West Malaysia." Graduation exercise presented to the University of Malaya in part fulfillment towards the Degree of Bachelor of Arts with Honours in Economics, 1971.

Commodities Research Unit, Ltd. *Establishment of a Tin Exchange in Malaysia: Draft Report Submitted to the Ministry of Primary Industries, Federation of Malaysia, August, 1977.*

Datuk Keramat Smelting Sdn. Berhad. "Guide to Straits Tin Market: The Operations of Straits Tin Market." No date.

de Koning, P. A. A. "The Achievements of the International Tin Council." Mimeographed. Address before a meeting of the Tin and Lead Smelters Club, The Hague, May 6–8, 1979.

International Tin Council. *Notes on Tin*. London: International Tin Council. Photocopied.

International Tin Council. Press Communiques. London: International Tin Council. Mimeographed.

Jindarah Phangmuangdee. "Analysis of United States Tin Disposals." Master of Economics thesis, Thammasat University, Bangkok, Thailand, 1975.

Landon, J. W. "Prospects for a Kuala Lumpur Commodity Exchange." Mimeographed. City of London Seminar, Kuala Lumpur, 1979.

Page, William. "One Attempt at Taking a Long-term View to Assist Decision-Makers in the World Tin Industry." Mimeographed. Brighton: Science Policy Research Unit, University of Sussex, September 1977.

Puey Ungphakorn. "The Economics of Tin Control." Ph.D. dissertation, University of London, 1949.

Straits Trading Company Limited. "Penang Tin Market." Form 100. No date.

Straits Trading Company Limited. "Tin Sales and Conditions Relating to Tin Bids." Form 101. No date.

Tan Teong Hean. "A Study of the Sources and Nature of Credit in Chinese Tin Mining in Malaya." Graduation exercise presented to the University of Malaya in part fulfillment towards the Degree of Bachelor of Arts with Honours in Economics, 1965.

Wit Sakyarakwit. "Tin: A Comparison of Gravel Pump and Dredge Mining." Master of Economics thesis, Thammasat University, Bangkok, Thailand, 1971.

Index

William Lee Baldwin is Professor of Economics at Dartmouth College, where he has taught since 1956. He has served as Chairman of Dartmouth's Social Science Division and Economics Department. He received his B.A. from Duke University in 1951, and a Ph.D. from Princeton University in 1958.

In addition to his continuing affiliation with Dartmouth, he has been Visiting Assistant Professor at Princeton in 1961–62, Research Professor at the Brookings Institution in 1963–64, Visiting Professor at Thammasat University in Bangkok, Thailand, in 1968–70, member of the Rockefeller Foundation's Special Field Staff, again attached to Thammasat University, in 1974–75, and consultant to the Federal Trade Commission in 1980.

In 1978–79, Professor Baldwin spent a sabbatical year in Thailand, Malaysia, and England while engaged in preparation of this book. During that period, he was given an appointment as Visiting Researcher at Universiti Sains Malaysia in Penang, Malaysia.

Professor Baldwin is the author of two earlier books published by Duke University Press: *Antitrust and the Changing Corporation* (1961), and *The Structure of the Defense Market, 1955–1964* (1967). He was co-editor of *The Role of Foreign Assistance to Thailand in the 1980s*, and has contributed a number of articles to professional journals.

Of related interest

Duke Press Policy Studies

The Making of Federal Coal Policy
Robert H. Nelson

Global Deforestation and the Nineteenth-Century World Economy
Edited by Richard P. Tucker *and* J. F. Richards

Modeling Growing Economies in Equilibrium and Disequilibrium
Edited by Allen C. Kelley, Warren C. Sanderson, *and* Jeffrey G. Williamson

A Society for International Development
Prospectus 1984
Edited by Ann Mattis

Duke University Commonwealth Studies Center Publications

Economic Imperialism in Theory and Practice
The Case of South African Gold Mining Finance, 1886–1914
Robert V. Kubicek